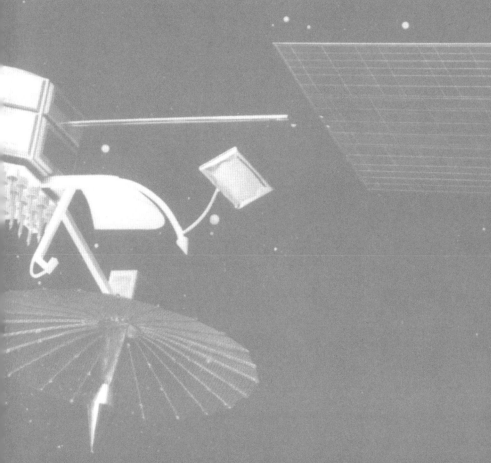

"READ YOU LOUD AND CLEAR!"

"READ YOU LOUD AND CLEAR!"

by Sunny Tsiao

The Story of
NASA's Spaceflight
Tracking and
Data Network

The NASA History Series
National Aeronautics and Space Administration
NASA History Division
Office of External Relations
Washington, DC   2008

NASA SP-2007-4232

Library of Congress Cataloging-in-Publication Data

Tsiao, Sunny, 1963-
  Read you loud and clear! : the story of NASA'S spaceflight tracking and data network / by Sunny Tsiao.
       p. cm. -- (NASA SP-2007-4232)
  Includes bibliographical references and index.
 1. Space vehicles--United States--Tracking--History. 2. Ground support systems (Astronautics)--History. 3. United States. National Aeronautics and Space Administration--History. 4. Project Apollo (U.S.)--History. I. Title. II. NASA SP-2007-4232.
 TL4026.T785 2007
 629.47'8--dc22
                               2007007826

*To my beloved Father and Mother,
Tom and Ellen*

ISBN 978-0-16-080191-4

For sale by the Superintendent of Documents, U.S. Government Printing Office
Internet: bookstore.gpo.gov Phone: toll free (866) 512-1800; DC area (202) 512-1800
Fax: (202) 512-2104 Mail: Stop IDCC, Washington, DC 20402-0001

ISBN 978-0-16-080191-4

CONTENTS

Preface ix
Foreword. xvii
Acknowledgments. xxvii
Introduction: Invisible Network. xxix
1 The Early Years. 1
2 Evolution of a Network 29
3 The Mercury Space Flight Network 65
4 Preparing for the Moon. 105
5 The Apollo Years 143
6 Era of Change 199
7 A Network in Space 243
8 The New Landscape 303
9 A Legacy 329
Endnotes. 343
Bibliography 383
Appendix 1: Acronyms, Abbreviations, Glossary 403
Appendix 2: Maps. 415
Appendix 3: Radio Frequency Chart 425
Appendix 4: Honeysuckle Station Log for Apollo 11 427
Index 455
NASA History Series 475

PREFACE

Much of what has been written on the topic of the National Aeronautics and Space Administration's (NASA) tracking and data networks has been on the Jet Propulsion Laboratory's Deep Space Network, the DSN. This is perhaps understandable as the DSN has played and continues to play a central role in many of America's most high-profile exploration missions. These have included the early Pioneer probes, the Mariner missions of the 1960s and 1970s, Viking and Voyager, and most recently, Galileo, Cassini-Huygens, and the new generation of Mars explorers that will prepare the way for eventual human voyages to the Red Planet.

The intent of this volume is to present a history of NASA's "other" network, the one established and run by the Goddard Space Flight Center (GSFC). The Spaceflight Tracking and Data Network, or STDN, was—in its various incarnations throughout the years—the network that tracked the first artificial satellites around Earth. It tracked Apollo astronauts to the Moon and back. Today, a network based in space called the Space Network, along with a much reduced Ground Network, work together to support the United States and international partners in all near-Earth space communications and spaceflight activities. The history of the STDN is not unlike a microcosm of the history of NASA itself. It spans 50 years. It has seen its share of triumphs and tragedies, and it is playing a major role in setting the pace for space exploration in the twenty-first century.

When considering sources for this history, the author searched for scholarly works that have been published on the subject of NASA's STDN.

There has been some coordinated effort to document NASA's human spaceflight and near-Earth communications networks throughout the years. Starting in the late 1960s, the GSFC saw the need to begin documenting this history. From that start, most of the literature on the subject has been independently commissioned reports sponsored by the Center itself.

The seminal work was by William R. Corliss in 1969 called *History of the Goddard Networks*. Corliss updated it five years later, expanding the subject to the end of the Apollo program. These initial works were relatively general, based primarily on secondary research—assimilation of information put out by the GSFC in the form of information pamphlets, brochures and public affairs news releases.

NASA historian Alfred Rosenthal drafted, in 1983, an unpublished work titled *Vital Links: The First 25 Years of NASA's Space Tracking, Communications and Data Acquisition 1958–1983*. *Vital Links* was a new work on the subject, taking the timeline up through the early part of the Space Shuttle program. A key part of Rosenthal's work was his interview of some 20 people whom he identified as principals on the subject. These were not formal oral histories but rather, topic-specific statements made by people who were major contributors on the STDN over the years. The interviews were done in 1982 and 1983 but never published. Many of these people are now deceased. The author drew heavily on these interviews and many of the quotes are used throughout this book. The draft of *Vital Links* was turned over to the NASA History Division where it is archived. It was a primary reference for this work.

In 1992, GSFC published the Contractor Report *Keeping Track: A History of the GSFC Tracking and Data Acquisition Networks 1957 to 1991*. It was edited by Kathleen M. Mogan and Frank P. Mintz. The report listed them as editors rather than coauthors because *Keeping Track* was, for the most part, edited together from Corliss's 1974 and Rosenthal's 1983 works. A final chapter was added which provided a very brief overview of the STDN in the 1980s. Mogan and Mintz included excerpts from about half of Rosenthal's interviews at the end of the report in a section called "Personal Views." They, in addition, conduct half a dozen interviews with key network personnel who were at Goddard in 1990. The material was then published by the GSFC as a reference report where it has circulated since. Copies have also been distributed to other NASA locations. Therefore, through Rosenthal, Mogan, and Mintz, over two dozen people who shaped NASA's STDN from 1960 to 1990 were interviewed. This was an invaluable resource. Taken together, *Vital Links* and *Keeping Track* served as the primary reference, the updating of which was an impetus for this book.

The most important archival materials were undoubtedly the Historical Reference Collection at the History Division of NASA Headquarters. There, correspondences and memoranda on the subject were found dating back to 1958. Letters between key Administrators and managers like Edmond C. Buckley, Gerald M. Truszynski, and Ozro M. Covington—early movers of the program—explained why certain decisions were made and how they arrived at those decisions. Files were found all the way to the present. Organized by subject and key words, files addressing a specific topic with regard to the Networks were found in the form of Congressional testimonies, news articles, technical presentations and photographs. These included those published by the Agency over the years as well those from open literature such as newspapers, magazines and professional journal articles.

Technical briefings and presentations provided to the author by NASA was another good source of firsthand information. Most of these came from the Office of Space Communications in Washington and some from the GSFC. The author also interviewed current and former NASA managers and others who worked with the Network, both inside the government and contractors, both U.S. and foreign. They brought unique, personal perspectives that answered questions or helped clarify ambiguities uncovered in the course of the research. These personal perspectives also provided stories which simply could not have been found anywhere else. NASA Oral Histories—most of which are now available online—provided much insight to this end. They complimented, and in certain cases, supplemented the histories obtained from other sources. Most of these Oral Histories were conducted by the Agency as part of the ongoing Johnson Space Center (JSC) Oral History Project started in 1998. Interviews from three groups of people were reviewed: NASA Administrators and high level managers, flight controllers and operations personnel, and astronauts. This produced a broad and diverse set of viewpoints from which a story could be weaved together, supplementing the material uncovered from other sources.

Books on the subject of tracking networks, space communications, space exploration, and political history in general provided good background information within which the topic was framed. Even though these generally lacked the details necessary to delve in-depth into the actual history of Networks, they provided the general science and historical background which complemented this history, and which in fact, allowed the story to be told from a popular point of reference. Other books of a more technical nature provided the material needed to explain some of the finer scientific points in layman terms.

Transcripts such as Administrator testimonies before Senate and House subcommittees and White House letters were examined. The World Wide Web provided convenient material, but they required verification from other sources. Most could be validated by authenticating or tracing back to the original Web site. Those from dubious sources were either not used at all or if used, clarified in more detail in the Endnotes section of this book. Internet sites were especially helpful for general historical information on topics such as history of a country, the climate of an island, or the evolution of a country's governing body and its political relationship to the United States.

A diverse range of photographs and pictures are included in this book. When discussing the topic of tracking stations and communication networks, it is easy to fall prey to reproducing pictures of big antennas and buildings page after page. After all, those are probably the most prominent components of a ground station. Some of those are included. But what paints a more interesting and complete picture are the people involved and the environment that they operated in. Therefore, a deliberately chosen mix of seldom before seen photographs along with better-known, more frequently reproduced pictures are included.

The history of NASA's STDN is a 50 year story, from its birth in the late 1940s to where it stands today in the first decade of the twenty-first century. The story is told here from a historical and not a technical point of view, although explanatory passages are present when warranted. The author hopes that in doing so, the chapters will appeal not only to the trained subject expert but also to a lay audience or just the space enthusiast at large who may simply have always been curious as to how NASA tracked the Lunar Rover on the Moon, for instance.

Since the beginnings of the STDN are inextricably tied to the formation of NASA as America's civilian space agency, a brief synopsis of how the Agency came about—the nuances of which affected how it would operate its tracking networks for years to come—is included in Chapter 1. For a comprehensive look at the establishment of NASA, the author recommends starting with *NASA's Origins and the Dawn of the Space Age*, Monographs in Aerospace History number 10, by David S. F. Portree, 1998.

Similarly, the history of the STDN is tied to the Space Shuttle. After the Apollo lunar missions, the Network turned its focus on supporting the reusable Shuttle. Much has been written on the history of the Space Transportation System (STS), so much so that in 1992, NASA's History Division published a monograph summarizing the vast amount of literature on the subject. The reader is encouraged to reference that work for more information. (Launius, Roger D. and Gillette, Aaron K., *The Space Shuttle: An Annotated Bibliography*, Monographs in Aerospace History, Number 1, 1992)

Finally, this book does not address the Deep Space Network DSN except where it overlaps and affects the STDN. Works on the Deep Space Network include *A History of the Deep Space Network* by William R. Corliss from 1976, and the 2001 NASA Project Histories publication SP-2001-4227, *Uplink-Downlink: A History of the Deep Space Network 1957-1997* by Douglas J. Mudgway. The latter work is a very comprehensive description of the DSN from a technical perspective. Mudgway followed this volume with a less technical but more historical look in his 2005 publication *Big Dish*. Since it is difficult to fully segregate the two networks, the following is an overview of the DSN which the reader may find beneficial as he or she delves into the history of the STDN in the coming chapters. Much of this description can be found in the 1989 Jet Propulsion Laboratory publication JPL400-326 *Goldstone Deep Space Communication Complex:*

Scientific investigation of the planets and interplanetary space has been carried out for over four decades by NASA mainly through the use of automated space probes and robotic vehicles. Although engineered to operate independently in the far reaches of space, these intricate and highly autonomous craft are dependent upon Earth-based DSN for guidance, control and reception of the vast amounts of scientific information they acquire and transmit back to Earth.

NASA's DSN is today among the largest and most sensitive scientific telecommunications and radio navigation networks in the world. Its principal responsibilities are to support unmanned planetary missions and radio and radar astronomy in the exploration of the solar system and the universe. The DSN is managed, directed and operated for NASA by the Jet Propulsion Laboratory (JPL) of the California Institute of Technology in Pasadena, California. It performs several, wide-reaching functions, all relating to the deep space environment: 1) Receive telemetry signals from space probes; 2) Transmit commands that control the spacecraft; 3) Generate radio navigation data to locate and guide the spacecraft to their destinations; 4) Perform radio science, radio and radar astronomy; 5) Perform highly sensitive Earth geodynamics measurements; and 6) Participate in the Search for Extraterrestrial Intelligence.

The history of the DSN dates back to 1958 when JPL deployed three ground communication stations to help receive telemetry and determine the orbit of Explorer 1, the first successful American satellite. Before that, engineering research into the tracking requirements for lunar and planetary spaceflight had been in progress for some time at JPL. Following Explorer 1, JPL developed a ground communications and navigation network for the Pioneer 3 and 4 missions, the first U.S. spacecraft to be launched beyond Earth orbit. On 3 December 1958, shortly after the establishment of NASA, JPL was officially transferred to the new space agency from the U.S. Army. Since then, it has grown with the U.S. space program into a world network consisting of 12 deep-space stations located at three communication complexes on three continents: Goldstone in Southern California's Mojave Desert; at Robledo near Madrid, Spain; and at Tidbinbilla on the outskirts of Canberra, Australia. These three locations allow the DSN to maintain around-the-clock communications with spacecraft traveling anywhere in the solar system.

The network consists of several key components. One is the Network Operations Control Center (NOCC), which controls and monitors the three complexes. Located in Pasadena, the NOCC is the operations hub for the DSN. Its functions include monitoring operations at the three complexes, validating the performance of the network for spaceflight project users, providing information for configuring and controlling the network, and planning and participating in network testing and mission simulations.

The Ground Communications Facility provides the necessary communication circuits that link the complexes to the control center at JPL and to the remote flight project centers in the United States and overseas that manage and operate the space probes. "Comm" traffic between these places can be sent in various ways, such as via land lines, ocean cables, terrestrial microwave, and communication satellites. These circuits are leased from commercial carriers by the NASA Integrated Support Network, commonly referred to as NISN—jointly managed by the Goddard and Marshall Space Flight Centers—which provides circuits as needed to all NASA centers and facilities.

The most visible arm of the DSN is of course the three Deep Space Communication Complexes that are dominated by a suite of large parabolic dish antennas. At each complex are two 34-meter (111-foot) diameter antennas, one 26-meter (85-foot) antenna, and one gigantic 70-meter (230-foot) antenna. A centralized Signal Processing Center remotely controls the 70-and 34-meter antennas, generates and transmits spacecraft commands, and receives and processes the spacecraft telemetry. These antennas form separate subnets according to their respective sizes, and each subnet has a different communication capability. For example, the 70-meter antenna subnet—which is the most sensitive—supports deep space missions while the 26-meter subnet supports spacecraft in near-Earth orbit. The two 34-meter antenna subnets support both types of missions.

The 26-meter systems were originally part of the Goddard STDN. They were consolidated into the DSN in 1985. The merger led to significant operational efficiencies by consolidating like maintenance, operations, and field support activities for tracking stations located geographically close to each other. From a mission's perspective, it also allowed the DSN to now track spacecraft in highly elliptical, high apogee Earth orbits.

In the United States, the DSN is staffed by JPL personnel, assisted by contractor engineers and technicians who are primarily responsible for operating and maintaining the Goldstone complex, the NOCC and the Ground Communications Facility. The two overseas complexes are operated entirely by the local government agency (the National Institute of Aerospace Technology, or INTA, in Spain and the Commonwealth Scientific and Industrial Research Organization, or CSIRO, in Australia) and their contractors under NASA funding. The international staff totals over 1,000 people.

Since its beginning, the DSN has provided principal tracking, telemetry, and command (TT&C) support for every unmanned lunar and planetary spacecraft that NASA has launched (plus secondary support on Apollo). The resumé is quite impressive. Moon exploration began with the Pioneer 3 and 4 probes in 1958 and 1959. This was followed in the 1960s by the Ranger 1 through 9 lunar television missions (1961–1965); the Lunar Orbiter 1 through 5 photographic surveys for Apollo landing sites (1966–1967); and Surveyor 1 through 7 lunar landers (1966–1968). Support for inner-planet exploration began in 1962 with JPL's Mariner series of missions to Venus, Mars and Mercury that encompassed six planetary flybys and the first planetary orbiter around Mars in 1971. Concurrent with these Mariner flights were the NASA Ames Research Center Pioneer missions, Pioneer 6 through 9 between 1965 and 1968. Mars exploration culminated in 1975 with the much-heralded landings of Viking 1 and 2.

The first JPL probe to explore the outer planets was Pioneer 10, which flew by Jupiter in 1973. Six years later, Pioneer 11 became the first spacecraft to flyby Saturn. Two Voyager probes performed a grand tour of the outer planets following their launch in 1977, passing Jupiter in 1979 and Saturn in 1980 and 1981. Voyager 2 then went on to make close approaches to the seventh planet, Uranus, in 1986 and three years later Neptune. In both cases, stunning close-up photographs of these distant planets were beamed back to Earth. They are now far beyond Pluto's orbit and astronomers anticipate they will venture beyond our known solar system in the next few years.

More recently, the DSN is supporting the Cassini-Huygens mission, exploring Saturn and its largest moon Titan. The latter is one of the most intriguing objects in the entire solar system. The second largest of any of the planetary moons (only Jupiter's Ganymede is larger), Titan is actually larger than Mercury and Pluto. It, in fact, has a planet-like atmosphere which is denser than those of Mercury, Earth, Mars, and Pluto. Composed predominantly of nitrogen with other hydrocarbon compounds, such elements are the essential building blocks for amino acids necessary for the formation of life. This has led some scientists to postulate that Titan's environment may be similar to that of Earth before life began putting oxygen into the atmosphere.

Finally, the DSN is once again supporting new NASA missions to our nearest planetary neighbor, this time with the Mars Reconnaissance Orbiter along with the Spirit and Opportunity robotic rovers. This is a prelude to eventual human exploration of the planet that is slated for the 2030 to 2040 timeframe. Pluto remains the only planet (dwarf planet) not having been visited by a manmade spacecraft. But that will change by the year 2015 as the New Horizons spacecraft—launched on 19 January 2006—is scheduled to encounter our solar system's most distant planet. It will then make an extensive study of as many Kuiper Belt objects as it can, a mysterious disk shaped zone beyond Neptune containing thousands of planetary-like objects, before permanently leaving the confines of the solar system to travel into interstellar space.[1]

FOREWORD

"Invisible" is an adjective often used to describe NASA's space communications and tracking programs. Behind this is the implicit characterization of the associated systems as collectively providing "infrastructure:" necessary support capabilities essential to the achievement of primary objectives. Infrastructures are by nature seldom noticed, with recognition normally reserved for those unfortunate occasions when it falls short and limits accomplishment of desired objectives.

The electronic highways linking orbiting spacecraft with their associated control and information handling capabilities on Earth were (and remain) absolutely essential to the spectacular successes of NASA's space programs. Yet many, commonly including the spaceflight programs themselves, take the existence and performance of NASA's communications and tracking infrastructure for granted. The thousands of dedicated and capable people who have devoted much of their careers to making this capability a success for nearly five decades understand and accept the fact that their success is measured by remaining invisible.

Those of us who have participated in this program are proud of our individual contributions and collective accomplishments, and greatly welcome this history being documented. We also hope a broader understanding of the unique challenges we faced, and of our solutions for these challenges, will prove helpful to those blazing similar new trails.

Infrastructures are notoriously tricky to manage, due in large part to funding issues associated with "taken for granted" resources. Customers, or users, of infrastructure assets and capabilities are usually detached from capital, operating and maintenance funding concerns, and seldom get excited about contributing! Those managing NASA's space communications program very successfully surmounted these challenges.

Supplying NASA's communications and tracking infrastructure also presented unique challenges. Unlike more traditional infrastructures such as roads and bridges that rely on established technologies, the technologies needed for space communications and information handling were in their infancy and needed to be appreciably advanced. NASA's leaders had to anticipate the direction and pace of the underlying technologies—predicting future capabilities, risks, and costs.

This evolving technology added yet another challenge: vaguely defined requirements. As these same technologies were integral to the development of user spacecraft and instrumentation, the development period for providing new space communications and tracking capabilities has been comparable to the design periods for its customers' spacecraft. The net result was that the tracking and data organization needed to evolve its enabling capabilities in advance of having defined user requirements.

The early leaders of NASA's space communications program deserve special recognition for their success under these conditions. In retrospect, they were extremely successful, synergistically integrating NASA's needs with those of the broader society. The tremendous advances in communications and information handling experienced over the four decades of NASA's space program were not coincidental. It is quite a tribute to early leaders such as Edmond Buckley and Gerald Truszynski that essential tracking and data capabilities have been there when needed, enabling NASA's impressive accomplishments!

A general characterization of NASA's space communications organization may help explain the associated program. This program relied heavily on support service contractors—people with needed skills provided under level-of-effort contracts—along with foreign nationals at overseas stations. Contracts were usually competitively awarded. The direct use of U.S. government civil service personnel was quite limited. Although numbers varied over the years, in general (prior to the 1996 reorganization) only about 1 percent of program personnel were civil service personnel at NASA Headquarters. About 10 percent were civil service (or corresponding JPL employees) at the Field Centers—primarily GSFC and JPL—and approximately 90 percent were support contractors reporting to the Field Centers.

The focus of the program was operations, and images of control rooms crowded with consoles suggest a large staff directly participating in operations. The reality is that only a small percentage of the people occupied console positions. The majority were devoted to supporting roles—planning and scheduling, maintenance, logistics, preparing to meet new mission requirements, management, and overhead duties.

As demonstrated by results, NASA has successfully and cost-effectively met the space communications needs of the Agency's ambitious space programs throughout its history. This did not happen by accident, but was effectively orchestrated by leaders grasping both the needs and possibilities. Regrettably slighting numerous dedicated people who made significant and vital contributions to NASA's space communications program over the years, I mention three leaders whom I feel merit special recognition: Edmond C. Buckley at Headquarters, Ozro M. Covington at Goddard, and Murray T. Weingarten at Bendix. Not just very competent managers, they were leaders—able to envision future possibilities and inspire others, especially subordinates, to pursue their visions. Personally, I feel very fortunate to have been able to work with and learn from them.

From NASA's establishment in 1958, Edmond Buckley headed NASA's network operations at Headquarters. As described in this history, Buckley's approach of consolidating management and funding in a single organization reporting directly to the Administrator prevailed. This feat must be recognized in the context of convincing very competent, territorial-minded peers to agree his approach would best serve NASA. NASA's networks—STADAN, MSFN and DSN—were consolidated under him to form the Office of Tracking and Data Acquisition (OTDA) in November 1961. Consolidating tracking and data activities also took better advantage of the limited pool of personnel knowledgeable of relevant technologies.

NASA's Centers have notoriously acted as independent fiefdoms, the drumbeat of "one NASA" periodically coming from various Administrators notwithstanding. Many at the Manned Spacecraft Center (renamed Johnson Space Center in 1973) especially resented GSFC being given responsibility for the Manned Space Flight Network. Ozzie Covington deserves special recognition for developing a cooperative relationship between Goddard and JSC, and especially with Chris Kraft, during the Apollo era. Ozzie strongly promoted a culture in which JSC was recognized as the MSFN's customer. Although this legacy was lost in later years, one must not underestimate its contribution to Apollo's successes.

Similarly, by force of his personality Murray Weingarten created a cooperative and professional relationship with NASA as his customer. Murray was especially attentive and responsive to changes in NASA's needs. BFEC was a dominant contractor in the early days of the Network, and its culture permeated much of the contractor community. After Murray departed and BFEC ownership transitioned to Allied Signal and then Honeywell, the contractual relationship with NASA became especially bottom-line oriented and more contentious. Again, the program had been well served by Murray's contributions.

Although much of the history of NASA's Spaceflight Tracking and Data Network focuses on the execution of the program, primarily by GSFC and its contractors, NASA Headquarters was also a key participant. Headquarters not only provided leadership within the Agency, but was also responsible for external coordination, program advocacy and—especially essential—securing needed funding. Thus, the history of the Headquarters component is inextricably intertwined in the history of the program.

NASA Headquarters normally has had only a half-dozen or so Program Offices—offices responsible for funding and executing Agency programs—with OTDA/OSC (Code T/O) being one. In 1990 the Associate Administrator for Space Operations (Code O) was designated as fifth in line of succession to the Administrator.[1] The roles and responsibilities of most Headquarters Program Offices changed over the years, but that of OTDA/OSC remained largely unchanged until 1996, organizational name changes notwithstanding.

For example, a post-*Challenger* organization review headed by Sam Phillips observed that ongoing operations were fundamentally different than occasional operations done as part of an R&D program. Phillips recommended NASA recognize differences, and manage operations in an organization experienced with continuing operational activities even while evolving in response to multiple changing requirements, technical upgrades, and continual maintenance. NASA's only organization experienced with such continual operational activity was OTDA (Code T), and in response to the Phillips' recommendation NASA renamed OTDA as the Office of Space Operations (Code O)—in preparation for moving all on-going operational activities, including STS operations, into this new organization.

However, the Shuttle program had no intention of transferring its operations to Code O—whatever its name—and Space Operations was in reality only a continuation of OTDA. Thus Code T became Code O in the mid 1980's, with essentially no change in its roles and responsibilities. It was again renamed in 1990, this time as the Office of Space Communications (OSC).

Gerald Truszynski, who followed Buckley as OTDA Associate Administrator in 1968, was equally capable and continued Buckley's philosophies for another decade. Truszynski very eloquently described the importance of Buckley's approach during an interview with Alfred Rosenthal in 1982. This interview is included herein (in Chapter 9), and the reader's attention is especially directed to it.

This structure endured and served NASA well for thirty-five years, cost-effectively providing the similar communications services necessary to the conduct of most agency missions. By understanding this fast-changing technology and the evolving industry associated with it, OSC has successfully delivered these essential services at costs which have decreased almost linearly from some 10 percent of the Agency appropriation down to about 5 percent over the first three decades.

History has repeatedly shown that societies are far more successful at adapting technologies developed for other purposes for general benefit, often military, than in envisioning these broader purposes first. For example, the Romans first paved roads for military purposes to be able to control the empire with a smaller army by being able to redeploy it rapidly. The significant boost to commerce of being able to rapidly move goods and services to distant markets over these paved roads was totally unanticipated.

Located at the intersection of the infant computing and communications technologies, NASA's space communications needs drove many communications and information handling advances—advances which have since been widely adapted, changing our daily lives extensively. NASA's leadership was not only in identifying needs and advancing necessary technologies, but also in knowing when to back off and capitalize on industry-funded advances.

A characteristic of synergistic relationships: it is seldom possible to isolate and credit contributions of individual participants. Advances are a collective product of all. As noted in Chapter 9, OSC's participation in advancing relevant technologies has served NASA and the nation very well. For example, global communications by satellite became a reality when it did partly because my predecessors chose to become anchor tenants, rather than providing a NASA infrastructure. Enabling capabilities such as the DSN 64-meter antennas and TDRSS came on-line ahead of defined requirements. TDRSS was implemented as a cost-effective alternative to expanding the ground stations to meet STS requirements, but greatly increased productivity of low-Earth orbiting science programs by enabling essentially "any-time" communications.

TDRSS was an early user of a technique for spectrum sharing, enabling simultaneous support of several spacecraft transmitting on a common frequency. This technique (Direct Sequence Spread Spectrum Code Division Multiple Access, or DSSS CDMA) now enables every teen and his or her parents in the United States to have their own cell phone. CDMA has now evolved to become essential in the U.S., and NASA's early role in fostering CDMA was substantial. (A different sharing technique is used in other parts of the world.)

The globe-spanning nature of NASA's tracking and communications activities led the program to develop relationships with counterparts internationally. For example, OSC-led delegations visited Soviet tracking stations in Eupatoria (Yevpatoria, on the shores of the Black Sea) and Ussuriysk (near Vladivostok) even before the Soviet Union collapsed. Such experiences both facilitated international understanding in themselves, and provided experience and models later adopted for broader space relationships.

Other allied responsibilities, such as spectrum management, were also vested in OTDA. Radio spectrum is a finite natural resource, spawning increasingly fierce battles over allocations as technology greatly expands utility. The International Telecommunications Union (ITU) had been established in 1865, and recognizing radio waves do not respect political boundaries expanded its charter to include allocating radio spectrum globally in 1927. The Federal Communications Commission (FCC) serves a similar role for U.S. domestic commercial uses, while the National Telecommunications and Information Administration (NTIA) coordinates U.S. government uses.

OTDA was designated NASA's spectrum management representative, routinely participating in government, commercial and international forums seeking equitable allocation of this resource. The Lewis (now Glenn) Research Center in Ohio has been NASA's center of expertise for this very specialized activity. An example of NASA's contributions: OSC played a major role in obtaining the international spectrum authorizations for commercial low-Earth orbit satellites at the World Administrative Radio Conference in Spain in 1992, invaluable to the commercial communications industry.

The world's space agencies understood that the use of standardized approaches for handling space data would be beneficial to all. A primary objective of standards is to reduce costs and enable interoperability by adopting compatible systems and procedures. Although space data formats were somewhat similar, the standards activity for space data is more complex, involving designing or adopting systems and procedures that can utilize these standardized data formats.

On its own initiative, NASA OTDA became a founding member of the Consultative Committee for Space Data Systems (CCSDS), which is now supported by more than 30 space agencies and their associated industrial bases distributed across the world space community. Acting as a technical arm of the International Organization for Standardization (ISO), CCSDS generates world standards in the field of space data and information transfer systems.

In recent years, NASA has sought to play an anchor tenant role and facilitate development of private, commercialized remote tracking and communications assets. Success has been limited, as a major commercial space industry has yet to materialize. But even in the absence of significant private revenues, costs have been reduced by sharing assets with other space agencies.

A less appreciated element of President Kennedy's challenge to send humans to the Moon and back was his rapid time scale: "before this decade is out." Not only did this schedule add an inspiring sense of urgency for the Agency, it also saved costs! Had he said "two decades," the cost of the endeavor would undoubtedly have doubled. However, this also spawned a culture of "technical excellence at any cost," an attitude extending well beyond priorities such as human flight crew safety.

Although many STDN contracts included incentives, these incentives were primarily used to reward technical performance, and occasionally, for meeting schedules. Seldom have they been used to motivate cost savings. In my view, failure to develop an effective means to balance contract performance with costs throughout the program has been a shortcoming of OTDA/OSC leadership. Adequate funding was generally obtained, and controls to prevent waste, fraud, and abuse were effective. But stewardship of public funds could have been improved by adopting lower cost approaches.

Progress towards reducing costs was steady but neither smooth nor assured, with social challenges greatly dominating technical ones. Although early on the STDN program was on the leading edge of relevant technologies, the program was often surprisingly slow keeping pace with these fast changing technologies. For example, NASA was not only one of the first customers for IBM's "big iron" mainframe computers, it was also one of the last mainframe customers as the world moved on to powerful mini-computers, workstations and desktop computers. Altruistically, program managers were hesitant to change from ways that had proven successful.

Other factors certainly came into play. Mid-level civil service managers often equated their own importance to the size of their budgets, and in turn the size of "their" support contractor staff. Some contractors were particularly effective resisting changes. As they generally held "level of effort" contracts, adopting more efficient systems and streamlining operations would reduce the size (and revenues) of their contracts. I certainly recall contractors making end-runs to influential members of NASA's appropriations committees in Congress with greatly exaggerated stories of how many voters' jobs could be adversely affected by proposed changes.

The "contractor marching army" working new mission needs also had an insidious side: these people were basically inventors, and often tended to invent new solutions rather than adapt old approaches or increasingly available commercial alternatives.

The point is that constraints on modernizing and streamlining were much more complex than simply taking advantage of advancing technology to reduce costs. Obtaining capital funding for modernization was only a minor impediment. There were two ways to view progress in this environment. On the plus side, operating costs have been continually reduced even while output metrics increased substantially. On the other hand, technology offered additional cost savings opportunities, accompanied by improved operational performance from streamlining and modernizing the systems architecture. OSC pressures to do better resulted in acrimonious relations between Headquarters and the GSFC/contractor communities.

OSC fell into a trap that plagues many confronting change—not appreciating the full dimensions of inertia and resistance. Technically educated managers naively assume the predictable behavior of the laws of physics also apply in the arena of human behavior. When egos, profits, and even jobs are threatened with change, resistance can be formidable.

In 1995, NASA sought more effective ways to operate, rejecting Buckley's model in favor of a "new" management and funding arrangement. As described in Chapter 8, OSC responded unreservedly to this call, and prepared to shift major responsibilities to JSC and a newly created office: SOMO (Space Operations Management Office). Operating costs had already been reduced by one-third over the first five years of the 1990s, even while output metrics such as the number of passes supported and data delivered had more than doubled as NASA's spaceflight programs grew rapidly. By early 1996 the OSC organization had been reduced substantially. After I retired, OSC was abolished, with management and funding fractionated across Headquarters.

Certainly money factored into this excursion—it is easy to idealistically underestimate effort needed to successfully provide "invisible" services and conclude doing better is a "slam dunk." The SOMO staff was extremely naïve as to the challenges of managing the space communications program, but little interested considering old ways. The consolidated contracting structure which became CSOC (Consolidated Space Operations Contract) was advocated directly to the Administrator by a contractor very familiar with the space communications program.

This recommendation was that by consolidating all contracts into a single large contract and giving the contractor overall responsibility, NASA could save $1.4-billion. One can only speculate as to why a contractor very aware of the dynamic nature of STDN support requirements—space missions lasting past projected lifetimes, launch delays, spacecraft degradations—would recommended an approach certain to result, as it did, in massive, *cost-increasing* change orders.

In retrospect this excursion was ill-conceived and short-lived, with SOMO and CSOC lasting barely five years, and the old structure restored within a decade. Generally, performance suffered and cost savings failed to materialize. Undoubtedly a major factor in the SOMO/CSOC fiasco was the same failure to comprehend the full nature of change that had initially plagued OSC. In October 2006 NASA restored Buckley's organization and funding structure, although reporting in at a slightly lower organizational level—a de facto reaffirmation of Buckley's wisdom a half century earlier.

An independent review of NASA's space communications program by the National Research Council in 2006[2], in explicitly addressing the SOMO/CSOC excursion, recommended future major realignments in top-level management, funding and contracting structures be preceded by a transition plan outlining objectives, ensuring past corporate knowledge is considered.

The bottom line: as I am quoted herein, I believe our nation has been well served by the STDN program. Not only have all mission support requirements been met, but advanced enabling capabilities have been there when needed. Being at the juncture of communications and computing technologies, the program has played an influential role their amazing advances over the past half-century—benefiting all. Our performance, speaking from a Headquarters perspective, has not been perfect; the primary shortcoming has been failure to appreciate social impacts of changes, and effectively counteract resistance to streamline, modernize, and reduce costs even further.

I am especially pleased to see this history documented, and congratulate both the NASA History Division and the author for an excellent, objective, and scholarly work. Well done!

Charles T. Force
February 2007

ACKNOWLEDGMENTS

Publication of this book would not have been possible without the generous contribution of many inside and outside of the space agency. I would like to first recognize the Office of Space Communications for sponsoring the work and for their interest in preserving the history and accomplishments of the Spaceflight Tracking and Data Network. The OSC's recommendations in general, and that of James Costrell in particular, provided invaluable guidance to help steer a clear path through the nearly 50 years of this history. I would like to acknowledge, along these lines, previous contributions by William Corliss, Alfred Rosenthal, Kathleen Mogan, and Frank Mintz on documenting this history. Their works served as an impetus for the current volume. Sincere appreciation also goes to the people I interviewed who provided first hand accounts of their experiences, perspectives, and reflections. They provided oral and written material that were not only substantial in nature but also unique and fascinating. These, in many ways, helped set the tone of the story. I also appreciate the team of reviewers for sharing their candid insight and knowledge, without which corrections and improvements to the original draft could not have been made. I am especially grateful to David Harris for his comments on the early years of the history and similarly to Charles Force on the later years. Working with Charlie was espeically bittersweet as he sadly passed away six months after having graciously written the foreword.

A cadre of historians, archivists, and professionals at the NASA History Division made this work possible. Colin Fries was instrumental in identifying and finding the numerous documents used from the Historical

Reference Collection, as was Jane Odom with the archival materials. Nadine Andreassen provided administrative assistance. Special thanks certainly go to Steve Garber who oversaw this project from its beginning to end. The library staff at the Goddard Space Flight Center and NASA Headquarters were also helpful in recommending reference sources and in locating otherwise hard to find reports and publications.

In the Communications Support Services Center at NASA Headquarters, Shelley Kilmer-Gaul designed and laid out the book and cover art. Meredith Yuskiw assisted with the layout as well as the re-creation of many of the images contained in this book. Stacie Dapoz performed the copyediting and proofread the layout, David Dixon handled the printing, and Gail Carter-Kane and Cindy Miller assisted in the overall process. Thanks are due to all of these fine professionals.

Thanks are also due to John Saxon who volunteered as my guide in Australia. He also provided the pages of the Honeysuckle Station Log for Apollo 11, an enlightening eyewitness documentation of the technical operation of a historical ground station on the epochal mission. I am also indebted to the hospitality of Mr. and Mrs. Mike Dinn; Glen Nagle and Candy Bailey, my hosts at CDSCC; Neal Newman who facilitated my research with CSIRO; and Kathryn Zaraza for enthusiastically providing logistics assistance. It was a delightful experience.

Credit also goes to Eleanor Ellis for transcribing the hours of interview material. Johnny Tsiao and John Hackney critiqued the manuscript from the lay readers' perspective, providing honest feedback from outside the space community. A goal of this book was to present the history in a way that would appeal to a broad range of readers. Their commentary was appreciated.

And finally, to my dear family—Lisa, Jordan, Marissa, and Andrew—who gave me their unwavering encouragement and support throughout: Thank you.

Sunny Tsiao
Monument, Colorado
March 2007

INTRODUCTION

INVISIBLE NETWORK

	Capcom:	3, 2, 1, 0.
00 00 03	Glenn:	Roger. The clock is operating. We're underway.
00 00 07	Capcom:	Hear loud and clear.
00 00 08	Glenn:	Roger. We're programming in roll okay . . . Little bumpy along about here.
00 00 16	Capcom:	Roger. Standby for 20 seconds.
00 00 19	Glenn:	Roger.
00 00 20	Capcom:	2, 1, mark.
00 00 23	Glenn:	Roger. Backup clock is started . . . Fuel 102-101, oxygen 78-100, amps 27.
00 00 39	Capcom:	Roger. Loud and clear. Flight path is good, 69.
00 00 43	Glenn:	Roger. Checks okay. Mine was 70 on your mark . . . Have some vibration area coming up here now.
00 00 52	Capcom:	Roger. Reading you loud and clear.
00 00 55	Glenn:	Roger. Coming into high Q a little bit; and a little contrail went by the window or something there.
00 01 00	Capcom:	Roger.

00 01 03	Glenn:		Fuel 102-101, oxygen 78-101, amps 24. Still okay. We're smoothing out some now, getting out of the vibration area.
00 01 16	Capcom:		Roger. You're through max Q. Your flight path is . . .
00 01 19	Glenn:		Roger. Feels good, through max Q and smoothing out real fine . . . Cabin pressure coming down by 7.0; okay; flight very smooth now . . . Sky looking very dark outside . . . Cabin pressure is holding at 6.1 okay.
00 01 46	Capcom:		Roger. Cabin pressure holding at 6.1.
00 01 49	Glenn:		Roger. Have had some oscillations, but they seem to be damping out okay now. Coming up on two minutes, and fuel is 102-101, oxygen 78-102. The g's are building to 6.
00 02 07	Capcom:		Roger. Reading you loud and clear. Flight path looked good. Pitch 25. Standby for . . .
00 02 12	Glenn:		Roger. BECO, back to 1.25 g's. The tower fired; could not see the tower go. I saw the smoke go by the window.
00 02 21	Capcom:		Roger. We confirm staging on TM.
00 02 24	Glenn:		Roger . . . Still have about 1.5 g's. Programming. Over . . . There the tower went right then. Have the tower in sight way out. Could see the tower go. Jettison tower is green.
00 02 48	Capcom:		Roger.
00 02 50	Glenn:		1.5 g's.
00 02 53	Capcom:		Roger, Seven. Still reading you loud and clear. Flight path looks good.
00 02 56	Glenn:		Roger. Auto Retro-Jettison is off; Emergency Retro-Jettison Fuse switch, off; Retro-Jettison Fuse switch, off.
00 03 03	Glenn:		UHF/DF to normal.
00 03 19	Capcom:		Flight path looks good; steering is good.
00 03 22	Glenn:		Roger. Understand everything looks good; g's starting to build again a little bit.
00 03 30	Capcom:		Roger.
00 03 32	Capcom:		Friendship Seven. Bermuda has you.
00 03 34	Glenn:		Roger. Bermuda standby . . . This is Friendship Seven. Fuel 103-101, oxygen 78-100. All voltages above 25, amps 26.
00 03 48	Capcom:		Roger. Still reading you loud and clear. Flight path is very good. Pitch -3.
00 03 53	Glenn:		Roger . . . My pitch checks a -7 on your -3.
00 04 00	Capcom:		Roger, Seven.

00 04 08	Glenn:	Friendship Seven. Fuel 103-101, oxygen 78-100, amps 25, cabin pressure holding at 5.8.
00 04 20	Capcom:	Roger. Reading you loud and clear. Seven, Cape is Go; we're standing by for you.
00 04 25	Glenn:	Roger. Cape is Go and I am Go. Capsule is in good shape. Fuel 103-102, oxygen 78-100, cabin pressure holding steady at 5.8, amps is 26. All systems are Go.
00 04 44	Capcom:	Roger. 20 seconds to SECO.
00 04 47	Glenn:	Roger . . . Indicating 6 g's . . . Say again.
00 04 53	Capcom:	Still looks good.
00 04 54	Glenn:	Roger . . . SECO, posigrades fired okay.
00 05 10	Capcom:	Roger, stand . . .
00 05 12	Glenn:	Roger. Zero-g and I feel fine. Capsule is turning around . . . Oh, that view is tremendous!
00 05 21	Capcom:	Roger. Turnaround has started.
00 05 23	Glenn:	Roger. The capsule is turning around and I can see the booster during turnaround just a couple of hundred yards behind me. It was beautiful.
00 05 30	Capcom:	Roger, Seven. You have a go, at least 7 orbits.
00 05 35	Glenn:	Roger. Understand Go for at least 7 orbits . . . This is Friendship Seven. Can see clear back; a big cloud pattern way back across towards the Cape. Beautiful sight.
00 05 54	Capcom:	Roger, still reading you loud and clear. Next transmission, Bermuda.
00 05 58	Glenn:	Roger. Understand next transmission, Bermuda . . .[1]

Bermuda, Canary Island, Kano, Zanzibar. The list goes on: Muchea, Woomera, Canton Island, Guaymas. Reading like a who's who from National Geographic, these obscure spots from around the world became astronaut John H. Glenn, Jr.'s only links back to Earth on that historic day in 1962 as he circled the globe at 27,350 kilometers per hour (17,000 miles per hour) inside *Friendship 7*. Completing an orbit every 90-minutes, ground stations that the National Aeronautics and Space Administration (NASA) had constructed at these and other places around the world anxiously tracked the progress of Glenn in his tiny, Mercury capsule. Not unlike runners in a relay, as *Friendship 7* passed from station to station, each Capsule Communicator—known as Capcom—handed him off to a waiting colleague down the line.

Even though the momentous flight of 20 February 1962 lasted only 4 hours and 55 minutes, the United States had prepared for it since before NASA even opened its doors as America's space agency some three and a half years earlier. By the time Glenn became the first American in orbit, NASA had in fact already established no less than 30 ground stations on five continents and several islands, along with ships in the Atlantic, Indian, and Pacific oceans. At the time, about half of these stations were devoted to tracking science and application satellites while the other half played a unique role in the then-burgeoning arena of human spaceflight, an arena which seven years later, would culminate with mankind's first steps on the Moon.

Connected by over two million circuit miles of land and ocean-floor cables, America's spaceflight network spanned just about every corner of the globe, from desolate volcanic atolls like Ascension Island to metropolitan, capital cities like Madrid and Canberra.[2] In the United States, stations could be found spanning the cold north of Alaska to the lush hillsides of North Carolina. Like the deep-sea diver who relies on his harness and cable to communicate with and find his way back to the mother ship, NASA's spaceflight tracking and data network was—and is—that electronic link to the satellites, spacecraft, and astronauts in space.

More than that, it is also an electronic link to the past.

Through the ages, communications have been the key to discovery. Whether by foot messengers or highspeed electronics, communications have been essential to the exploration of new frontiers. Take the 1960s, for example, where radio communication with probes preceded man to the Moon. Regardless of how sophisticated it may be, no spacecraft is of any value unless it can be tracked accurately to determine where it is and how it is performing. Only in doing so can the data it is collecting—whether pictures of celestial objects or television broadcast signals—be transmitted, received and used on the ground. This data, reduced into useful information by computers and electronics on the ground, enable the user here on Earth to analyze data from space. In the case of human spaceflight, the stakes are much higher. A failed communications link potentially compromises not just the success of a mission but also puts the lives of astronauts at greater risk.

These electronic downlinks—called telemetry—carry everything from astronauts' pulse-rate to so-called "housekeeping" data, which give an indication of the health and status of the spacecraft. Conversely, uplinks transmit commands from mission controllers, scientists and engineers up to the spacecraft. These radio frequency or RF links bridge the expanse of space, tying the spacecraft and Earth to each other. Without tracking, telemetry, command, and control, satellites would merely be inanimate objects in space and astronauts would be beyond the reach of the thousands who support their mission back on the ground.[3]

William C. Schneider, NASA's former Associate Administrator for Tracking and Data Systems from 1978 to 1980, commented on the vital role that space communications played, for example, during tense moments on Apollo. "Very few people really understand and appreciate the importance of reliable tracking and communications to all of NASA's programs, particularly when it comes to manned missions. Sitting in Mission Control in Houston during a lunar flight, there was a communications blackout when the spacecraft was behind the Moon. This also happened to be a period when the spacecraft has to fire its engines to start its descent to the lunar surface. Not having communications during this critical phase made us all feel very nervous and anxious, and the relief, when the capsule was able to confirm that all was well, was almost indescribable."[4]

In the early days of space exploration, prominent engineers, scientists, and astronauts were certainly at the forefront of public attention and acclaim. These professionals in a way became the new Magellans and the new Columbuses, the new breed of modern-day explorers and map makers. Instead of crossing oceans at the mercy of the wind, astronauts now traveled the vast expanse of space in craft designed by engineers. Whether a mission was the flight of a communications satellite or the exciting journey of an astronaut, *all* were supported by a unique team of men and women on virtually every continent who operated an intricate system of ground stations, computer facilities, and communication centers. These, tied together, made up NASA's Spaceflight Tracking and Data Network (STDN).

John T. Mengel, one of the "founding fathers" of America's spaceflight tracking networks, once commented on this behind-the-scenes criticality:

> True, tracking and data acquisition were support functions, but this support, going back to the early days of the space effort, was critical to success in the competition-charged arena of the sixties and seventies, when the ability of the United States to succeed in the field of space travel and space research was being questioned. We proved it could be done.[5]

To operate a network, teamwork and cooperation were essential. This was true not just for those working at a station on some remote island but also on the much larger, agency-to-agency level. For instance, over the years, cooperation among the triad of NASA, the Department of Defense (DOD) and the Department of State, has been crucial. To NASA, the Air Force, with its launch facilities and worldwide network of radar installations, was a huge asset. Help from the Navy was important too, providing tracking and recovery ships in all three major oceans. However, like most big projects where different organizations have to depend on each other to get the job done, the NASA/DOD cooperative was not without its share of problems. One example was that the space agency's desire to work with the Air Force was tempered sometimes by the fear that the DOD might try to "elbow in" on the fledgling

Edmond C. Buckley was NASA's first Associate Administrator for Tracking and Data Acquisition, leading the new Office of Tracking and Data Acquisition (OTDA) at Headquarters from 1961 to 1968. In this role, Buckley dealt directly with Congress, the DOD and the State Department to oversee the space agency's tracking needs. Prior to becoming the Associate Administrator, Buckley was NASA's Assistant Director for Space Flight Operations. (Rensselaer Polytechnic Institute photograph, copy in Folder 10/1/1 NASA Australian Operations Office, Yarralumla, ACT)

Agency's new, and in many ways, more glamorous programs. Congress, too, was sensitive to the NASA/DOD relationship. Questions about duplication of facilities, for instance, often came up during budget discussions. Influential members on the Hill played the role of watchdogs to make sure that NASA remained true to its civilian charter. Despite constant budget and political battles, Congress, nevertheless, generally understood that NASA and other government offices simply had to work closely together in order for America to succeed in space.

To maintain communications with an orbiting spacecraft, a tracking network has to be global in nature. Therefore, international cooperation is not only important but absolutely necessary in getting stations established at optimal locations around the world. Whether dealing with strong allies like Australia and Great Britain or venturing deep into Africa for the first time, the role of the State Department was indispensable as a facilitator to help the space agency intermediate discussions at the highest levels of government. With this liaison, officials could more effectively manage NASA operations in the face of geopolitical unrest around the world. For example, during the pioneering flights of Project Mercury, the Guaymas tracking station in Mexico often had to be surrounded by troops to protect it against unruly mobs espousing anti-U.S. sentiment.[6]

The different persuasions of different governments sometimes require delicate diplomacy on the part of the United States. In particular, the ability to cater to the sensitivities of diverse cultures is important. NASA has found that the single, best way to accomplish this is to invite foreign nationals to join in the operation of the tracking network, providing them with a sense

The 85-foot tracking antenna which received signals of mankind's first steps on the Moon on 24 July, 1969 still operates today at the Canberra Deep Space Communication Complex in Tidbinbilla, Australia. The antenna was moved from its original Honeysuckle Creek site some 30 kilometers (20 miles) away in 1985. Clearly visible are the two large yokes which allow the antenna to be moved in two "X-Y" axes as it tracks a spacecraft. (Photograph by the author)

of ownership to become partners in the space effort. This "nationalizing of the stations" from the early days of the network would leave an indelible legacy around the world that continues even today.

Such international cooperation could in fact make or break a station. Putting a site in Havana, Cuba in the late 1950s presented a volatile situation. Conversely, other places such as Australia—where as many as 10 sites were active during the 1960s—enthusiastically embraced the opportunity to participate in this new frontier, adopting the American space program as their own. Australia's Parkes Observatory was selected by NASA (with great national fanfare), to help receive the Apollo 11 moonwalk telemetry. Indeed, the Canberra Deep Space Communication Complex (CDSCC) on the outskirts of the Australian capital remains active to this day as part of NASA's DSN. In fact, the 85-foot antenna at nearby Honeysuckle Creek (HSK) that actually received video of Neil Armstrong and Buzz Aldrin's first steps on the Moon is still used at the CDSCC in Tidbinbilla, having been moved from

HSK when it closed in 1984. The years have not changed this. International cooperation is just as vital today as it was then, as the space agency routinely works with the Europeans, Russians and Japanese on a diverse range of programs including the International Space Station, Earth science research, and planning of future space communication needs.

As much as the story of NASA's spaceflight tracking networks is about the stations and the technologies, it ultimately boils down to the individuals, the workers in the trenches who made it happen. From Administrators, to Station Directors, to the teams of contractors, all did their jobs because they believed in it and put their hearts into it. Even though space is often associated with "high-tech," it is still the people involved who were movers of the program, who ran the day-to-day operations of the facilities, and who left behind their legacy. This emphasis on people especially characterized the early years of the Agency's tracking networks. There were no notebook computers in 1957 when the first satellites were launched. In fact, there were no digital computers of any kind. Instead of clicking on a "mouse" to retrieve data, technicians would interrogate lines drawn on graph paper from mechanical strip chart recorders. Whether one was the head of a station or just feeding teletype printouts to the engineers, everyone had a job to do.

While those working with the communication networks of today certainly get their share of excitement, those who were around for the early human flights and satellite launches fondly remember the "glory days." As one former official at the Goddard Space Flight Center (GSFC) recalled:

> I didn't geographically know the world then in the detail I do today. Not only was it educational and exciting just to learn the names of all the remote tracking stations, but there was also the excitement of talking to the people at all the different tracking stations and ships.[7]

Former Flight Director Christopher C. Kraft—who wrote the Foreword to Hamish Lindsay's book *Tracking Apollo to the Moon* on the Australians' perspective of working on the Apollo Network—recalled fondly:

> I have a very soft spot in my heart for the network and the people who operated it. These people were as much a part of the success of our efforts as were the flight controllers and the other people at Houston and the Cape. Whatever we did was in large measure dependent on the reliability of the worldwide tracking and communications network. . . . This is truly one of the unsung accomplishments of the space program.[8]

As the network expanded in the 1960s and early 1970s, so did the team of people who ran it. It was during this time that contractors from the aerospace industry, like Bendix Field Engineering Corporation (BFEC), made an indelible mark on the history of the spaceflight network. So close was the relationship that station workers were often called the "badgeless controller" because it didn't matter if they were government civil servants or contractor employees. The two could work side by side and one would not know which was which unless you looked at his badge.[9] For many, working the networks became a way of life. For some, entire careers were spent working in remote places around the world.

An examination of the history of America's spaceflight tracking network is not unlike driving down a long and winding road—with many turns and twists. Much of its early fate was tied to the Cold War and the Space Race borne from it. In the mid-1950s, even before NASA was formally established, the United States began building a network to track what it hoped was to be the world's first artificial satellite. However, that network—called Minitrack—ended up as its first test tracking Sputnik 1, which the former Soviet Union shocked the world with on 4 October 1957.

Minitrack next set the stage for a greatly expanded network called the Satellite Tracking and Data Acquisition Network, or STADAN. Led by the new Goddard Space Flight Center (GSFC) in Greenbelt, Maryland, this much more expansive network soon became the centerpiece of all near-Earth tracking and space communications activity for the United States. The STADAN supported not only NASA's own science and application satellites but also the world's first commercial satellites which were just then being launched, including the COMSAT Early Bird, the world's first electronically active, commercial, communications satellite.[10]

However, even as the sophistication of satellites was rapidly evolving and launches were taking place on a more regular basis, something even more dramatic was about to change the very fabric of the Goddard network. This something was the world of *manned* or *human* spaceflight.

To support this new national priority, engineers at Goddard, along with the Manned Spacecraft Center in Houston, Texas, developed a network that ultimately tracked astronauts to the Moon and back. Initially called the Mercury Space Flight Network, or MSFN (the word "Mercury" would later be replaced with the word "Manned"), it quickly became just as expansive as the STADAN, even surpassing the latter in terms of capability, locations and new technology. Much of its early requirements were based on experience gained from aircraft and missile testing and from the tracking of lunar and planetary probes such as the unmanned Surveyor spacecraft which preceded man to the Moon. From this, cadres of network planners worked out solutions to the high data rate requirements needed to send humans into space.

As both the STADAN and the MSFN matured, the ever increasing demand for data in the right amount (bandwidth), at sufficient speed (data rate) and at the right time (timing accuracy) drove their evolution. Real time operations at ever-increasing data rate became the prime consideration. "To know accurately where the spacecraft was at any given moment was extremely important to us," said Kraft. "A review of the early history, particularly of the Mercury flights, reveals that we often had to make some very delicate decisions whether to continue some of these flights or not. Determining the orbital trajectory and providing tracking and communications support to meet these parameters from launch to reentry was certainly no easy job for the Goddard team."[11]

To this end, Kraft recalled a specific problem that the network had to deal with involving the critical "Go/No Go" decision as whether to continue a flight or initiate an abort. During the launch of an Atlas rocket—an Air Force Intercontinental Ballistic Missile which NASA used to launch the Mercury astronauts and the Agency's early, large, satellites—there was only a 30- to 120-second window after its main engine had cutoff to decide whether to continue or to abort. This was all the time controllers (or the astronaut) had to turn the spacecraft around and fire its retrorockets so that its trajectory would allow the spacecraft to reenter in the Atlantic recovery area before its impact point reached the African coast. Kraft recalled:

> Initially, there were very few people who believed that this would be possible.... We called it the 'short arc' solution. People like John Mayer, Bill Tindall, Lynwood Dunseith and Goddard's Jim Donegan deserve credit for giving us the know-how to solve this critical problem, a problem which many mathematical and trajectory experts believed could not be solved. By analyzing the data generated by the Cape, Goddard and the Bermuda tracking station, we were able to get the answers, even within the very limited timeframe available.[12]

During NASA's first man-in-space program Project Mercury, the network was not well centralized and relied on the not always reliable teletype communications. For mission assurance, flight controllers were dispatched to most of the primary tracking stations so that they could maintain immediate contact with the spacecraft from the ground on a given pass. Astronauts too were sent as capsule communicators or Capcoms to the various sites. Here, something interesting happened within the NASA culture. For unmanned science and application satellites, there was no debate as to who was in charge. Goddard was clearly the lead and the STADAN was centralized at the Network Operations Control Center (NOCC) in Greenbelt. However, things were not so clear-cut when it came to the "manned" side of the house. Here, both Goddard and Houston's Manned Spacecraft Center (MSC) had a stake.

Despite the fact that there was not always full agreement even within Goddard's own top management (who thought the center should be devoted to scientific exploration of space and questioned whether it should be in the business of tracking human missions), Headquarters in due time delegated Goddard with the full responsibility for running the MSFN. This decision by Headquarters was fairly logical as MSC (renamed the Lyndon B. Johnson Space Center, or JSC, in 1973) certainly had enough on its plate as NASA's lead center for human spaceflight to be worried about also having to run a worldwide tracking network. Nevertheless, the decision did not eliminate Houston's concerns about the network; the "Not Invented Here" mindset was not something that easily went away.

As history has shown, JSC did not have to worry about the Goddard-run network, as not a single mission—manned or unmanned—has ever been compromised due to network failure. This "invisible network" as

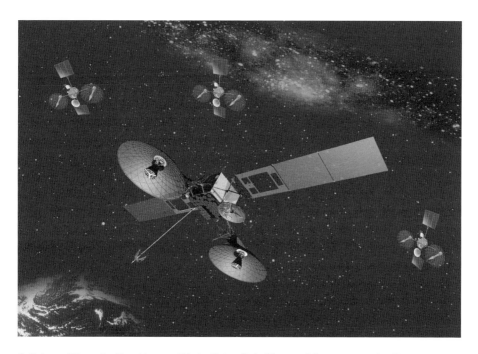

Artist rendition of a Tracking and Data Relay Satellite providing communications support from Earth orbit. The long and winding road has taken NASA's Spaceflight Tracking And Data Network from a handful of ground stations in the late 1950s to the Space Network of the twenty-first century. (NASA Lithograph LG-2001-8-033-GSFC)

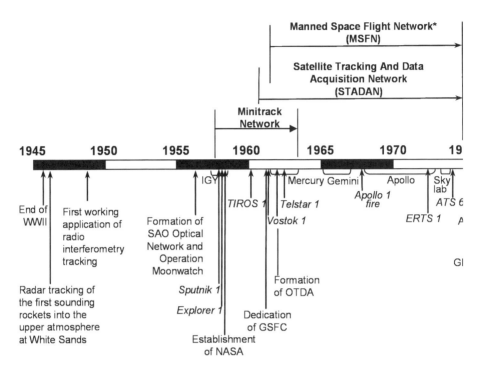

A chronology of NASA's near-Earth Spaceflight Tracking and Data Network. (Figure by the author)

it has been called, operating behind the scenes, has always been there. Like a light switch, it's there when needed. In the final analysis, much of the success on what could have been a very divisive issue was due in no small part to some very effective personal relationships between people who understood what was needed to get the job done at both centers, people like Ozzie Covington, Bill Wood, Tecwyn Roberts, and Chris Kraft.

Following the incredible journeys of Apollo to the Moon and America's first space station Skylab, as well as the world's first international spaceflight in the Apollo-Soyuz Test Project, human spaceflight became suddenly very quiet in the late 1970s. Although the GSFC was busy supporting Earth science and application satellites, in the cost-conscious days following Apollo, the wisdom for having a separate network specifically for human spaceflight was questioned. It was in this era of change that Goddard merged its two networks (the STADAN and the MSFN) in 1975 into a single, leaner, and ostensibly more cost-effective network called the Spaceflight Tracking and Data Network, or STDN.

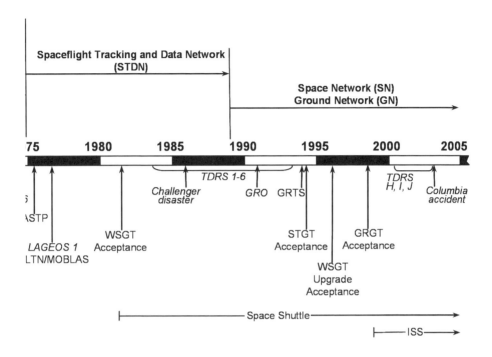

But no matter how large or small a ground network is, due to "line-of-sight" limitations imposed by the curvature of Earth, it is still located on the surface of Earth and can thus only provide coverage for some 15 percent of an Earth-orbiting spacecraft's ground track—limited to brief periods when it is within the line-of-sight over a given tracking station. To overcome this, NASA's network for tracking and communicating with near-Earth spacecraft (that is, the STDN but not the DSN) had to move into space. Thus, the Tracking and Data Relay Satellite System, or TDRSS, was born.

With the advent of TDRSS in the 1980s, America's spaceflight tracking and communications network evolved into a full-fledged Space Network (SN) constellation of relay satellites and ground terminals along with a greatly reduced Ground Network (GN). Today, many Earth-orbiting spacecraft rely primarily on the SN for their tracking, telemetry and control needs with many missions that do not require continuous contact with the ground opting, mainly for cost reasons, to transmit their spacecraft science data via the GN. Digital and telecommunications technology have also evolved significantly in recent years to impact space communications. The ever increasing demand for higher bandwidths (traffic capability) and lower "bit-error rates" (higher accuracy) in the modern information age explosion have led to the complete transformation of how space communications are done. Such is the trend in spaceflight communications that started in the 1990s in which NASA historically set the precedence but is now heavily influenced by the commercial sector. In this new age of information technology and international teamwork, the Agency's role is no longer the same as it was when the Space Shuttle first flew into space in 1981 or even when the International Space Station was first being built in the 1990s.

Aside from geopolitics, the root cause driving such change can almost always be traced to two factors: technology and cost reduction. Few would be surprised that the demand for better technology is always a driver. But in addition to better technology, as space moves from the realm of government sponsorship to a commercial commodity, cost reduction in today's world of real-time global communications and data demands is perhaps more important than ever before.

In the 1990's, "Faster, Better, Cheaper" also drove much of the way business was conducted in the space industry, both commercially and at NASA. This approach impacted the space program in ways ranging from economics to performance and, some would say, safety. New ground and space communication networks such as Universal Space Network (USN) entered the playing field. This network provided multi-mission ground terminals which offered users the advantage of low cost services: pay only for what you use. They targeted not only commercial users of satellite services but also government users, including NASA. As we broach the twenty-first century, satellite users now routinely rely on the internet for data access and file transfers. This access extends to such remote locations as the South Pole, where NASA was instrumental in providing internet access to researchers.

When one speaks of space exploration, "high technology" usually comes to mind. To this end, the aphorism "If NASA can put a man on the Moon . . ." has become a part of the English vernacular, perhaps even to the point of being somewhat trite. Satellite and spacecraft—some human, some robotic—sent on their missions into space, are there to make the unknown known. But whatever data that astronauts and instruments record, the information is of no value if it cannot be returned to Earth. At its height, over 6,000 men and women—from network engineers to field technicians to NASA managers—were committed to this vast undertaking. As former Station Director and later Associate Administrator for Space Communications (1989 to 1996) Charles T. Force says of this invisible network, "The nation was well served."[13]

This is the story of that invisible network.

CHAPTER 1

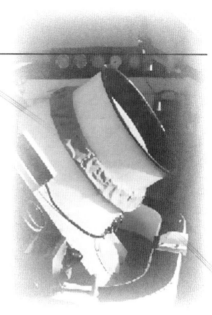

THE EARLY YEARS

Tracking, in the context of spaceflight, refers to the collection of spacecraft position and velocity measurements so that its motion on orbit may be determined. The roots of modern day tracking and data acquisition can be found in the immediate years following the Second World War when the United States entered into a period of intense research to develop a viable ballistic missile technology. Science, and in particular, the development of sounding rockets (small rockets launched into the upper part of the atmosphere for research and experiment), played a part in this advancement. However, its pace was clearly driven by national security.

World War II had just forced the United States and the Soviet Union to become reluctant allies. With the war now over, the potential for an armed conflict between the two superpowers was something that both governments took careful steps to avoid while at the same time, and in no uncertain terms, prepared for. Even in the last days of the war, both countries postured to shape the political landscape that would soon emerge after Nazi Germany and the Imperial Japanese governments had been defeated. It was in this Cold War atmosphere that a major concern arose in the American military leadership, one that fueled the perception that the United States not only trailed, but trailed badly, the Soviet Union in the area of ballistic missile technology.[1] Much of this concern could be traced to that fact that much larger missiles were then needed by the Soviets to carry their nuclear warheads which were, at the time, bigger and much heavier than their American counterparts.[2] Also capitalizing on this state of apprehension was the very effective use of propaganda by the Soviets, as the usually secretive society openly and routinely paraded these large rockets in Red Square during times of state celebrations.

As the pace of missile testing accelerated, the Department of Defense (DOD) converted several former War Department weapon proving grounds in the continental United States into missile ranges. Key among these was the Whites Sands Missile Range (WSMR) in southern New Mexico. Here, captured German V-2 rockets were brought and flown after the war. In a move called Operation Paperclip, German rocket scientists, under the direction of Wernher von Braun, brought over 300 boxcar loads of missile materials to White Sands where they were tested as sounding rockets equipped with atmospheric sampling devices and telemetry transmitters.[3] It was in New Mexico where America's first suborbital rockets were launched above Earth's atmosphere in 1946 when the WAC Corporal ventured some 80 kilometers (50 miles) up, almost into the ionosphere.[4]

Other pioneering work at the Army range included the "Albert monkey flights" and the Navy's Aerobee sounding rockets.[5] In 1950, the Air Force converted the Long Range Proving Ground at Cape Canaveral, Florida into the Atlantic Missile Range, which it later renamed the Eastern Test Range (ETR). With a corridor that extended all the way from the Florida coast, over the Caribbean to Ascension Island in the South Atlantic, ETR was perfect for testing the early long-range intercontinental ballistic missiles (ICBM) in the Air Force arsenal such as the venerable Atlas. Located on the coast of Florida, easterly launches out of "The Cape" also enjoyed the enormous benefit of having the velocity from Earth's rotation "kick start" a rocket's journey into space. On the other coast, the Navy established the Pacific Missile Range (PMR), tracking missiles launched out of Point Mugu, California and later, Air Force launches from Vandenberg Air Force Base, on intercontinental trajectories to Kwajalein in the Marshall Islands some 10,000 kilometers (6,200 miles) away in the South Pacific.

In those days, there were essentially two proven methods for tracking vehicles into the upper atmosphere: optical and radar. *Optical tracking* has been around since humans first studied the stars and sailed the seas. As the name implies, visual sightings from ground-based, high-power telescopes were used to provide measurements of a satellite's position against the background stars. In April 1956, the Working Group on Tracking and Computation (WGTC), as part of the National Academy of Sciences' Space Science Board, approved a plan by the Smithsonian Astrophysical Observatory (SAO) in

Cambridge, Massachusetts, to establish an optical tracking network to photograph very small objects in anticipation of tracking the International Geophysical Year (IGY) satellites. The IGY was an 18-month (July 1957 through December 1958) scientific undertaking sponsored by the International Council of Scientific Unions designed to promote and stimulate a broad, worldwide investigation of Earth and the near-Earth cosmic environment. At the time, it was quite the watershed event in terms of fostering scientific interest on an international basis when East-West tensions were dominating the news.

To prepare for U.S. IGY activities, the SAO was given $3,380,000 by the National Academy of Sciences (NAS) to implement an optical network. This network, proposed by Harvard astronomer Fred L. Whipple and Ohio State University professor J. Allen Hynek, featured sophisticated Baker-Nunn high precision telescopic cameras that had unusually large apertures. These high-resolution cameras were named after its principal creators, James G. Baker, consultant to the Perkin-Elmer Corporation of Norwalk, Connecticut, and Joseph Nunn of South Pasadena, California. Built by the Boller and Chivens Company of South Pasadena and instrumented with optics fabricated by Perkin-Elmer, the Baker-Nunn was basically an f/1 Schmidt camera with a 20-inch (50.8-centimeter) focal length and a wide 5° by 30° field-of-view so as to accommodate star-field photography for the purposes of optical tracking. (A Schmidt camera is a high-powered telescope with a film plate holder instead of eyepiece at its focus. It is typically used as a survey instrument in which a large amount of sky is covered.) Another key component of the camera was its film. Manufactured by the Kodak Corporation, film for the Baker-Nunn used high speed Royal X-Pan emulsion, standard for high grain black and white resolution at the time. The 55.625-mm film size was, however, unique to the needs of the SAO. To achieve the required resolution, a single frame was rather large, measuring 2.2 inches by 18 inches (5.6 centimeters by 45.7 centimeters). It required close to an hour to process and dry the film in preparation for making position measurements.[6]

The camera operated by alternately tracking a satellite and the star background. Superimposed on the same strip of film was the image of a crystal-controlled clock from a separate optical system which was periodically illuminated by strobe lights. This was done to establish a precise time reference. From the photographs, the position of a satellite could be accurately determined by comparison with the position of the background star field.[7] The instrument was sensitive enough; under favorable lighting conditions, it could photograph a 16th magnitude object, which corresponds to something that is 25 times dimmer than the faintest star visible to the naked eye.[8]

By the late 1950s, the Smithsonian's optical network was global in scope and was concentrated in a geographical band of about 40° north and south of the Equator. There were 12 stations located at sites selected, among other things, for their year-round mild weather:

>North America
> Haleakala, Maui
> Jupiter, Florida
> Organ, New Mexico
>South America
> Arequipa, Peru
> Curacao, Dutch Antilles
> Villa Dolores, Argentina
>Australia
> Woomera
>Africa
> Olifantsfontein, South Africa
>Europe
> San Fernando, Spain
>Asia
> Mitaka, Japan
> Naini Tal, India
> Shiraz, Iran[9]

    Although optical tracking could be effective, it had its drawbacks. In particular, its usefulness was limited by something known as the acquisition problem. Simply put, in optical tracking, there was no way to initially find what it was supposed to track. Before the optical network could be put to use, the SAO had to come up with a plan to address the acquisition problem. Here, Whipple had an idea. To encourage participation and popular interest in this new field, Whipple's plan was to use a network of amateur volunteers worldwide to literally serve as eyes to visually find—or acquire—a satellite. The strategy was rather simple. After surveying the night sky and finding the satellite, these participants would pass the "acquisition data" on to one of the 12 Baker-Nunn stations that was closest to them.

    Not surprising, the intention to use amateurs in this fashion was initially ill-received by the NAS Space Science Board's technical working group who thought the work "too technical" for just anyone to perform. But just as Whipple had anticipated, the announcement of the formation of Operation Moonwatch in early 1956 brought an enormous response. Visual observation teams sprang up in North and South America, Africa, Europe, Asia, in the Middle East and even at such remote places as Station C and Fletcher's Ice Island T-3 in the Arctic Basin. In the Washington, DC area, the first Moonwatch station equipped with 12 telescopes was set up in an

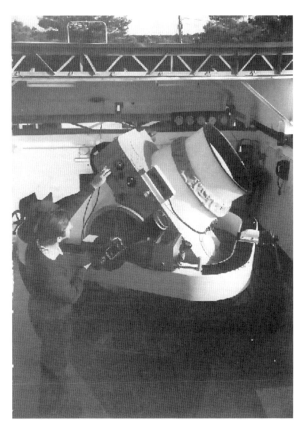

An optical satellite tracking network using Baker-Nunn cameras established by the Smithsonian Astrophysical Observatory was in place by the late 1950s. The camera would alternately track a satellite and the star background to obtain a fix on its location. Baker-Nunn cameras were also used extensively by the Air Force to track military satellites. (*http://history.nasa.gov/SP-4202/p9-151.jpg*)

apple orchard in Colesville, Maryland. The orchard was owned by a Mr. G. R. Wright, a Weather Bureau employee who ended up chairing the national advisory committee in charge of planning the unusual network.

Dry runs to test the ability of spotters to actually sight a tiny sphere hurling through space began in June of 1957. An imitation of the proposed IGY satellite flew over Fort Belvoir, Virginia. This test satellite was definitively "lo-tech," a makeshift device made up of a tiny flashlight bulb fastened to a plumber's suction plunger that was towed at the end of a clothesline behind an airplane flying at 1800 meters (6,000 feet). Light output from this object was calculated by Whipple to be about the same as that reflected off a metal sphere on orbit from a rising or setting Sun. The contraption worked though, and the experiment went off without a hitch. Soon, variations of the technique were used to train Moonwatchers from around the world. Two years later, some 250 teams with approximately 8,000 members were functioning under Moonwatch. Teams came from all walks of life, organized by universities, public schools, government agencies, private science clubs, and amateur astronomers. The United States alone accounted for 126 groups.[10]

The endurance of Operation Moonwatch was a testament to its true popularity, as more than 5,000 volunteers equipped with telescopes and binoculars continued scrutinizing the skies over the next two decades (long after the original need had been met). More than 400,000 satellite observations were recorded. Whipple would praise the infectious enthusiasm of these early space buffs, saying, "Quite a number have gone into science or space programs because of their amateur involvement in Moonwatch."[11] After the National Aeronautics and Space Administration was formed in 1958, it would continue to fund Moonwatch at a low level through the SAO—at about $30,000 per year—until the organization was officially disbanded in July of 1975.

Despite its relatively good accuracy, optical tracking had inherent disadvantages in addition to acquisition. For instance, in order to get good photographs, the weather must be favorable and lighting must be correct. The latter condition meant that a site must be in darkness while the satellite being tracked was in sunlight. The problem with that though, is a satellite spends a significant portion of its orbit either in Earth's shadow where it is not illuminated, or in sunshine where there is not enough contrast to see it (like trying to see stars during the day). Thus optical tracking could only be used during a short window just before and after sunrise or sunset. Therefore, from any given station, it was usually impossible to get more than a few good observations of the orbit per pass.[12]

Besides optical systems, there was *radar*, an electronic system. Radar—or Radio Detection and Ranging—in which the location of an object is precisely determined by measuring returns on electromagnetic waves reflected off the object back to the transmitting station, had been a topic of research by scientists and engineers since the late 1800s when German physicist Heinrich Hertz first began experimenting with radio waves in his Frankfurt laboratory. Research continued throughout the first half of the twentieth century on both sides of the Atlantic. But it was not until World War II when the allies first used it to great advantage for early warning—in particular during the Battle of Britain—did radar come into its own. By the end of the war, it was standard operating equipment on Navy ships and aircraft on both sides.

Towards the end of the war, missile testing activities soared, and along with it, the need for reliable tracking techniques. In the fall of 1944, Major General G. M. Barnes, then Chief of Research and Development Service, U.S. Army Office of Ordnance, recognized the urgency for the Army to establish a research program in the field of guided missiles. To this end, he was convinced that as a vital part of such a program, it was necessary to have somewhere within the limits of the continental United States, a range where test firings of such missiles could be conducted routinely and safely. With this

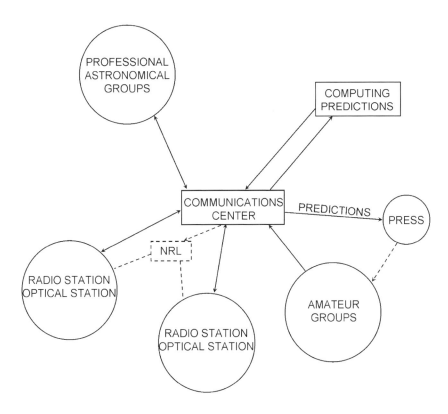

Diagram of Operation Moonwatch showing the vital role of amateur star gazers in this schematic from an original set of NRL plans. (Constance McLaughlin Green and Milton Lomask, *Vanguard: A History*. [NASA SP-4202, 1970])

in mind, military and civilian engineers from the Ordnance Department and the Corps of Engineers conducted surveys of all open areas in the United States to find a place suitable for such firing activities.

Certain fundamental requirements were set forth, such as: extraordinarily clear weather throughout the preponderance of the year; large amounts of open, uninhabited terrain over which firings could be conducted without jeopardy to civilian population; accessibility to rail and power facilities; and to whatever degree possible, proximity to communities to provide for the cultural needs of personnel to be employed at such an installation. The result of these studies indicated that a relatively desolate area around the White Sands Desert and the Alamogordo Bombing Range in southern New Mexico nearly filled all of the specifications.[13] While it was recognized that both the length and width of the range were not as large as might be desired, it could be utilized efficiently for the early types of missiles that were being developed and for the reduced-scale prototypes of the later and larger types that the Army then had in mind. In this way, White Sands quickly became the hub of activity for U.S. missile development, and with it, tracking technology.[14]

Throughout the late 1940s and early 1950s, radar was increasingly used to supplement optical tracking. This work, at a time when the accepted practice was that optics were the primary way for data recovery, was not received with open arms by all. In spite of this initial reservation, it did have the momentum of technical advancement on its side.[15] It was at White Sands during this time that a young engineer named Ozro M. Covington began developing the concept of centralized networks for tracking and communications, honing skills which would later be used to lead NASA's human spaceflight network. After working in the field of communications and radars as a civilian in Europe and at Fort Monmouth, New Jersey during World War II, Ozzie Covington became the Technical Director of the Army Signal Missile Support Agency at White Sands in 1946 (a position he would hold until leaving for NASA in 1961.) Here, Covington and range engineers tested and refined networking communications and data processing techniques and developed what was, in essence, the forerunner to the mission control concept.

The idea of a worldwide communication complex to support spaceflight was not a new concept to many at White Sands. By the time Covington came onto the scene, it had already been the subject of study and a number of possible approaches had been considered.

> We supported the firing of V-2 rockets brought to the U.S. from Germany after World War II and, of course, our own U.S. developmental missiles, [said Covington] to monitor these flights, which could reach 100 miles (160 kilometers) in altitude and some 100 miles downrange. We established a chain of five radar tracking stations. These were linked to our 'C' Station, which also contained early computer capabilities to give us a real-time data system. We really had an early mission control center from which we not only monitored these flights but also sent guidance signals to the missiles. While we developed our electronic capabilities, we also used the real-time data to guide an array of optical instruments to keep an eye on these firings. After a V-2 went south to Juarez, Mexico, instead of north, these radars and their associated display equipment became the primary source of data for the range safety officer.[16]

Among these first networks was the very successful tracking system developed at the range using an elaborate chain of FPS-16 C-band, single-object tracking radars made by the Radio Corporation of America (RCA) to track missiles launched on extended horizontal flights. The units were networked to transmit data to a control center where consoles displayed information on the radar returns for the test engineers. Operating either on commercial line power or generators, these transportable radars could be moved from one place to another or to undeveloped sites using prefabricated transportable pads so as to provide support from different locations all around the range. A modernized version of this radar tracking system continues to be used at White Sands today.[17]

Still, a third approach for tracking emerged in the late 1940s: *radio interferometry*. Like most electronic tracking techniques then in use, radio interferometry required the presence of a signal source, or transmitter beacon, on the object being tracked. Despite its somewhat intimidating name, the basic operating principle of radio interferometry is relatively straight forward. Since radio signals—like sound—travel in waves, separate antennas will receive the waves at slightly different times. Knowing that electromagnetic waves propagate at the speed of light, measurements of the different arrival times of the waves, called phase differences, can then be used to calculate position solutions for an object. This technique had the advantage of yielding highly accurate tracking angles and could be used under virtually any atmospheric condition, even underwater. Given its potential, a number of groups were involved in its research after World War II. A significant milestone was reached in 1948 when engineers of Consolidated Vultee Aircraft Corporation (Convair) created the first working application for the U.S. Army on the Azusa missile tracking system.

To understand how this technique was utilized and eventually developed into what would become the Minitrack Network—the progenitor to the STDN—one has to go back to the year 1955. That summer, plans were drawn up by the United States to launch the first artificial satellite as President Dwight D. Eisenhower announced on 29 July that the United States would participate in the upcoming IGY by orbiting a 9-kilogram (20-pound) satellite. In keeping with the spirit of the IGY, Eisenhower announced that the American satellite program would be civilian in nature. Despite the announcement, it was well understood by the science community that assistance from the military was going to be required, since at the time, it alone possessed the capability to launch a rocket large enough to orbit a satellite.

Following this announcement, a Committee on Special Capabilities was appointed by the DOD to select the appropriate organization to lead this effort. This committee would end up selecting a proposal submitted by the Naval Research Laboratory (NRL) as the one which seemed best qualified to promote the civilian nature of the project while at the same time support military interests in space technology development. The NRL was a good choice. It was well qualified, having already conducted research in this field for nearly a decade, even developing an underwater tracking system using sound interferometry. By the early 1950s, the NRL detachment at White Sands had built and fielded a rudimentary tracking system for the development of the Viking short range guided missile.

In this way, the effort selected by the committee, called Project Vanguard, thus became the United States' official IGY entrant in September 1955. The project was led by Milton W. Rosen, Chairman of the NAS Advisory Committee on the IGY, and was managed jointly by the Office of Naval Research and the NAS Committee on the IGY.[18]

Once Vanguard was given the go-ahead, a significant portion of the early work was devoted to solving the problem of just how to track such a small object in space, since the Vanguard satellite would only be 15 centimeters (6 inches) in diameter. Most atmospheric research scientists were advising Rosen's team to rely on optical tracking; it was, after all, the tried and proven method. In addition, camera observations of meteoroids entering Earth's atmosphere had demonstrated that modern terrestrial optical instruments could indeed spot small objects weighing only a few pounds that moved at high velocities in the upper atmosphere.

Rosen was well aware of the limitations of optical tracking—in particular, the acquisition problem—and wanted something better. In other words, optical instruments could see a satellite, but could they find it? To get a better handle on the problem, he had Richard Tousey at the NRL verify visibility computations that had been performed earlier. Tousey's answer confirmed what Rosen had already suspected. In his opinion, the probability of successful optical acquisition of a Vanguard-sized satellite on the first visual pass was only 1×10^{-6}, that is, literally one in a million.[19]

Thus, convinced from the beginning that his team must look elsewhere for a solution, Rosen asked John T. Mengel and his NRL Tracking and Guidance Branch to develop an electronic system for use in conjunction with an optical one. For ballistic missile guidance, Mengel had led the development of the mentioned Azusa system at White Sands. For tracking a satellite, however, that system was out of the question as it required an onboard transmitter weighing 13.6 kilograms (30 pounds), far too large for a small satellite. Mengel's team immediately went to work and within a year came up with an interferometer system that, although based on the techniques developed for Viking, required only a 0.37-kilogram (13-ounce) transmitter using a different operating frequency and antenna pattern. It was, in essence, a pioneering new system.[20]

Hired by nuclear submarine pioneer Hyman G. Rickover a year after the Japanese attack on Pearl Harbor, Jack Mengel, who later became the first Director for Tracking and Data Systems at the Goddard Space Flight Center, began his career for the NRL working on antisubmarine devices. "At first, I was responsible for telemetry and radio control systems in the Rocket

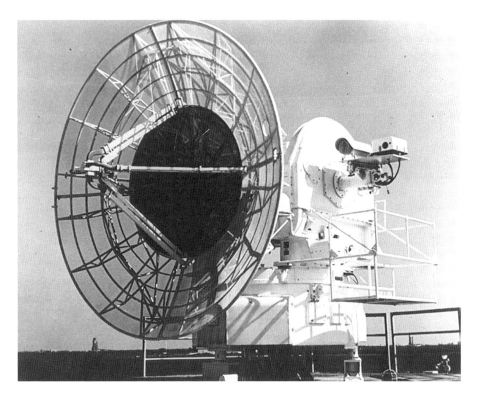

The RCA FPS-16 radar has been widely used at missile ranges since the 1950s. It is a transportable, Single-Object Tracking Radar which operates in the C-band of the electromagnetic spectrum. The photograph shows a FPS-16 at Patrick Air Force Base used in 1961 to track launches out of Cape Canaveral. (Photo courtesy of Patrick Air Force Base Office of History)

[Sound] Branch, which used V-2 rockets of World War II fame for scientific observations in the near-Earth regions. This was followed in 1950 by work on a Viking missile project and its possible use in submarine warfare."[21] Here, his work in developing an X-band interferometry system that operated at the centimeter (one inch) wavelength provided the high precision missile guidance that the U.S. military was looking for.

Expounding on this initial work, he and his team then converted this system into a 108 megahertz (MHz) system for tracking satellites. This new radio interferometry satellite tracking system had the catchy name Minitrack, which Mengel derived from "minimum-sized tracking system." The name was fitting, as tracking was accomplished using a small oscillator onboard the satellite that illuminated pairs of antennas at a ground station with which angular positions of the satellite was derived using the radio interferometry technique. "What apparently had sold the Defense Department on our proposal was the fact that it consisted of a good radio system which would be infallible in picking up the satellite in orbit," explained Mengel.[22] This pioneering tracking network was to become operational in 1957 at a cost of only $13 million.[23]

Despite the low price tag, America's first spacecraft tracking network had to overcome several key technical hurdles in order to achieve a level of tracking accuracy never before attempted. Angle measurements and accurate timing were vital. One had to know precisely at any given instant in time where a satellite was in orbit. From the very beginning, the ability to ascertain precise timing, in particular, was crucial. Chesley H. Looney, Jr., the former Head of the Time Measurement Branch for Minitrack at the GSFC said:

> In the early days of the space program, prior to the establishment of NASA, we had to develop a reliable, low-frequency phase-measurement system to track our Vanguard satellites. While a high frequency interferometer system had been used at White Sands Missile Range, New Mexico, to track military rockets, this system was inadequate to meet the needs of low-powered satellites then being planned. Thus, a new system at a different frequency and with a different antenna layout to provide the proper precision and signal resolution had to be designed . . . In those days, our budgets were very tight and we were forced to use low cost hardware. Yet, the equipment had to be capable of phase-angle measurements with an accuracy not attempted before. In the process, we developed an elegantly simple, filter-inductor system. In our related time measurements, we achieved an accuracy of 10-milliseconds per day and broke new ground in phase-angle tracking later embodied in commercial designs.[24]

As eventually implemented, Minitrack used a 10-milliwatt, 370-gram quartz crystal controlled oscillator transponder aboard the satellite which operated at the fixed frequency of 108 MHz. It had a battery life of 10 to 14 days, sufficient for a planned Vanguard mission. The new tracking network was considered quite revolutionary for its time and was thus subject to much scrutiny by the general scientific community. In a series of papers and speeches over the next two years, Mengel and his colleagues—notably Roger L. Easton, his assistant at NRL, and Paul Herget, Director of the Cincinnati Observatory and a consultant on Project Vanguard—responded to this scrutiny by presenting the difficulty of tracking and its importance in the U.S. satellite program.

In a March 1956 paper, Mengel explained the challenge of tracking an artificial satellite to fellow scientists and engineers. In it, he said:

The final realization of man's efforts to place a satellite in orbit about the Earth will immediately pose a new series of problems: How to determine the precise orbit that it is following and how to measure what is happening within the satellite from the vantage point of a ground station. The immensity of the first of these programs, how to prove that the satellite is in fact orbiting—the acquisition phase—can be realized by an analogy. Let a jet plane pass overhead at 60,000 feet (18,000 meters) at the speed of sound, let the pilot eject a golf ball, and now let the plane vanish. The apparent size and speed of this golf ball will approximate closely the size and speed of a satellite three feet in diameter at a height of 3,000 miles (4,800 kilometers). The acquisition problem is to locate the object under these conditions, and the tracking problem is to measure its angular position and angular rate with sufficient accuracy to alert nonacquiring tracking stations, those trying to follow the satellite by optical means, as to the time and position of expecting passage of the object.[25]

In describing how Minitrack worked, Mengel and Herget likened the antennas of a station to human ears. In short, it used stereo-phonics to pinpoint the location of a given signal source. For instance, a person locates the source of a sound by virtue of the phase difference in the sound waves which arrive at slightly different times at each of his ears. The antennas of a Minitrack station functioned much like ears. The electronic receivers of the Minitrack system were connected to pairs of receiving antennas set at a known distance apart, which indicated the direction of the signal by phase differences in the radio wave that it received from the satellite. The direction to the satellite, measured relative to the baseline vector between the two antennas, could then be calculated from the phase difference using triangulation.

For example, if the waves arriving at two antennas were out of phase by one-third of a wavelength, the extra distance they traveled to the farther antenna could be deduced and the angle to the baseline calculated using trigonometry. However, in order to find the actual length (that is, magnitude) of the extra distance, it must be known whether this phase difference was only one-third of a wavelength or some multiple thereof. To resolve this ambiguity, other pairs of antennas were set up but spaced closer together at distances of less than one wavelength. These arrays provided shorter fractional phase differences, and on the basis of whether the resulting directions fell within the reception pattern of the antenna system, the number of wavelengths involved could be determined.

A Minitrack antenna array was quite large, consisting of a 150-meter (500-foot) cross of eight linear dipole antennas (metal frame antennas with many elements like the traditional rooftop, television antenna) mounted above 18-by 3-meter (60-foot by 10-foot) ground screens which served to eliminate multipathing, or interference, from ground reflection. In the north-south direction, three pairs of antennas spaced 150, 20 and 4 meters (500, 64 and 12 feet) apart, respectively, were used; in the east-west direction, two pairs with spacings of 150 and 20 meters were used. The resulting reception pattern was a fan-shaped beam spanning 100° north-south by 10° east-west, measured at the 6 decibel (dB) reception power point, a common demarcation used to measure antenna reception strength.[26]

In this setup, the north-south and east-west determinations provided the two angles needed to locate the actual direction to a satellite. However, more information was still needed. To verify precise antenna alignment (that is, its orientation with respect to a fixed celestial coordinate system) and timing, calibrations had to be done. This was done periodically by having an airplane equipped with a Minitrack transmitter and flashing lights fly over the station at night at an altitude of 6,100 meters (20,000 feet). An optical system called MOTS—Minitrack Optical Tracking System—was used for this purpose.

MOTS was one of the lesser known tracking systems implemented early in the network. Developed under the direction of Edmund J. Habib at the GSFC, stations used it to calibrate and determine position misalignments in the Mintrack system. It worked something like this: A time-coded lamp on the belly of the airplane, operating simultaneously with an onboard 108-MHz Minitrack transmitter, allowed for precise observation of the airplane's position against the background star-field as the Minitrack array on the ground recorded the movement of the Minitrack beam. In this way, a precise comparison between the optical position (aircraft belly light) and the radio frequency position (aircraft transmitter) could be made for calibration purposes.[27] Precise optical and radio fixes could then be made to correct for any misalignment of the antenna. Using this setup, Minitrack was sophisticated enough to record phase differences to an accuracy of one-thousandth of a wavelength and the time of observation to one millisecond, which in turn produced a direction fix to the satellite accurate to within 20 arc-seconds (1/180th of a degree). Such was the kind of resolution that was needed in order to track Vanguard.[28]

As implemented, the Minitrack stations were all very similar. Their principal antennas consisted of the large arrays for angular tracking, one fixed antenna array for telemetry reception, and a single rhombus (diamond shaped) antenna for communications. Three trailers housed the ground station core electronics consisting of telemetry recording equipment and communications hardware. Site selection had to consider many factors. Due to the size of the main array, a station required a large plot of land, at least 23 acres in an area that was relatively smooth, with a gradient of less than 1°. To maximize the observation time during each pass, Minitrack specifications called for the terrain adjacent to a station to not exceed an elevation angle of 10° for at least 800 meters (one-half mile), and 20° for 8,000 meters (5 miles). Finally, to minimize interference, engineers selected locations away from electric power plants and even airports (although this recommendation was not followed at many of the sites eventually constructed).[29]

The final decision by the IGY selection committee on the U.S. satellite project came down to two proposals: the Navy's Project Vanguard and a competing concept called Project Orbiter submitted by Army scientists—including Wernher von Braun. The latter, however, originally contained no provision for electronic tracking. While it may be too simple to say that the NRL proposal was selected because it had provisions for electronic tracking, it is clear that the inclusion of a Minitrack Network in the NRL proposal undoubtedly had a direct bearing on the committee's decision in August 1955. Mengel later supported this claim, saying, "What apparently sold the Defense Department on our proposal was the fact that it consisted of a good radio system which would be infallible in picking up the satellite in orbit."[30]

In addition, NRL also described in their proposal a somewhat less elaborate version of the Minitrack system as eventually implemented. Developed by Roger Easton, this toned-down system became known as the "Mark II," or unofficially, the "Jiffy." Some in the halls of NRL referred to it as the "Poor Man's Minitrack." Using less-sophisticated radio frequency phase-comparison (instead of interference), the Mark II, nevertheless, went on to become the nucleus of Project Moonbeam, a program sponsored by NRL to encourage amateur radio enthusiasts and their clubs to build their own ground stations to track satellites.[31]

With its selection accomplished, design and construction-preparation of the Minitrack Network began in the fall of 1955. From its inception, Minitrack was a project that had to have cooperation amongst all the stakeholders. Considering all the technical and administrative requirements that were needed to coordinate the work of various military units, university laboratories, individual experts and private contractors, its progress was remarkably smooth. Years later Mengel was able to recall "a few personality clashes" but in his opinion, those were "par for the course for a program that made use of some of the best astronomers in the country."[32]

During initial development and construction, the U.S. military played the decisive role. Three Army agencies—Corps of Engineers, Map Service and the Signal Corps—were responsible for most of the actual construction. The Air Force's main role was the installation of tracking radars at Patrick Air Force Base in Florida and on Grand Bahamas Island. Overall program management was under the Naval Research Laboratory. In addition, the Navy's Bureau of Yards and Docks obtained use of the land needed to construct the prototype station at Blossom Point, Maryland. Outside of the military, the Department of

Minitrack antenna arrays were large, spread over an area one and a half times the size of a football field. (Constance McLaughlin Green and Milton Lomask, *Vanguard: A History*. [NASA SP-4202, 1970 file 6589 NASA Historical Reference Collection, NASA History Division, Washington, DC 20546])

State provided the much-needed assistance to the Navy as the executing agent for foreign affairs, conducting negotiations for land leases on foreign soil. This cooperation between the various departments of the U.S. government was a testament to the priority of Project Vanguard. There was with it a definite sense of urgency. Chesley Looney recalled years later that, "While our budget was low, . . . [we] had high priority which allowed us to push ahead without administration hurdles. All of us were anxious to succeed."[33]

Things moved quickly. In the spring of 1956, a project advance team led by Commander Wilfred E. Berg, surveyed 17 Latin American sites from which six were selected. They were:

> Batista Field in Havana, Cuba
> Paramo de Cotopaxi near Quito, Ecuador
> Pampa de Ancon at Lima, Peru
> Salar del Carmen at Antofagasta, Chile
> Peldehue Military Reservation near Santiago, Chile
> Rio Hata in the Republic of Panama.

One of the lesser-known tracking systems developed by GFSC was the MOTS—Minitrack Optical Tracking System—which was used to obtain precise, position calibration and misalignments in the Minitrack system. The photograph shows David W. Harris (who later became a Deputy Associate Administrator at NASA Headquarters) operating the camera at Blossom Point in 1963 after it was modified to photograph passive satellites such as Echo. Note the photographic plate at the bottom of the camera. MOTS was used until the 1970s. (NASA Image Number G-63-826)

Shortly after Berg's group returned in early May, NRL eliminated the Panamanian site based on studies indicating that a station at San Diego would be more useful for early orbit determination. This decision would make the station on the coast of California the first to receive confirmation from Vanguard that it indeed had successfully attained orbit.[34]

As the network began to take shape, there emerged basically three categories of stations. The first type involved three tracking units located immediately downrange of Cape Canaveral on the islands of Grand Bahamas, Antigua, and Grand Turk, networked with a radar installation at Patrick Air Force Base, Florida. These monitored the launch, ascent, and orbital insertion

The antenna system of a Minitrack station was made up of eight individual antennas connected to form three pairs in the north-south direction and two pairs in the east-west. The outermost pairs in each direction were 152 meters (500 feet) apart while the inner north-sourth pairs were separated by distances of 19.5 and 3.7 meters (64 and 12 feet). The spacing of the inner east-west pair was 19.5 meters. (Mengel and Herget, Tracking Satellites by Radio. Folder 8800, NASA Historical Reference Collection, NASA History Division, NASA Headquarters, Washington DC)

("early ops") of the Vanguard vehicle. In the second category were the so-called primary stations located roughly on a north-south line along the east coast of North America and the west coast of South America so as to form a "picket line" across the expected path of satellites. To preclude over-flight of the Bahamas, launches out of the Cape were limited to orbital inclinations of 35° or less (in other words, the ground track of satellites were bound between 35° in latitude north and south of the Equator). With this picket line, radio signals from a satellite could be intercepted as it crossed the 75th West meridian on each orbit. Seven primary stations formed this line:

> Blossom Point, Maryland
> Fort Stewart, Georgia
> Antofagasta, Chile
> Santiago, Chile
> Quito, Ecuador
> Lima, Peru
> Havana, Cuba[35]

Lastly, there were other stations outside of the picket line in locations around the world at San Diego, California; Hartebeestpoort outside of Johannesburg, South Africa; and Woomera, South Australia. This last group of stations was needed to help determine the initial orbit of the satellite. With this arrangement, Mengel estimated that there was a 90 percent chance of capturing every pass of a satellite which was at an altitude of 500 kilometers (300 miles) or higher.

The first station established was on a 23 acre NRL managed site in the U.S. Army Blossom Point Proving Ground, Maryland, some 60 kilometers (40 miles) south of Washington, DC. The Army had specifically permitted this portion of the range to the Navy in 1956 for use as a Vanguard support site. Bordering the Potomac River, Blossom Point provided an ideal setting for a ground station, with horizon-to-horizon look angles along with an interference-free, low-noise environment. As the network developed, this prototype station functioned somewhat like a test station for the rest of the network. Here, engineers, scientists, and station operators received their training, and procedures were developed for system tests, simulations, and checkout, including use of the MOTS cameras.

When it was first conceived, NRL engineers envisioned using only four sites, along with the test-bed at Blossom Point. By the time Vanguard was ready to issue its first full progress report in December of 1955, however, Mengel's team was thinking in far more elaborate terms. By the time it was completed, the Minitrack Network would consist of 14 sites, including the DOD launch support radar stations. With the obvious need to place ground stations in other countries, the ability to secure foreign cooperation became paramount. In fact, the ability to work effectively with local governments and foreign nationals would play directly into how well stations could accomplish their missions in the years to come. This NASA hallmark began in the days of Minitrack. As B. Harry McKeehan, Chief of International Operations at the GSFC from 1963 to 1980, explained:

> Very early in our planning for the tracking network, there emerged a definite commitment that our foreign tracking stations were to be conceived carefully and operated with the full support and active cooperation of the foreign governments involved. Thus, even in times of political turmoil, we were usually able to continue our operations without serious interference. . . . [In] instances when stations had to be closed because of political conditions within the country, even in those cases, we continued to recognize and honor our commitments to the host government and our local employees.[36]

McKeehan soon found that one of the best ways to secure international cooperation was to involve local nationals in station operations. "A major stabilizing factor in our foreign operations was our policy of maximum utilization of local people in station positions. This proved to be cost effective as well as giving the local government and population a feeling of having a stake in the NASA missions." McKeehan explained how this "nationalizing of the stations" was done.

> Our efforts normally started after the technical requirements for a specific program had been identified and the tracking coverage determined. Then, we explored with our colleagues in NASA Headquarters and the appropriate officials of the Department of State, the country to be approached for a 'tracking partnership'. We considered the country's geographic features, accessibility, political stability and available logistic support. Also, we looked into possible sources for local employees to reduce the number of U.S. personnel. Then, we would work through the U.S. embassy, establishing contacts with representatives of the foreign governments and explaining our needs to their leaders. This approach, together with the worldwide excitement of the U.S. space effort, opened doors and usually assured us of the support needed.[37]

To meet technical standards, NASA provided training and other educational programs for local employees which ultimately ended up also broadening the scientific base in those countries. "With concern for the sensitivities of other countries, we learned to manage responsibly the resources and manpower needed to assure our success in exploring space. It helped create a great reservoir of goodwill throughout the world for the U.S. space effort and . . . proved to be a valuable future investment," said McKeehan.[38] Even today, as the space agency continues to promote international goodwill—albeit in a very different global environment—the lasting effect of this policy can still be felt, the roots of which date back to this time.

As site construction finished, all three military branches also contributed to their operation and maintenance. Army personnel operated the five prime Minitrack stations in Latin America plus the one at Fort Stewart, Georgia. Meanwhile, the Naval Electronic Laboratory operated the station at San Diego, Blossom Point, and the tracking units on Grand Bahama Island, Antigua, and Grand Turk. The Air Force operated their radar installations at Patrick and Grand Bahama. In Australia, the method of operation was somewhat different. There, the Weapons Research Establishment (WRE), an agency of that commonwealth's Department of Supply (DOS), constructed and operated the station just outside of the village of Woomera. Under agreement with the Australian government, the DOS supplied the land, power, facilities, and workers. In return, the United States furnished all the technical equipment, trained the WRE personnel, and installed the initial equipment. Half way around the world, an agreement was reached with the South African government allowing for the National Telecommunications Research Center to staff the station just outside of Johannesburg with South African nationals.

Finally, the NRL contracted with Bendix Radio in Towson, Maryland, a Division of Bendix Aviation Corporation, to build the station electronics. Being the equipment supplier for the DOD, Bendix was a logical choice to provide this service. (After NASA was formed and operation of Minitrack was turned over by the DOD to the new space agency, Bendix would quickly assume operations of the stations in addition to just providing the hardware.)

One of the early challenges of Minitrack was the need for rapid data processing in order to compute and determine a satellite's orbit. To this end, NRL began working to implement a solution very early on in the network design process. As early as September 1955, Paul Herget met with Cuthbert C. Hurd, then Director of Electronics Data Processing Machines, and other officials of International Business Machines (IBM), to discuss the computational requirements of the Earth satellite program. In March of the following year, the Office of Naval Research issued a request for proposals for the renting of computer facilities and the furnishing of mathematical and programming services for Project Vanguard. IBM and two other companies responded with IBM winning the contract at a bid of $900,000.[39]

Under the contract, IBM was to supply personnel to operate, around-the-clock, its 704 computer for six weeks after installation and checkout, and then to provide, free of charge, orbital computations over the life of the first three successful satellites or lifetime of the Minitrack Network—whichever occurred first. In July 1956, NRL's new Vanguard IBM Computing Center opened in downtown Washington, DC. John P. Hagen, Director of Project Vanguard, appointed a committee led by Joseph W. Siry, Head of the Theory and Analysis Branch at NRL, to oversee the orbital computation work. Joining Siry on the committee were Gerald M. Clemence and R. L. Duncome of the Naval Observatory, and Paul Herget from the University of Cincinnati.[40] Since this was a new center, Siry had IBM network the facility via a transceiver to a backup site located at the company's Research Computing Center in Poughkeepsie, New York. This ensured computational redundancy and reliability that the NRL felt it needed.[41]

Network operations could be summarized as follows: Within minutes after a satellite pass, observations from a station were transmitted by teletype to the Vanguard Control Center. After being (manually) inspected for errors, the data was recorded onto reel-to-reel magnetic tapes where they were taken by courier to the Computing Center. There it was transferred onto punched cards and fed into an IBM-704 analog computer (better known at the time as an "electronic calculator"). The machine was programmed with mathematical calibration formulas and correction factors for each station to account for certain systematic errors such as atmospheric conditions and refraction of the radio signal as it passed through the ionosphere. Distances and directions from the given ground station to the satellite were then computed.

As observations on various points along the trajectory were collected, the calibration formulas were updated and the satellite path calculated (a process known as differential correction and orbit determination). Adding higher order corrections to account for parameters such as atmospheric drag and nonuniformity of the Earth gravitational model—for example its nonspherical "pear shape"—and variations in geographical mass concentrations, the IBM-704 was able to compute (and predict) a satellite's position and velocity 150 times faster than the actual progress of the vehicle. This was certainly no small feat in an era before the digital computer. In this manner, the orbit of the satellite was steadily refined as it made more and more revolutions around Earth.[42]

By September 1957, Minitrack was in place. To the disbelief of its builders, however, the network would begin its operations by tracking not an American satellite but a Russian one. Just three days after completing a checkout the network, on 4 October 1957, the Soviet Union shocked the world by announcing the launch of Sputnik 1 (the name meant "satellite" or "fellow traveler"). Once they were sure that their 45-kilogram, beach ball-sized sphere had indeed made it into orbit, the Soviet news agency Tass announced from Moscow that radio signals from the satellite could be heard by receivers set at 20 and 40 MHz. Amateur radio operators from all over the world scrambled to pick up the steady beep of the transmission.

The unexpected Soviet announcement disrupted Minitrack preparations as all the equipment had been set up to operate at the planned American satellite frequency of 108 MHz. In the words of one space pioneer, the United States was "caught with its antennas down."[43] Stations had to quickly convert their Minitrack receivers to operate on the new frequency and to put as many stations as possible into operation before Sputnik's transmitter gave out. Unlike today where receiver frequencies can be changed by the mere turn of a knob, entire racks of equipment had to be changed out. Instructions sent by teletype and equipment flown in by the Army enabled stations at Blossom Point, San Diego, Antofagasta and Lima (and later at Santiago and Woomera) to convert to operating at 40 MHz within a few days.

Although the Central Intelligence Agency (CIA) had been fully aware that the USSR. was planning to put something into space, the actual launch nevertheless still caught American space planners off guard. Chesley Looney illustrated just how much so:

> When the Soviet Union launched its Sputnik in 1957, we were totally demoralized initially. This lasted for about 24 hours. Then we pulled up our socks and quickly put together a 40 megahertz system and three boxes of antenna gear which we flew to Lima, Peru, and to Antofagasta and Santiago, Chile, and quickly started to track this new bird. Unfortunately, it was not ours, but while we were not happy, we gained some valuable experience for the first Vanguard vehicle tests expected to follow in December 1957.[44]

Before its batteries died some 14 days later, the Naval Research Laboratory was able to collect some very useful tracking information on Sputnik 1 (and later Sputnik 2). Despite the obvious American disappointment, the whole experience gained tracking the first two Soviet satellites ended up being very useful for Minitrack, in essence providing a "real world" opportunity to check out communication lines, time references, clocks and data transmission accuracy. The United States had chosen a frequency of 108 MHz because it would give a more accurate indication of direction than the lower frequencies. For example, a radio signal transmitted on a carrier frequency of 40 MHz is bent approximately seven times more sharply than one at 108 MHz; for 20 MHz, it is 29 times. Therefore, a scientific fallout from tracking Sputnik was the U.S. realizing a large amount of data from which new knowledge about the electrified layer of the ionosphere and its effect on radio wave propagation was gained.

Two months after the stunning blow of Sputnik 1, the United States was ready to attempt orbiting its own satellite. On 6 December 1957, the Navy's Vanguard rocket, with a tiny 1.5-kilogram (3.3-pound), grapefruit-sized satellite inside the nosecone, rose a few feet off its launch pad at Cape Canaveral, and then to the dismay of those watching, promptly crashed back down on itself erupting into a huge ball of fire. Unlike the Soviet Union's delayed announcement of Sputnik, the launch attempt of Vanguard was seen on American television. The psychological impact of this spectacular failure could not be understated. Just one month earlier, the Soviets had succeeded in orbiting a dog on the 450-kilogram (990-pound) Sputnik 2. Five days after the failed attempt, President Eisenhower, who up until then had been publicly endorsing the Navy's Vanguard as the official U.S. satellite program, reluctantly (and quietly) granted the Jet Propulsion Laboratory under the direction of the Army the approval to attempt a satellite launch.

Three months later on 31 January 1958, the Army came through. Explore 1 was launched into orbit atop a Jupiter C ICBM to become America's first artificial satellite. In addition to a 108 MHz transmitter (which operated for a remarkable 113 days), the 68-kilogram (31-pound) satellite had onboard an experiment devised by Professor James A. Van Allen of the University of Iowa to detect charged particles in Earth's magnetic field. In time, telemetry downlinked from the first four Explorer satellites led to the discovery of the Van Allen Radiation Belts, hailed as the single most important discovery of the IGY.

Seven weeks later on 17 March, the Navy—on its third attempt—finally followed Explorer 1 into space with the launch of Vanguard 1. Although its success was delayed, the satellite and its all-important accompanying tracking network were to pay great dividend almost immediately. Utilizing the precise tracking capability afforded by Minitrack, telemetry provided by this tiny sphere, which when analyzed, showed Earth's geode to be slightly pear-shaped rather than a perfect sphere—a fact verified some 10 years later in beautiful color photographs taken by Apollo astronauts on their way to the Moon.

* * *

Spurred on by the sudden urgency brought about by Sputnik, President Eisenhower in early 1958 revisited an initiative that up until then had not exactly occupied high priority on his administration's agenda. That is, a national governing agency whose charter was to develop, direct and coordinate all American space-related activities.

NASA, as it turned out, was not to be a completely new creation, but was instead, transformed from a predecessor organization, the National Advisory Committee for Aeronautics, or NACA. Founded in 1915, NACA was a civilian U.S. government agency whose original charter was to advise on and coordinate research that was being conducted by other organizations in the (then) new field of aeronautics. However, it quickly established itself as the leading research organization in the field, pushing back the envelope of flight, from supersonic flight in 1947 to high altitude experimental aviation that would set the foundation for eventual human flight into space.

After Sputnik—throughout the remainder of 1957 and into 1958—the status and progress of existing U.S. space activities was reexamined by Congress, Eisenhower, and the military. The nature, scope, and organization of long-range space goals were debated. The principal issue at hand was not *whether* there should be an organized space program but rather *the extent of military involvement*. To settle the issue, Congress conducted several hearings by the Military Preparedness Subcommittee of the Senate Committee on Armed Services from November to January. Eisenhower himself advocated civilian control of existing space related activities but showed little progress in resolving the broader, long range issues. However, Congress soon forced the administration's hand.

In January 1958, it introduced a bill authored by New Mexico Senator Clinton D. Anderson on behalf of the Joint Committee on Atomic Energy. The bill amended the Atomic Energy Act by giving the Atomic Energy Commission a major share of the nation's space program. Not surprisingly, the Administration decidedly opposed this bill. But it was put in a position where it had to propose an alternative. On 4 February 1958, President Eisenhower announced that he was assigning James R. Killian, Jr., the responsibility of developing a definitive solution. (Killian, who was at the time President of the Massachusetts Institute of Technology, took a leave of absence between November 1957 and July 1959 to serve as the first presidential science advisor.) Following the Sputnik crisis, Eisenhower established the President's Science Advisory Committee (PSAC) of which he appointed Killian chairman.

A month later, on 5 March 1958, President Eisenhower received the answer he had wanted, approving the recommendations of the PSAC that the "leadership of the civil space effort be lodged in a strengthened and redesignated National Advisory Committee for Aeronautics and that legislation be enacted to give NACA the authority and flexibility" to carry out its expanded responsibilities.[45] Politics aside, the recommendation was not unexpected. NACA had begun posturing itself in January of that year by proposing an interagency space program made up of leadership from its own organization, the DOD, the National Science Foundation (NSF) and the NAS. Meanwhile,

NACA's leaders—in particular Hugh L. Dryden—had close ties to many on the President's advisory committee, including Killian himself and World War II aviation hero, General James Doolittle.

Killian called the President's 5 March endorsement the conclusive act with respect to building a civilian space agency upon the NACA structure. Several factors favored into this decision: 1) NACA was already an established research Agency with a large technical staff and sizable research facilities; 2) NACA had been progressively moving into space research; 3) If NACA would not be allowed to move further into the space field, its whole future would be in jeopardy; 4) NACA had a long history of close cooperation with the DOD; and 5) NACA's liabilities could be overcome.

This last point was important as the committee specifically identified what these liabilities were: 1) NACA did not have an across-the-board space competence nor did it have much experience in the administration of large-scale developmental contracts; 2) Most of the nation's space work had been done by or for the DOD, and NACA would have had to tap this competence without impairing the military space program; 3) NACA was not in a position to push ahead with the immediate demonstration projects which may be necessary to protect the nation's world prestige; 4) NACA was limited by the somewhat inflexible hiring and pay provisions of civil service regulations; and 5) NACA's organization and procedures were geared to a much lower level of expenditure than would be the case after its expansion. Nevertheless, the advisory committee concluded that these liabilities could be overcome—or at least be mitigated—by enacting appropriate legislation.

Towards this end, four specific recommendations were proposed. First, NACA should be renamed the National Aeronautics and Space Administration. Second, it should be permitted to establish pay rates in excess of those of the Classification Act of 1949. Third, the Agency Head should be appointed by the President with the advice and consent of the Senate.[46] Finally, the composition of the 17-member NACA governing committee should be changed.[47]

By far the most important point to come out of the 5 March 1958 Presidential memorandum was the clear emphasis on the civilian nature that the new U.S. space agency should have. Language to the effect permeated the memorandum, such as "entail increased expenditures and the employment of important numbers of scientists, engineers, and technicians," and that "an aggressive space program will produce important civilian gains in general scientific knowledge and the protection of the international prestige of the United States," and straight to the point that the "long-term organization for the federal space program . . . should be under civilian control."[48]

In making the recommendation, the PSAC cited several other options which were considered but in the end did not have the advantages that transforming NACA would have. For example, the DOD was not recommended because its deep involvement in the ballistic missile program conflicted with the desired civilian scientific emphasis. The Atomic Energy Commission was not recommended because its charter was not the space program. Finally, putting the civil space program under a proposed Department of Science and Technology was not recommended due to the delays and the process that would have been required with creating a brand new government department, something that the Administration did not want to have to go through.

On 2 April, one month after President Eisenhower approved the plan, the White House set the process in motion. A draft legislation establishing NASA was sent to Congress by the Bureau of the Budget, and a White House directive was issued to NACA and the DOD instructing them to take certain preparatory actions pending congressional approval on the draft legislation. This draft to Congress was accompanied by a personal message from Eisenhower to congressional members urging them to enact appropriate legislation in order to promptly establish a national space program which he deemed was essential to the general welfare and security of the nation. He added that the space program should be given high priority and, of course, be soundly organized.

In summarizing its civilian nature, the President said: "I recommend that aeronautical and space science activities sponsored by the United States be conducted under the direction of a civilian Agency, except for those projects primarily associated with military requirements. . . . The civilian Agency should be a new one and include aeronautical activities as well."[49] In anticipation of congressional approval and to pave the way for the new Agency, the White House initiated a series of preparatory actions under the supervision of Dr. Killian and the Bureau of the Budget.

In May 1958, it instructed NACA to prepare a full explanation of the proposed legislation for presentation at congressional hearings. NACA was to also make plans to reorient its programs. Internal organizations and management structure were changed to carry out the functions to be assigned to the newly formed National Aeronautics and Space Administration, and as such, to lay the groundwork for whatever expansion might be necessary in order to implement the legislation when enacted. Along these lines, the White House directed the DOD and NACA to jointly review current Defense space programs to determine which ones should be transferred to the new Agency, and of those transferred, which would require continued military support. Finally, Killian wanted NACA officials to meet with the NAS, the NSF and the country's academic community so as to ascertain how one might best secure their participation at large.

By the time the draft legislation reached Congress, the ground work had been done. It was received with enthusiasm on both sides of the aisle. On 6 February 1958, the Senate created the Special Committee on Space and Astronautics. In a statement confirming its priority, Lyndon B. Johnson, the Senate Majority Leader from Texas, was named chairman. The House mirrored the move when on 5 March, it created a Select Committee on Astronautics and Space Exploration, naming the Speaker of the House John W. McCormack from Massachusetts as the chairman. Hearings got underway in April and three months later on 16 July, Congress passed the National Aeronautics and Space Act of 1958 (Public Law 85D568) that NACA would become NASA after 90 days, unless the transition was proclaimed sooner by the new NASA administrator. Thus, less than six months after signing off on the proposal to Congress, the piece of legislation establishing the National Aeronautics and Space Administration as a legal United States government organization returned to Eisenhower's desk where it was signed on 29 July 1958.

Thus in one move, NACA and its 8,000 employees were absorbed intact into the new space agency. Its three research Centers (Ames Aeronautical Laboratory, Langley Aeronautical Laboratory and Lewis Flight Propulsion Laboratory), along with the Army's Jet Propulsion Laboratory, formed the space agency's first Field Centers.[50] NACA's meager $100 million annual operating budget would soon pale in draconian fashion to the funding that the new Agency would begin receiving as America committed itself to the space race.

On 1 October 1958, just three days shy of the one year anniversary of Sputnik, NASA officially opened it doors as the nation's space agency.

CHAPTER 2

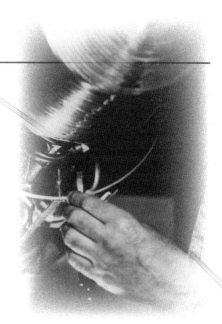

EVOLUTION OF A NETWORK

The formation of NASA had an immediate, galvanizing effect on United States space activities. One major effect was bringing together organizations and programs that up until then had been functioning as separate entities. Consolidated along with the NACA were: Earth science satellite and lunar probe programs from the DOD's Advanced Research Projects Agency; the Army's JPL in Pasadena, California;[1] the Army Ballistic Missile Agency's Development Operations Division in Huntsville, Alabama including the important Saturn launch vehicle project; the IGY Vanguard and Explorer satellite programs, and with it, the Minitrack Network.[2] Within weeks, Minitrack shifted from military control to NASA management. As part of this reorganization, contracts to industry were expanded to allow contractors to begin assuming the day-to-day network operations that had previously been performed by the military.

One such contractor was Bendix. Murray T. Weingarten, President and Chairman of the Board for Bendix Field Engineering Corporation (BFEC) from 1973 to 1989, would recall that in this arrangement,

The contractor forces were able to respond quickly, yet under control, to the many needs of NASA, whether they involved manning a far-off tracking station or chartering and flying a Douglas DC-3 aircraft to calibrate these tracking stations. . . . The word went out, if you had a problem, whether it involved calibrating a tracking station or a need for unique equipment, call 'Hunter 6-7700', which was Bendix's local phone number in Maryland. It was a fast and cost effective means to operate, a fact confirmed many years later through numerous congressional oversight hearings.[3]

NASA depended on a large contingent of contractors to operate its tracking networks. This government/contractor "marriage" was a crucial ingredient as to why its tracking and data network has been as successful as it has been. Weingarten would attribute this long-term success to an infectious, esprit de corps of those who made the network happen.

Whatever needed to be done, we did in an efficient and cost effective manner. The spirit was there. We were there to assist NASA in helping the United States become the first nation to land a man on the Moon. In those early days, NASA had many talented, strong, dedicated people who were decision makers gathered from other government agencies. We in industry became part of that team and were able to participate in the planning and the execution of those plans. The program had top priority and the team concept prevailed. Of course, there were problems, but we did not allow them to fester. They were discussed openly and resolved on a timely basis and the program moved forward. The program was a pioneering effort, and new program techniques, such as award fee concepts, were developed, all helping to move the program forward effectively and efficiently. This enabled us to break new ground, establish new policies, practices and procedures to have effective operations in all parts of the world, including such exotic places as Kano, Nigeria; Zanzibar, and Madagascar.[4]

To see how this "can do" attitude shaped the network, one needs to go back and examine how things began to change soon after NASA was established.

Soon after the Agency's formation in the fall of 1958, many of the people directly responsible for the Minitrack Network transferred from the Office of Naval Research and the White Sands Missile Range to what would become the Goddard Space Flight Center (GSFC). Even before NASA officially opened, Congress authorized in August of 1958, $3.75 million for a

"Space Projects Center" to be located near the Capital. On 1 August, Senator J. Glenn Beall of Maryland announced that the new Center would be located in Greenbelt just outside of Washington, DC on land that was already owned by the Agriculture Department's Agricultural Research Center. Plans for the new Center were approved by NASA's first Administrator T. Keith Glennan in November. To accomplish this transfer with as little disruption as possible to Project Vanguard, an agreement was reached between NASA and the Navy that provided for the continued use of Naval Research Laboratory facilities until the new Greenbelt facility could be completed.

It was around this time that Wesley J. Bodin, a sonar researcher with NRL, became curious about the new space program, enough so that he transferred into the first group at the new Goddard Center. "When I say transferred, it was more than people," said Bodin. "We transferred people, equipment, and functions. We had a network already built with the Army Corps of Engineers [Minitrack].... This operating network was transferred—completely—into NASA."[5]

Initial personnel transfer took place on 30 November with 157 members from the NRL being designated to what was temporarily the Vanguard Division of the yet unnamed space center. A month later another 46 persons were moved to form the core of the Space Sciences Division. Over

American rocket pioneer Robert H. Goddard in a 1932 photograph. (NASA Image GPN-2000-001336)

the next three months, 73 additional people transferred to various new divisions. Major construction in Greenbelt soon began in April 1959.[6]

On 1 May, NASA officially announced that the Center would be named the Goddard Space Flight Center (GSFC) in honor of American rocket pioneer Robert H. Goddard who on 16 March 1926 launched the world's first liquid propellant rocket. Construction of the new Center continued over the next 18 months. Although the Center bustled with activity as soon as NRL personnel were transferred, NASA waited until 16 March 1961—the 35th anniversary of Goddard's pioneering launch—to officially dedicate the Center. In the meantime, NASA found office space wherever possible in the area to house the growing contingent of engineers and scientists. To maximize work space, Jack Mengel's people in the Space Communications Branch, even turned barracks into offices at the nearby Naval Station in Anacostia.[7]

Organizational changes soon followed. From the two groups that came over to NASA—the Vanguard group and a Space Sciences group—Jack Mengel formed a new networks division while John Townsend headed up the sciences division. (Mengel served as the first Director for Tracking and Data Systems at GSFC until retiring in 1972. Townsend became the Goddard Center Director from 1986 until his retirement in 1990.) Mengel would recall that:

> The excitement of the space program became a virtual magnet which attracted the very best people to handle the many new functions which had been assigned to the Goddard Center. This, of course, included tracking, data processing, computer operations, and worldwide communications which were needed to tie the system all together. Fortunately, we were able to assemble a truly priceless team of very dedicated and talented people.[8]

Part of the new Center's early organization included the Space Task Group (STG), a cadre of engineers and scientists out of Langley Research Center who were responsible for planning NASA's "manned satellite" activities. Although located at Langley, the STG was actually part of Goddard's early responsibilities as it assumed the lead for NASA's new space projects. It was a busy time. The Center grew so rapidly that most of the people working human spaceflight—commonly known as manned spaceflight—also had to move off-Center for a while into a rented facility some five miles away because the Greenbelt Center literally ran out of space.[9] In its multirole function, the Center was assigned the broad scope of managing scientific, communications, and meteorological satellite projects, and developing sounding rocket and spacecraft experiments. To support these activities, NASA's first real-time computer complex was built at the Center. The building that housed this new computer facility was among the first to be constructed at Goddard. Some

Dedication of the Goddard Space Flight Center in Greenbelt, Maryland on 16 March 1961. (NASA Image No. 022)

recall that it was still under construction and surrounded by a sea of mud even as IBM contractors installed interfaces to Cape Canaveral.[10]

In January 1961, four months before Alan B. Shepard, Jr. became the first American in space, some 670 people from the STG separated from the Goddard organization to form the nucleus of what went on to become the Manned Spacecraft Center (MSC) in Houston, Texas. In a move that would engender dissension between Houston and Greenbelt for years to come, Goddard still retained responsibility for managing and operating the tracking and data network for the human spaceflight program. It thus became the focal point from which all network activities (excluding the DSN which remained with the JPL) were coordinated and directed, and where data from the individual field stations were collected and processed.

While Minitrack continued to track Explorer and Vanguard satellites using radio interferometry, changes were being made during this time to accommodate technology advancements rapidly occurring in the field of RF transmission.[11] In 1960, Minitrack equipment was adjusted by network engineers to receive satellite transmissions in the Very High Frequency (VHF) part of the electromagnetic spectrum at 136 MHz. This modification greatly allevi-

A VHF double-stacked, yagi antenna, similar to common rooftop antennas used to receive over-the-air television signals, but much larger. The antenna consists of both vertical and horizontal dipoles which collect signals from a satellite, passing them to a preamplifier which boosts their strength for processing by a telemetry receiver. The photograph show an antenna now at the Canberra Deep Space Communication Complex in Tidbinbilla, Australia, that is used to support the Interplanetary Monitoring Platform 8 (IMP-8) satellite. (Photograph by the author)

ated low frequency interference at many of the sites and was implemented by NASA so as to conform with new international standards being recommended by the International Telecommunication Union in the burgeoning field of space communications. In addition, telemetry reception at many network sites was improved with the addition of nine and sixteen yagi antenna arrays.

As launch vehicles became steadily more reliable in the early 1960s and satellites were being launched into higher inclination orbits (the tilt of a satellite's orbit measured with respect to the Equator), their ground tracks went well outside the ±35° latitude window that Minitrack had been designed to support. For the first time, polar orbits, where the ground track of a satellite covered the entire globe (ideal for weather observation and surveillance), came onto the scene. For this reason, Minitrack stations were soon added in northerly and southerly locations outside the original "picket line," this even while some stations were being shut down. From 1959 to 1960, Minitrack reached a plateau in terms of the number of network stations.

By the early 1960s, a whole new generation of scientific and applications satellites boasting capabilities well beyond those of the IGY satellites sprang to the forefront. These ranged from weather, communication, surveillance, and Earth science spacecraft to lunar and planetary deep space probes. The orbit mechanics of these higher altitude satellites meant slower angular rates with respect to a given location on the ground as a satellite circled the globe. Elliptical orbits also meant that fewer stations could observe a satellite as it circled the globe. Some, like the Interplanetary Monitoring Platform (IMP), were on extremely high eccentricity orbits with apogees out to 240,000 kilometers (150,000 miles). Since interferometer effectiveness depended on measuring angular differences, the data that Minitrack could provide for precise orbit determination became increasingly inadequate. On top of that, modulation techniques also advanced to accommodate the ever increasing demand for higher communication data rates, more complex ground command operations, and higher bandwidth digital telemetry items such as television.

To accommodate this new generation of artificial satellites—soon to be followed by something even more ambitious in "*manned* satellites"—an evolution in NASA's tracking network had to take place. "While the Vanguard program was intended to plow new ground, show new rocket technology and develop new guidance and control systems," said Ches Looney when recalling the importance of these changes that were seemingly taking place overnight, "the manned space flight program, including its tracking system, required the use of proven, off-the-shelf hardware, to offer maximum reliability, because after all, human lives were [going to be] at stake."[12]

Looney, who was the Associate Division Chief for Advanced Development at Goddard from 1959 to 1966, added:

> We began to realize that while Minitrack was a marvelous system and absolutely essential in its time, the basic approach used was leading us into a blind alley. As things progressed, we were able to improve satellite trajectories and orbits as well as orbit calculations, which together with a new S-band ranging system [a higher frequency capable of handling higher data rates], allowed us to use large parabolic antenna dishes without independent acquisition aids such as Minitrack.[13]

In another sign of the changing times, computer telemetry processing facilities were installed at the field stations to directly feed into Goddard computers. This was a far cry from Minitrack, when tracking data was recorded and had to be shipped to Goddard by air or surface courier, a process that took anywhere from two weeks to a month. Although coarse determination of the orbit track could be done at a station, precise orbit computations would be unknown for a couple of weeks.[14]

The genesis of these changes can be traced back to December 1959 when NASA formulated its first comprehensive program plan as the country's new space agency in a document simply entitled "NASA Long-Range Plan," prepared by its Office of Program Planning and Evaluation. The report that the Agency came up with was essentially a forecast of what it could do within certain budgetary constraints and reasonable extrapolations of the state-of-the-art in launch vehicles—the pacing facet of space technology at the time. Plans such as this were being modified continually, particularly in the face of Soviet space "firsts," which at the time seemed to be occurring at an alarming pace.

Satellite launches over the coming two to three years were forecasted as part of this plan. These projections provided Goddard engineers with a set of requirements from which specific approaches for comprehensive tracking, telemetry, and command (TT&C) design were selected, developed, tested, and eventually turned into operational hardware. One of the first items of business began in 1960 when the GSFC started a process to upgrade the TT&C equipment at existing Minitrack stations. In addition to modernizing these old Minitrack sites, new stations were established. The network that emerged became known in the space community as the Satellite Tracking And Data Acquisition Network, or STADAN.[15]

Over the next two years, large aperture, high gain circularly polarized, parabolic antennas replaced the less efficient, linearly polarized, yagi antennas so as to accommodate communications using the more bandwidth efficient S-band frequencies (2,100 MHz). The Minitrack approach gave way to far better equipped 12-meter (40-foot) and 26-meter (85-foot) dishes. The large 26-meter parabolic dish antennas, in particular, were designed to support the Nimbus meteorological satellite program with its flood of high data rate telemetry cloud cover photographs. In addition to receiving downlinked telemetry, these antennas could transmit satellite commands from a single disc-on-rod uplink antenna mounted on the side of the dish connected to a high power amplifier at the base of the antenna. At the same time, a somewhat scaled down, 12-meter version of the same antenna type was installed at a number of stations to circumvent the cost of the larger dishes.[16]

A 26-meter pointing antenna system was a pricey item, costing some $910,000 (in 1961) to design, engineer, fabricate, and erect. Mengel was to recall that in 1961, NASA Administrator Keith Glennan said never to ask him again for any more such pricey equipment.[17] However, this was only a harbinger of things to come as more advanced spacecraft and elaborate missions soon called for the use of even more sophisticated systems in the STADAN.

Just as critical to the STADAN mission of downlink telemetry acquisition and uplink satellite commanding was the tracking function. It had become a great deal more difficult as NASA's programs broadened to embrace satellites in high eccentricity, high apogee orbits all the way to geosynchronous orbits. In these orbits, angular rates with respect to a ground station changed very slowly (virtually none in the case of geosynchronous). Because of this, interferometry effectiveness was greatly reduced and sometimes eliminated altogether.

These tracking problems were anticipated by NASA early on. Addressing the issue, Edmund J. Habib at Goddard and Eli Baghadady of Adcom, Inc., devised a solution called the Goddard Range And Range Rate (GRARR) system. GRARR differed from Minitrack in that it was an *active* rather than a *passive* tracking system. In Minitrack, a beacon aboard a satellite continuously sent out a signal which was received on the ground when it came into range of a station. With the GRARR, a concept called "side-tone ranging" was used. In side-tone ranging, range tones were sent from the ground station up to the spacecraft on a VHF carrier separate from the S-band uplink carrying the command signal. An onboard transponder then returned these VHF tones along with the downlinked S-band telemetry. At the ground station, the precise time delay between the transmitted and received tones was measured to determine the range (and its rate of change) from the site to the spacecraft. With similar information from other STADAN stations from around the world, the orbit in which a satellite is traveling could thus be determined very precisely. When the system was first set up around the network, ranges out to 400,000 kilometers (250,000 miles or approximately the distance of the Moon from Earth) could be measured. Typical ranging accuracy was ±25 meters (82 feet), impressive for its time; range rate accuracy was just as good at ±15 centimeters (6 inches) per second.[18]

In describing the complexity of ranging systems, George Kronmiller, former Goddard GRARR Project Manager explained:

> During the early VHF years, our commanding and telemetry units were the same as the spacecraft's. We never took telemetry through the ranging system, not because we couldn't receive the signal, but because we didn't have the data handling gear to capture telemetry data and transmit it to NASCOM [NASA Communications Network] lines. Actually, until the STDN [Spaceflight Tracking and Data Network which Goddard formed in 1975], commanding and telemetry were kept separate from range and range rate, even with S-band. With the development of USB [Unified S-band], ranging and commanding could be handled on the same carrier. After the MSFN [Manned Space

Women operators, such as Melba Roy at the Goddard Space Flight Center in a picture taken in 1964, were a big part of the tracking workforce. (NASA Image GPN-2000-001647)

Flight Network] and the STADAN combined to form the STDN, VHF remained the same, but the S-band system combined into a unified command, ranging and telemetry system using the same NASA standard near-Earth transponder on the spacecraft.[19]

GRARR equipment was usually housed in two trailers, one accommodating receiving and the other transmitting equipment. It operated in two different frequency bands: VHF (Very High Frequency) at around 150 MHz and an Ultra High Frequency (UHF) band at about 2,500 MHz. Two antennas were used, one for VHF and the other for UHF. Each was used for both transmission and reception. Angular tracking measurements could also be made by this equipment, but its accuracy was only good to within ±0.1 degree in elevation and azimuth, sufficient for the tracking of distant spacecraft but not always good enough for following spacecraft whose angular rates changed more rapidly. Thus, the angular information was often used as pointing or acquisition information for other instruments. Each of the two uplink transmitters radiated at 10,000 watts of power—about 5,000 times the power of an outdoor "walkie-talkie" hand radio. Since it continuously radiated while the equipment was on, station staff had to be cautious so as to avoid exposure, particularly at the microwave frequencies. A system of flashing red warning lights could be found around a station so that people knew when the transmitters were on.[20]

By 1962, the initial GRARR ground elements were installed in trailers at Rosman, North Carolina; Tananarive, Madagascar; and Carnarvon, Wester Australia. The system was continually improved and used throughout the 1960s, supporting numerous Goddard satellite programs to provide range and range rate data whose accuracy would not be surpassed until the use of lasers a decade later. On Explorer 35—a so-called "Anchored Interplanetary Platform" orbiting the Moon—its range was measured to within 1,500 meters (4,900 feet) and the range rate accurate to within 65 centimeters (25.5 inches) per second at the lunar distance. Goddard engineers eventually considered the system reliable out to 1.3 million kilometers (800,000 miles), or three times the distance from Earth to the Moon. This was quite an accomplishment for 1960s technology.[21]

Yet a third improvement had to do with the dramatic increase in the data transfer volume that the Goddard network began to experience in the early 1960s. The observatory class of satellites such as the Infrared Space Observatory (ISO 1 in 1962) and the Orbiting Geophysical Observatory (OGO 1 in 1964) ushered in an era of information explosion. Meteorological satellites, like the Television Infrared Observatory Satellite (TIROS 1 in 1960) and Nimbus 1 in 1964, also required large amounts of telemetry to be downlinked to the ground during any given data pass over a ground station.

To meet these high "data dump" requirements, Satellite Automatic Tracking Antennas (known by the unflattering acronym SATAN) were installed throughout the network so as to supplement the telemetry reception workload of the parabolic dishes. These were essentially smaller, automatic tracking or "auto-track" replacements for the fixed Minitrack yagi antenna arrays. Operating in the VHF-band, SATANs were less sensitive (lower antenna gain) than their larger S-band counterparts, but were much less expensive to purchase, install and operate than the large dish antennas. Eventually, 30 SATAN antennas were installed throughout the STADAN during the 1960s.[22]

As the technology of NASA's global tracking network migrated from one of using radio interferometry and optics of the late 1950s to the more sophisticated STADAN of the 1960s, requirements on the locations of the stations changed with it. When the original Minitrack Network was developed, site selection was driven (and constrained) primarily by the short range of the tiny Vanguard transmitter, the technology of the receiving electronics and the general lack of precision tracking available during the satellite's early orbit. In other words, no one was exactly sure just where a satellite might appear over the horizon. Thus, Minitrack stations were originally spaced rather closely along a longitudinal line roughly following the 75th meridian (the "picket line"). Later, with downrange tracking ships and much better ascent and orbit injection tracking, rough ephemerides (position-time information from which orbital elements are derived for orbit determination) were available even before booster burnout.

As a result, the three original "early ops" stations in the British West Indies—Antigua, Grand Bahama and Grand Turk—just downrange of Cape Canaveral were shut down in 1961. In some cases, actions by the DOD were a factor. A year earlier, for example, the Chula Vista station on Brown Field near San Diego had to be moved to Mojave when the Navy closed down that air station.[23] (It was at this new location that the Mojave Station went on to become the STADAN side of NASA's Goldstone Communication Complex.)

The picket line also changed. Take the Minitrack station in Chile for example. It had been located since 1957 at a spot called Salar del Carmen in the Pacific port town of Antofagasta on the northern part of the country. Initially operated by the U.S. Army, it was operated by a joint NASA/University of Chile team under contract NAS5-1925 when the space agency was established in 1958. However, with the improved capabilities of the nearby stations at Santiago and Lima, Antofagasta became redundant and was closed in July 1963. Some University of Chile personnel were transferred to the improved Santiago site (actually at nearby Peldehue) while most of the American contractors returned to the United States for reassignment. Telemetry equipment was redistributed, largely to the stations at Santiago and Lima. However, heavy equipment like power generators and air conditioners were returned to the U.S. for use at other stations.

While the Antofagasta site was being shut down, the neighboring station at Santiago underwent a $1.2 million improvement to include the installation of a 12-meter (40-foot) medium gain antenna, a 370-square meter (4,000-square foot) operations building and a new collimation tower (a tower located a few miles from the main antenna equipped with an RF emitter used as an aim point to checkout and calibrate the automatic tracking capability of the antenna). Under the original contract with the University of Chile which allowed the United States to put a station on Chilean soil, the school had "agreed to provide . . . land as may be necessary for the effective operation of the station."[24] Since the entire reservation on which the station was located was used only for pasteurization, obtaining the additional 10 acres that was needed for modernizing the station did not present a problem.

The station staff at Santiago almost immediately doubled in 1963, from 38 (16 Americans and 22 Chileans) to 62. Under the contract with the university, NASA eventually replaced half of the additional 24 people with trained Chileans. Said Wes J. Bodin, the Associate Chief of the STADAN Engineering Division at the time, this was just an example of the rapid growth that the STADAN saw during this time. "When a requirement came in, we added to the network." Permanent or temporary sites were added as mission requirements called for. Bodin cited a couple of examples:

> When NASA started launching spacecraft into polar orbits, the second injection burn . . . was too far east of Johannesburg, so we added Tananarive. When we needed stations with large high-gain antennas, and this was a new requirement, we installed our first 26-meter (85-foot) antenna in Alaska. It wasn't too bad working there, so long as you were finished construction work by December when the temperature could hit 60 below![25]

This time period also saw what would be the first of several episodes in the political environment of the host country that would directly affect NASA's operation of its stations. Up until the 1950s, Cuba was a relatively stable independent island, enjoying strong agricultural trade with the United States and serving as a popular tourist destination in the western Caribbean. The U.S. had occupied this largest of the Caribbean islands in the Spanish-American War of 1898 until its independence was granted in 1902. After that, the United States continued to have a major influence in Cuban affairs, even occupying it briefly a second time from 1905 to 1909.

In 1940, Fulgencio Batista was elected president and over the next four years, began a series of idealistic reforms. Not surprisingly, the Cubans did not see eye-to-eye with his isolationist views and voted him out of office in 1944. In 1952 however, Batista regained power in a bloodless coup three months before the planned election. This time, he instituted a dictatorship. As a result, many factions and guerrilla groups began opposing him. One such group was led by Fidel Castro Ruz who had participated in a failed attempted to overthrow the Dominican Republic government in 1947. He was a staunch nationalist known for his opposition of American influence in Cuba. In 1953, Castro attacked the Moncada barracks, the main provincial garrison of Batista's army, but was captured and exiled to Mexico. However, he returned to Cuba in November of 1956 and over the course of the next two years, his "26th of July Movement" gathered strength and influence, eventually to the point where Batista finally had to flee the country on 31 December 1958. The next day, Castro established a one party communist state in Cuba, the first of its kind in the Western Hemisphere.

At the time of Batista's ouster, 75 percent of Cuba's farmable land was owned by foreign interest (mostly U.S.). The new government adopted land reforms and confiscated all the private property owned by upper class Cubans and foreign companies. Although Castro did not officially reveal his Marxist leanings until 1961, the United States had recognized the intentions of the new government, and as a result, relations rapidly deteriorated. On the day that Castro seized control of the Cuban government, the Minitrack station, which was located at Batista Military Airfield, was shut down even as the American Embassy scrambled to learn the fate of NASA Station Director Chester 'Chet' Matthes and his family, as well as that of the military personnel assigned to the tracking station.[26] By noon the next day, confirmation was received that all were safe. Station personnel were then told by the State Department to follow instructions from the American Embassy and that the station would reopen when the Embassy received word and considered it advisable. But that word never came. Instead, they were told to close out the station, remove the assets and leave the country.[27]

In events unrelated to Cuba during this time, operations at the Fort Stewart Station in Georgia underwent some reorganization. When the Minitrack station was established there in 1957, the site was primarily an Army anti-aircraft training facility. After the Korean cease fire, the Army recognized that it needed to maintain a ready and able armor force that could be deployed at a moment's notice to deal with threats that may suddenly erupt. As a result, Fort Stewart was turned into an Armor and Artillery Firing Center since its old anti-aircraft ranges and impact areas were better suited for this purpose than for the new age of missiles. Because of this realignment, the NASA tracking station no longer fit the bill for the post's new mission and the Fort Stewart Station was phased out of the Minitrack Network.[28]

With the abrupt shut down of two primary stations, a gaping hole materialized in the middle of the picket line. NASA had to quickly find a replacement. Attention focused on Fort Myers, Florida. Situated essentially midway between Fort Stewart and Havana, it was a logical choice, geographically. Equipment was immediately relocated to Florida from the two sites that had shut down and satellite pass responsibilities redesignated and combined at the new site.

From 1960 through 1963, most of the early Minitrack stations were refurbished and transitioned into the STADAN. For example, Johannesburg, Quito and Santiago received 12-meter (40-foot) antennas, and all acquired dual receive/command SATANs. Santiago was modified into a GRARR site, as well. The South American stations, in particular, continued as a centerpiece in the evolved network, providing NASA with a tracking and data acquisition capability in the southern hemisphere.[29]

With satellite ground tracks now extending well beyond the original ±35° latitude window, new stations at locations outside that latitude range had to be built. From 1960 to 1966, the GSFC oversaw the build up of eight additional STADAN stations on four continents. These were the most advanced stations to date. In North America, they were located at Fairbanks, Alaska; East Grand Forks, Minnesota; Rosman, North Carolina; and Saint John's, Newfoundland, the eastern-most point of Canada.[30] In Europe, the Winkfield Station was erected in Berkshire, England. The addition of these sites greatly expanded satellite coverage in the northern hemisphere, increasing the observation window from 35° latitude to 65°. In the south, the Johannesburg Station was joined by a new one at Tananarive in the Malagasy Republic on the island of Madagascar.

Fairbanks and Rosman soon became the most extensive stations in the system. In March 1962, Gilmore Creek, on the outskirts of Fairbanks, received the first of the large parabolic dishes, featuring both the 12- and 26-meter (40-and 85-foot) dishes, along with the GRARR and a full complement of SATAN equipment. In November 1966, in what was essentially a cost saving move, the STADAN site in College, Alaska, still operating Minitrack equipment, was consolidated with the nearby Fairbanks Data Acquisition Facility (DAF) at Gilmore Creek for an estimated annual savings of $292,000.[31] This was a relatively easy decision by Goddard as these sites in the vicinity of Fairbanks were less than 40 kilometers (25 miles) from each other. With this move, the last of the remaining Minitrack antenna arrays was transferred to Gilmore.

The Fairbanks Station—also known officially as "Alaska"—was located in the central part of the state 22 kilometers northeast of Fairbanks at a well-known landmark where Rose Creek joined Gilmore Creek. Being the most northerly of all the stations, Alaska provided crucial support for polar and high eccentricity, elliptical orbiting spacecraft such as the Nimbus weather satellites and the Alouette. (On 29 September 1962, Canada became the third country to orbit an artificial satellite with the launch of the Alouette, a science satellite to study the affect of aurora borealis, or the Northern Lights, on radio frequency (RF) signal propagation in the upper ionosphere and its effect on communications.) Conditions at the station were at times harsh, characterized by cold winters of below -60°F and summers in which the temperature could rise above 90°F. During the winter months for example, the station had to operate a steam boiler system to forestall the spread of glacier formations around the site.

Even though work conditions for personnel assigned to remote locations such as Alaska could at times be trying, severe injury or death in the field was not a common thing. A case in point, though, was recorded on 3 May 1969 at the Fairbanks Station. At approximately 4:30 p.m., a RCA Service Company (the station's maintenance and operations contractor) employee named George Matilla was found crushed to death beneath his car at the station's garage. Since it was a weekend, no one knew exactly what had happened. His death was later ruled as an accident by the Alaska State Patrol who concluded that the jack had slipped as he worked alone under his car, fatally crushing him.[32] Throughout the years, there have been other stories as well, particularly at the more desolate locations, of those who wandered out-

Orroral Valley (ORR), nestled some 50 kilometers southwest of Canberra in the Namadgi National Park, was one of 10 NASA stations in Australia at one time. The most comprehensive Satellite Tracking And Data Acquisition Network station in the Southern Hemisphere, ORR operated from 1965 to 1984. Clearly seen are the 26-meter antenna adjacent to the main Operations Building and the 2 SATAN antennas. Note the Minitrack array in the foreground. (Photo courtesy of CSIRO)

side the station compound to check on equipment or simply "went fishing" who were never heard from again.[33]

In the Southern Hemisphere, cooperation between Australia and the United States in the field of space research began in 1957 with construction of facilities at Woomera, South Australia, for the IGY. In a formal exchange of notes in February 1960, the two governments agreed on a cooperative space program:

> The object of such further and extended cooperation would be to facilitate spaceflight operations contributing to the advancement of our mutual scientific knowledge of man's spatial environment and its effects; the application of this knowledge to the direct benefit of man; and the development of space vehicles of advanced capabilities, including manned space vehicles.[34]

The Australian ground stations were established and run by the Department of Supply, or the DOS. In this agreement, construction and operation of the stations were financed by NASA while the DOS provided management and staffing, either from its own resources within the Weapons Research Establishment or through Australian industry contractors. In this way, Australia had the most autonomy and independence in running the NASA stations outside of the United States. Unlike most stations where the local staff reported to an onsite, NASA-appointed Station Director, the DOS in Australia appointed its own people. In this way, a great sense of ownership was promoted—one which prevails to this day.

The Australian government welcomed space activities as intrinsically good and providing a medium for strengthening diplomatic cooperation with the U.S. It viewed such activities as having strong spin-off potential to Australia in such fields as defense, communications, and astronomy. From the beginning, Australia was sensitive to the importance of retaining independence and having a substantial share in the control of network activities on its soil. Robert A. Leslie, who headed Australia's network operations—a "father figure" in the commonwealth's space work—stated its position very clearly in a 1974 letter to the Minister of Science in which he reiterated this right of Australian ownership:

1. The employment of foreigners where Australians could be employed would be undesirable.

2. Control of the activity in foreign hands might have been a matter for question on security grounds and would have been a matter of national controversy.

3. The maximum spin-off to Australia—for example, to industry, universities and government agencies—could best be affected by Australians.

In return for what could be viewed as granting of Australian autonomy, the DOS made an annual contribution to NASA, a subsidy of $140,000 for operations of the Australian Tracking Stations. Although insignificant in terms of absolute dollar amount to NASA's budget (measured in the billions of dollars), this token sum has nevertheless been viewed importantly by the U.S. as a sign of goodwill and good faith on the part of the Australian government.[35]

In August 1960, the Minitrack station at Satellite Tracking Centre just northwest of Woomera was moved to the nearby Island Lagoon dry lake bed 22 kilometers (14 miles) to the south and its operations combined with the new Deep Space Network (DSN) site there. A station was also opened at Carnarvon, Western Australia, collocated with a Mercury Space Flight Network station that was then being constructed. However, the largest STADAN site in

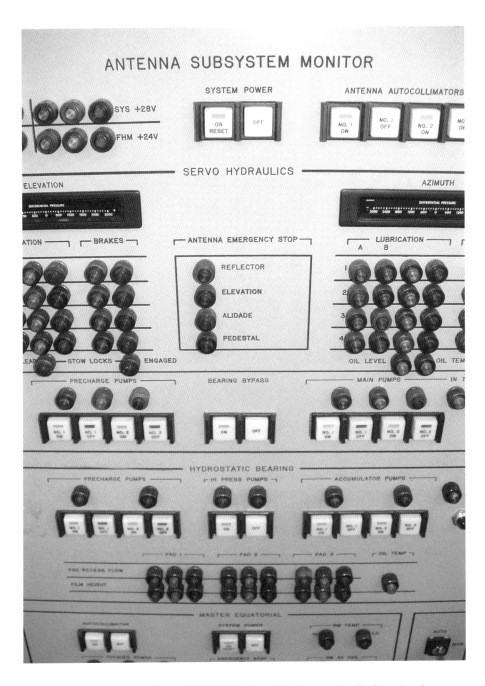

A vintage 1960s Antenna Control Unit (ACU). Graphical computer displays, touch screens and joysticks have, for the most part, replaced the lights and buttons of the old ACUs. Shown is a control unit on display at the Space Museum at the Canberra Deep Space Communication Complex in Tidbinbilla. (Photograph by the author)

the Southern Hemisphere officially opened on 24 February 1966 at Orroral Valley 50 kilometers (30 miles) southwest of Canberra.

The Orroral Station was needed as Woomera (officially closed on 22 December 1972) was no longer a cost-effective location for NASA. Situated in the quiet valley of Namadgi National Park, Orroral provided an RF-quiet area to meet requirements for the new 26-meter (85-foot) antennas that were being installed across the STADAN. The rolling hills surrounding the area made Orroral an ideal environment for a tracking station by naturally shielding the site from man-made radio frequency interference. When the station site was chosen, the isolated valley located deep in the "Land of the Echidna and Platypus" was accessible only to light and off-road vehicles. Even today, one needs to cross the 100-year-old Tharwa Bridge just to reach the site. Before construction could start, suitable roads had to be paved for the passage of heavy construction equipment and antenna hardware. As at sites around the world, NASA provided this as part of its construction.

Six months after the station became operational, network responsibilities at Island Lagoon were phased out and what was left of the Minitrack equipment was moved to the new location. The net effect of this growth was to more than double the time that high inclination orbiting satellites could be in contact with the ground. Orroral became the most comprehensive STADAN site in the Southern Hemisphere. Construction of the station began in November 1964 and was completed in July of the next year at a cost of £1,000,000 (about $1.5 million U.S. dollars in 1964).[36]

The layout of the station was fairly typical of the new, larger DAF across the STADAN. Equipment was contained within an area of about 40 acres and laid out so as to minimize interference to the receiving systems from buildings, support structures and other antennas. The centerpiece of the station was the 26-meter diameter parabolic tracking antenna. Weighing with its support structure and hydraulic gimbal-drive mechanism some 400 metric tons (882,000 pounds) and standing 36 meters (120 feet) high, it was easily the most prominent structure on site. From the Operations Building, the antennas were controlled and all tracking activities coordinated with the rest of the network.

During acquisition, the antenna—directed by data passed from Goddard—searched a specified area of the sky until it picked up the beacon signal transmitted by the satellite. The antenna data system then measured and encoded the look-angles and fed this information into a servo control system and an Antenna Control Unit (ACU) in the control center at the Operations Building. The information was also recorded on teletype punched tape and passed to the control center for orbit determination.

To receive telemetry downlink from a satellite, a command signal was encoded and transmitted using one of two 2,500 watt transmitters feed-

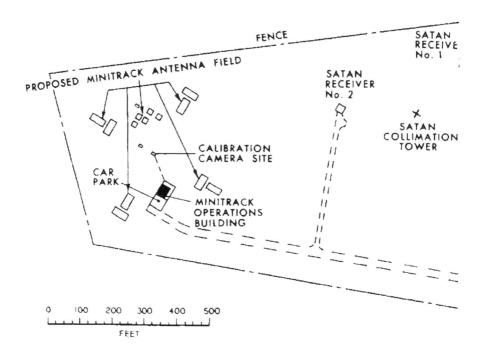

Layout of the Orroral Valley STADAN Station. Note that the Minitrack array transferred from Woomera was eventually placed further west of the SATAN receivers than shown

ing a small command antenna that was attached to the rim of the main dish. Signal from this uplink activated a transmitter in the satellite and data stored onboard the satellite could be transmitted to the ground station. Data and timing signals were then processed for recording onto 14-track reel-to-reel magnetic tapes and, in the case of time-critical information, for transmission by teletype back to Goddard. Backup receiving was usually provided by an onsite steerable Yagi antenna which could be directed by information from any of the primary tracking dish antennas.[37]

A continuous electrical power supply to a station was obviously vital to its operations. Even a brief outage could cause loss of data or compromise the mission if it were to occur at a key moment. If available, network stations used commercial line power backed up by its own diesel power gen-

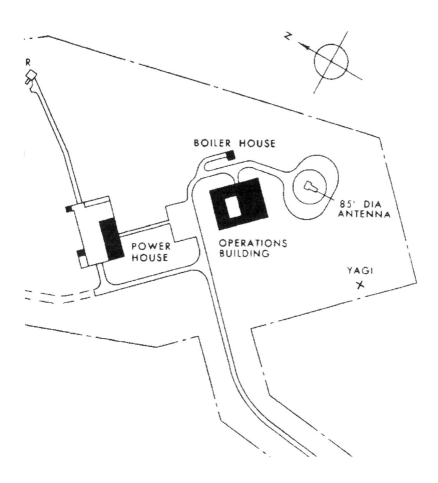

in the diagram. (STADAN Facility, Orroral Valley, ACT: Information Brochure. Folder 680/5/23 NASA Australian Operations Office, Yarralumla, ACT)

erators. If line power was not available—as was the case at ORR—then power was generated onsite exclusively using dual-redundant generators with sufficient capacity to provide full backup should one unit fail. At Orroral, for instance, two 250,000-watt and two 500,000-watt units were used. Sites usually had two separate power distribution systems, one to run the station core electronics and tracking equipment and the other for utilities, air conditioning, lighting, and facility electricity needs.[38]

Finally, a DAF was not cheap to operate, requiring a staff of about 150 technical, administrative and maintenance workers rotating in shifts. The operating cost was about $1.2 million a year in the mid-1960s, almost as much as the onetime cost to build a station. NASA spent over $140 million in Australia alone in the 1970s. Besides just getting the technical job done, such international cooperation also yielded social benefits. For example, by the mid-1970s, seven of the ten stations which had been built in Australia since 1958 had become popular tourist attractions, often offered as stops by touring companies. The Department of Science officially estimated that from 1965 to when it closed in 1974, the Carnarvon Station was visited by 7,000 to 10,000 people every year (this, in a continent of around 12 million in 1970). Similarly, the DSN site at Tidbinbilla on the outskirts of Canberra was drawing some 8,000 visitors annually.[39]

In addition to the permanent sites at Orroral and Carnarvon, several project-specific (temporary) stations were also added by Goddard in Australia. One reason that the Southern Hemisphere played host to most of the temporary sites was because that was where the critical Earth-injection phase of a mission to send satellites into high-inclination orbits usually took place. Key among these places was Darwin, a transportable station on the northern shores of Australia, selected specifically to support the Orbiting Geophysical Project missions, and Cooby Creek near Toowoomba in Queensland, which was added to support the Applications Technology Satellites (ATS). Kano, Nigeria in 1965 provided another temporary site for STADAN, briefly housing equipment supporting the International Satellite for Ionospheric Studies (ISIS). Goddard even ventured into Pakistan where a site was set up for a while to observe LANDSAT.[40]

As STADAN grew, one of its requirements was to receive high bit rate digital telemetry from a new observatory-class of large, science satellites. These included the Orbiting Geophysical Observatory (OGO), Orbiting Astronomical Observatory (OAO), Orbiting Solar Observatory (OSO) and the Nimbus weather satellites. While the facility in Alaska could support polar orbiting and high inclination satellites, it could not support this new class of observatories, which orbited Earth at lower inclinations.

For these missions, Goddard needed a wideband DAF at a latitude that was suitable for a high percentage of data recovery on these orbits. Dubbed internally as Project 3379, network engineers began looking for a suitable location for their newest station in 1962. Site surveys eventually whittled the number of possibilities down to two: Fort Valley, Georgia and Rosman, North Carolina. Rosman—the actual station site was some 15 kilometers northwest of the town itself—was eventually selected. The choice was due in part to Rosman being closer to Greenbelt which expedited experimenters with evaluation and control of their projects. Located 55 kilometers southwest of Asheville on the grounds of Pisgah National Forest, the site had an advantage of the natural setting of the surrounding hills which provided shielding from electromagnetic interference. Because it was on national forest land, the location also lacked high voltage transmission lines and commercial circuits in its vicinity

Chapter 2 \ Evolution of a Network   51

Two 26-meter tracking antennas dominate the skyline of the Rosman STADAN station in the hills of the Pisgah National Forest, North Carolina in this undated photograph. (Unnumbered photograph. Folder 8820, NASA Historical Reference Collection, NASA History Division, NASA Headquarters, Washington DC)

and, in the early 1960s, had an absence of commercial airline routes directly over the station.[41]

NASA invested $5 million in 1962 to build the station, filling a critical need as the network modernized from Minitrack to STADAN.[42] Rosman was established specifically as a full service Data Acquisition Facility and was among the first not to have the old interferometry tracking system. The site was well equipped to handle the new network mission, providing a full suite of the most up-to-date equipment for its time: telemetry reception with two 26-meter antennas that could autotrack at 1.7 and 2.29 GHz; a 3 kW command uplink system; a SATAN 16-element automatic yagi antenna array; S-band GRARR; a 1,200-square meter (13,000-square foot) operations building; a 420-square meter (4,500-square foot) power and service building housing four 200-kilovolt diesel generators, garage, and utilities; a 140-square meter (1,500-square foot) building to house the antenna hydraulic drive system; and a collimation tower with a small transmitter building located 2 kilometers (1.3 miles) west of the main antenna to serve as a boresight (beam center) for calibration and testing of the main tracking antennas. The first 26-meter system became operational in

July 1962 followed by the second unit in August 1964. These were immediately used to support S-band communications, which began earlier that year with the launch of the Hughes-built Syncom 1, the world's first geosynchronous communications satellite, on 14 February 1963.[43]

The Tar Heel state embraced its new scientific venture enthusiastically. On 26 October 1963, Rosman was officially dedicated with a ceremony replete with VIPs. Principal speakers included Governor J. Terry Sanford; U.S. Senators Sam J. Ervin, Jr. and B. Everett Jordan; Dr. George L. Simpson, Jr., NASA Assistant Administrator for Technology Utilization and Policy Planning; Edmond C. Buckley, Director of the Office of Tracking and Data Acquisition; and Dr. Harry J. Goett, GSFC Center Director. U.S. Representative Roy A. Taylor served as master of ceremonies.[44]

Although small compared to the 70-meter (230-foot) antennas used at the time by the DSN, the 26-meter (85-foot) dishes used in STADAN were nevertheless massive in their own right. Their movement was controlled by two large hydraulic motors that enabled three kinds of operation:

1. Manual operation in which the antenna was guided by an operator, who sitting at his control console, actually steered the antenna by sending electrical servo-signals to the motors.

2. Computer operation in which commands were sent by a computer that predicted the path of the satellite. Due to its large aperture, the 26-meter (85-foot) antenna had a rather narrow beamwidth (the direction in front of an antenna in which RF signals can be reasonably detected and focused). Consequently, it must be pointed near the horizon in the vicinity in which the satellite was expected to rise before the signal could be acquired (the acquisition process). By predicting the path of a satellite in orbit, the antenna could be pre-positioned by the computer thereby reducing the search time.

3. Automatic operation, or autotrack, in which a satellite's movement across the sky was automatically tracked by the antenna as it moved across its field-of-view. In this mode, the ACU converted the position of the satellite relative to the antenna boresight into electrical control signals which were sent to the antenna. In autotrack, the antenna was usually pointed just above the horizon at an azimuth where the satellite was expected to break horizon. Once acquired, autotracking enabled the dish to follow a spacecraft from horizon to horizon. During this process, the station computer converted the angular position of the

antenna into electronic code which was automatically punched on teletype tape. This data was then transmitted to Goddard via the NASCOM where it was processed along with pass data from other stations for use in orbit determination calculations.[45]

Output from each of these antennas was fed to a telemetry receiver located in a nearby control and operations building. The job of the receivers was to convert the radio frequency signal from the antenna into a lower frequency signal which could then be recorded onto reel-to-reel magnetic tape—a process known as down converting. The output of the receiver contained many individual pieces of information, not unlike, for example, the picture and sound for television received on a single channel. The magnetic tape recorders used to record the telemetry operated on the same principles as everyday-use home tape recorders but were much more robust in terms of the amount of data and speed they could record at. In addition to the telemetry itself, a precise time reference, the signal strength and operator voice annotations, if any, were all recorded simultaneously onto the tapes. Once packaged, the tapes were shipped to Goddard for processing.

In spaceflight communications, precise timing data, in particular, is of the essence. A spacecraft in circular orbit at 480 kilometers (300 miles) above the surface of Earth travels at a speed of 7,600 meters-per-second (25,000 feet-per-second), or roughly 27,300 kilometers-per-hour (17,000 miles-per-hour)! Therefore, accurate timing information is of the utmost importance so that computation of the spacecraft's position in space can be made and that timing of various scientific events as downlinked to the ground can be pinpointed. In the STADAN, DAF timing was synchronized to the U.S. Bureau of Standards Time Standard Radio Station—WWV in Boulder, Colorado—to an accuracy of ±0.001 seconds, or one one-thousandths of a second. While this level of accuracy may not seem like much in the age of the Global Positioning System, where timing accuracies are measured not in terms of milliseconds but *micro*seconds (one millionth of a second), it was quite the accomplishment in the analog era of the 1960s. These time-code generators produced electronically coded pulses which were recorded onto magnetic tapes simultaneously with the tracking and telemetry signals to provide the needed timing reference. Finally, to compliment the new, large, parabolic antennas, Rosman, along with the other DAFs at Gilmore Creek and Orroral Valley, continued to use the SATANs. Even though older, they still provided the much larger field-of-view needed to track the older generation satellites that still operated on VHF frequencies.[46]

During this time, the STADAN also saw the addition of one of the few network additions not in the English-speaking world. In 1964, a transportable station was set up on the island of Madagascar just off the east

African coast. A common language was obviously desirable from the standpoint of working with local nationals and with the local government. This was always one of the factors NASA considered when choosing locations for its ground stations. However, a station was required in the western part of the Indian Ocean for tracking highly elliptical, high apogee orbit injections that occurred over that part of the world. Also, NASA had to look for another location when it became apparent that the South African station would not be able to cover some of these critical events.

On 19 December 1963, the U.S. entered into a 10-year agreement with the Malagasy Republic allowing for the installation of a transportable ground station outside the port city of Majunga in northwest Madagascar. This agreement was reached in accordance with the spirit of a United Nations resolution calling for the application of results of space research to benefit all peoples. In addition to benefiting that region of the world by generating much-needed weather forecasts (especially during hurricane season), the station provided jobs for some 200 local residents in nontechnical positions for handling of day-to-day station maintenance. In reaching this agreement, NASA sent a delegation to the capital city of Tananarive where they were "received by the president, Mr. Philibert Tsiranana, most graciously in an office decorated with space memorabilia."[47] He soon gave the United States his enthusiastic support and permitted NASA to start bringing telemetry vans into Majunga.

The initial equipment consisted of five 9-meter (30-foot) trailers, each housing a 136.2-KHz and 400-MHz telemetry receiver that were geared to support the Nimbus, POGO (Polar Orbiting Geophysical Observatory) and A-12 Goddard satellite programs. Much of the equipment at Majunga came from the Australian sites of Muchea and Woomera, which had phased out at the conclusion of Project Mercury the previous May.[48] An MPS-26 radar was also temporarily deployed to support the EGO (Eccentric Geophysical Observatories) satellites.[49]

Explanatory literature handed out to familiarize station workers assigned to Madagascar described the environment as an area of mild winters and rainy summers, a relatively expensive but charming place to live. The handbook noted that the people of Madagascar were not politically minded and were predisposed to favor America and Americans. Harry McKeehan, who represented GSFC in negotiations with the Republic, called "our friendship with the president and the people of this island republic invaluable in building and operating this Indian Ocean site."[50] This cooperation was to play a pivotal role later when a political uprising in nearby Zanzibar created a tense situation, one in which American lives were put in jeopardy that required an evacuation to the Malagasy Republic (see Chapter 4).

Table 2-1: Stations of the Satellite Tracking And Data Acquisition Network (STADAN)

Station	Latitude Longitude	From	To	Equipment and Capabilities	Remarks
North America					
Alaska (Fairbanks)	64°59'N 147°31'W	1962	1984	GRARR, MOTS, Minitrack, SATAN, One each 40, 45, 85-foot dish	College site closed in 1966 and Minitrack equipment transferred to Gilmore. The most comprehensive of the STADAN stations.
Blossom Point, Maryland	38°25'N 77°06'W	1956	1966	Minitrack, MOTS, SATAN	The prototype Minitrack station. Used as a servicing station and training facility. NASA ceased joint operations with the Navy in 1966. Continued to be used by the NRL as a satellite control center.
East Grand Forks, Minnesota	47°56'N 97°01'W	1960	1966	Minitrack, MOTS, SATAN	Located near Grand Forks Air Force Base, a key ICBM station.
Fort Myers, Florida	26°33'N 81°52'W	1959	1972	Minitrack, SATAN, MOTS	Station formed when Havana and Fort Stewart shut down. Minitrack equipment transferred.
Mojave, California	35°20'N 116°54'W	1960	1969	40-foot dish, SATAN, Minitrack, MOTS	Moved from Brown Field near San Diego in 1960. Located in the Goldstone complex in valley adjacent to DSN site.
Network Test and Training Facility, GSFC, Greenbelt, Maryland	38°59'N 76°51'W	1966	1986	Minitrack, 59-foot antenna used for IUE support transferred to the Naval Academy. 30-foot antenna moved to Wallops in 1986	Until 1974, the NTTF was used to test equipment bound for the network and for personnel training. In 1974, it became part of the operational STDN. Station responsibilities were transferred to Wallops in 1986.
Rosman, North Carolina	35°12'N 82°52'W	1963	1981	Two 85-foot dishes, Three SATANs, GRARR, ATS	The most extensive STADAN site along with Alaska. Designed to receive high bit rate telemetry from observatory satellites. Turned over to the DOD in 1981.
Saint John's, Newfoundland	47°44'N 52°43'W	1960	1970	Minitrack, MOTS	First of the northern latitude stations to support tracking of polar orbiting satellites

continued on the next page

Station	Latitude Longitude	From	To	Equipment and Capabilities	Remarks
South America					
Lima, Peru	11°47'S 77°09'W	1957	1969	Minitrack, MOTS, SATAN	One of the original Minitrack stations. Turned over to their university under contract.
Quito, Ecuador	00°37'S 78°35'W	1957	1981	Minitrack, MOTS, SATAN, 40-foot dish	Equipment transferred to Dakar, Senegal after station phase out after STS-2
Santiago, Chile	33°09'S 70°40'W	1957	1988	Minitrack, MOTS, SATAN, GRARR, USB, a 30 and 40-foot dish	Most extensive of the South American stations. Operated jointly with the University of Chile who continues to run the station today.
Europe					
Winkfield, England	57°27'N 00°42'E	1961	1981	Minitrack, 14-foot dish, SATAN, MOTS	Operated by the British under agreement to the U.S.
Africa					
Johannesburg, South Africa	25°53'S 27°42'E	1957	1975	Minitrack, MOTS, SATAN, a 40 and 85-foot dish	Operated by South African nationals. Located with DSN site. Phased out under political pressure.
Majunga/ Tananarive, Malagasy Republic	19°S 47°18'E	1964	1975	45-foot dish, GRARR	Transportable site erected at Majunga which was moved a year later to a more permanent establishment at Tananarive to support the MSFN. Forced out due to political strife.
Australia					
Carnarvon, Western Australia	24°54'S 113°43'E	1964	1974	GRARR, C-band radar, 30-foot USB	Co-located with a MSFN site.
Darwin, Northern Territory	12°17'S 130°49'E	1965	1969	14-foot antenna to support OGO project	Temporary mobile OGO project station. Antenna transferred to Kauai MSFN station.
Orroral Valley, Australian Capital Territory	35°38'S 148°57'E	1965	1984	Minitrack, MOTS, two SATANs, 85-foot dish	The 85-foot antenna was relocated to the University of Tasmania after the station closed where it is still in use by the school
Toowoomba/ Cooby Creek, Queensland	27°24'S 151°56'E	1966	1969	SATAN, ATS	Temporary site used to support the ATS project
Woomera, South Australia	31°23'S 136°53'E	1957	1966	Minitrack, MOTS, SATAN	Closed out within 12 months after Orroral Valley became operational. Minitrack equipment transferred to Orroral.

Table 2-1 is a summary of the STADAN ground stations highlighting their capabilities and their roles in the network. Appendix 1 shows the network as it appeared throughout the 1960s into the mid-1970s.

☆ ☆ ☆

Network stations were operated by either American contractors, local nationals (if overseas), or a combination of both. To promote international goodwill, NASA used local nationals to the extent possible for non-technical positions. The staff usually worked under the direction of NASA civil service supervisors assigned to the station, led by a Station Director and sometimes a second-in-charge, the Deputy Station Director.[51] While Goddard contracted with several companies to perform station support requirements that ranged from facility maintenance to transportation to administration, it generally relied on two commercial contractors to run the network: BFEC and RCA.[52]

As the tracking network expanded from Minitrack to STADAN, so did the level of contractor support provided. In FY1963, Bendix as a corporation did $32.5 million of business with NASA. Over the next five years, as the STADAN expanded and became more complex, more stations were added. By FY 1968, Bendix was doing $123.8 million worth of work for NASA, making it the space agency's seventh top contractor in overall dollar amount obligated.[53] Of this amount, $39.1 million, or about one-third, were two prime contracts with Goddard (NAS5-9968 and NAS5-10750) to operate and maintain its two networks, the STADAN and the MSFN.[54]

Bendix employees received a number of incentives to work at overseas locations. By the late 1960s, for example, an array of seven allowances, each tailored to the specific conditions of the locale, was being offered. One, the Foreign Assignment Allowance, was a 10 percent of base pay "sweetener" applied to compensate for the inconveniences and cultural disadvantages at locations where employees had to learn a foreign language and assume additional assignments due to lack of trained local personnel. The Cost of Living Allowance was granted in circumstances where costs of subsistence, services and goods exceeded comparable expenses in the Baltimore-Washington area. Since that particular variable allowance did not cover housing expenses, a separate Housing Allowance was also established. For contractor personnel on location with families, many companies provided an Education Allowance that reimbursed the cost of providing educational services to children at local schools who would otherwise be tax-subsidized in United States public schools. As the years went by, many contractor employees ended up spending their whole careers overseas, doing very well for themselves, moving from station to station as the landscape of the tracking network evolved.[55]

A hallmark of STADAN was the training that its people received. Goddard and the contractor teams hired both degreed and nondegreed personnel in various disciplines, although most were in the field of electronics and electrical engineering. Technical training had been taking place at Blossom Point since the inception of Minitrack in 1956. In the fall of 1966, as NASA ceased joint satellite tracking operations with the Navy at Blossom Point, satellite training operations were consolidated with manned network training at Wallops Island. This was subsequently relocated to the Network Test and Training Facility (NTTF) at the GSFC itself.[56]

The NTTF employed about 40 contractor instructors who held either a Bachelor of Science degree or a teaching certificate in the appropriate technical field. The program was modeled along the lines of a vocational school, with a format for both classroom instruction and hands-on laboratory training. The instructional emphasis was decidedly pragmatic, fostering in students the capacity to quickly adjust and adapt to changing situations which might arise in the field. Instructors were sometimes dispatched to field stations themselves, providing the opportunity to conduct on-site training with foreign local nationals.

To ensure quality control and technical competency, Goddard, in the fall of 1967, instituted a formal job certification program for American citizens assigned to field stations throughout the network. All individuals studying at the NTTF were awarded primary certification in one technical area prior to his or her graduation. Certification was based on both academics and hands-on test performance in technical skill areas such as telemetry, communications, teletype operations, the IBM 1218 computer, and the FPQ-6 radar just to name a few. Certification tests examined the readiness of a technician in four fundamental areas: 1) Safety Procedures, 2) Operating Procedures, 3) Preventive Maintenance, and 4) Corrective Maintenance. Incumbent personnel (those veterans already serving on-site prior to the creation of the Certification Program) in some cases received certification by waiver based on their field experience; otherwise, they too had to take the test. Primary certification in a particular specialty meant that a person could practice that skill at any network location. The program also allowed personnel to obtain "secondary certification." This feature allowed station staff members to receive a number of secondary certifications in different technical skill areas outside their primary area. But these were generally limited for use only at the particular station where they worked.

The program was very successful. So successful, in fact, that three years after it was implemented, in a move to further promote foreign cooperation through the hiring of local labor, the Certification Program was extended to include skilled positions. Initial implementation began in 1970 at

Chile, Ecuador, and Madagascar. It eventually resulted in more than 200 local nationals being certified, mostly for Electronics Technician positions.[57]

☆   ☆   ☆

Ground stations and the people who operate them, however, do not make a network by themselves. A network requires a coherent and integrated communication system that enables the stations to talk to each other and to the GSFC. The NASA Communications Network, known as NASCOM, was that critical ground link which tied the system together. But when NASA first constructed its networks, NASCOM did not yet exist. Initially, each of the Agency's three networks—the Satellite Tracking And Data Acquisition Network (STADAN), the Manned Space Flight Network (MSFN) and the Deep Space Network (DSN)—all had their own communication system linking the ground stations to the control center. Even at Goddard, the satellite and human spaceflight communication organizations were in separate divisions within John Mengel's Tracking and Data Systems Directorate.

But by 1962, complexity of both the STADAN and the DSN had noticeably increased. With the new MSFN coming online, it became clear that running three separate communication systems at the same time was just not going to be very efficient, both technically and in cost. Technically, economically and logistically, a single system made much more sense. Studies were done to this effect which showed significant savings could in fact be achieved through circuit sharing and common use of field equipment. In July 1963, the Communications Division was formed at Goddard under the direction of Lavern R. "Vern" Stelter—who later became Chief of the NASA Communications Division—to coordinate all Agencies's ground communication activities. Up until then, this had been done by Headquarters.

In describing how the tracking networks worked, some have likened it to the human body. That is, if the control centers were the brain, then the tracking stations were the eyes and ears and the NASCOM lines the central nervous system. The flow of information between the worldwide network of stations and the operations center at the GSFC during a mission was quite extensive. The degree of mission success often hinged on the ability of a station to transmit critical information quickly and accurately. The NASCOM voice circuits enabled Goddard to manage, exercise, coordinate and brief all stations simultaneously during simulations and on actual missions. It also enabled the stations to discuss problems and procedures among themselves.

Even before the formation of NASA, John Mengel's Minitrack team at the NRL was faced with the job of getting tracking data back from the picket line and the other stations around the world. To develop and operate this system, Mengel recruited experienced range workers from Edwards (known as Muroc at the time) in Southern California, White Sands in Southern New Mexico, and from industry. As personnel from the military ranges and surveillance networks brought their knowledge to the NRL, its pool of communications knowledge quickly grew. This was vital since many of the then proposed Minitrack stations had to be in either undeveloped parts of the world or at great distances like Australia.

At first, the challenge was just getting data back to NRL. Speed was not too important. Similarly, science data was not time critical and were recorded on 14-track magnetic tapes and airmailed back to the United States. Within the continental United States, existing teletype lines were found to support Minitrack stations at San Diego, Blossom Point, and Fort Stewart, all of which were on established military installations. Not surprisingly, the real communication challenges were in South America where lines had to be added in uninhabited rainforests and undeveloped terrain. Throughout the 1940s and 1950s, the United States had worked with Latin American countries as part of its Inter-American Geodetic Survey (IAGS) to produce comprehensive maps of the entire Western Hemisphere. The U.S. government was able to cultivate important relationships through this endeavor which enabled the Army to go into some of these countries and lay communication lines that were needed by the South American stations. Stelter explained how it all worked out:

> Aside from the tremendous technical advances which made this network possible, we had the unqualified support and cooperation of every foreign country involved. Without their support, there would have been no successful NASA Communications Network. Indeed, if we had to pay for every person involved in this worldwide effort, we could not have afforded it. At that, our phone bill during the Apollo program was $50 million per year! On our part, we tried our utmost to make our foreign colleagues full participants and they in turn provided hundreds, possibly thousands of people, to assure the reliability and performance of the NASCOM network on which our success, and often the lives of our astronauts, depended.[58]

All of NASA—not just STADAN and the other tracking networks—relied on the NASCOM for communications, everything from everyday telephone conference calls to high rate telemetry transmissions. To this end, a combination of permanently rented circuits for teletype and voice were used. These included commercial landlines operated by telephone companies, ocean-floor cables, high frequency radio links and microwave relays, and finally, NASA and commercial communications satellites. By 1969, some two million miles of circuits in one form or another were being used.[59]

Data was routed using these circuits to key switching centers at various hubs in different parts of the world—Canberra, London, Madrid, Pasadena—which piped the data back to the United States.[60] Said Stelter:

> Initially, we had to rely on high frequency radio and on other rather primitive transmission technology which was anything but reliable. Somehow, we were able to make it work. . . . Later, we added the first computerized switching facilities, developed by UNIVAC [UNIVersal Automatic Computer], which was a significant advance. It assured us that all data from the spacecraft would get to the user in a reliable manner and outbound commands would in turn reach the spacecraft quickly. No easy task![61]

Other communication systems existed, but NASCOM was by far the world's largest, real-time communications system. It was very much a prototype, pioneering the broad-band communication systems of today.

As Minitrack steadily evolved into STADAN in the 1960s, the real-time link between satellites and controllers back at Goddard also improved. This was still at a time when very few commandable features were actually built into satellites. While things like tape recorders could be commanded to read out the data they had stored, little else could be directed from the ground. With the introduction of more complex satellites, particularly the OGO and OAO, by the mid-1960s mission controllers possessed the command capability for controlling many more satellite functions in addition to just the onboard scientific experiments. Goddard controllers, linked directly to STADAN stations, could now in effect "drive" a satellite via commands uplinked through a station as a satellite passed overhead.[62]

For Minitrack, transmission speed was a nicety but not a requirement. Now, with the advent of this new class of large observatory satellites and their high-rate data dumps, speed became not only desired but necessary. In addition, with Project Mercury on the horizon calling for real-time voice and tracking, telemetry, and command capabilities, a change was needed to bring the old NASCOM links literally "up to speed."

Preparations for the manned-satellite program, in particular, drove several requirements. It needed real-time, wide-band communications—the first for quick trajectory computations to support mission decisions and the second for the heavy data volume required to monitor spacecraft health and status and to talk with the astronaut. A most obvious requirement was that ground-based flight controllers, some of whom were stationed at tracking sites, had to be able to talk directly with the astronaut during a pass. Voice

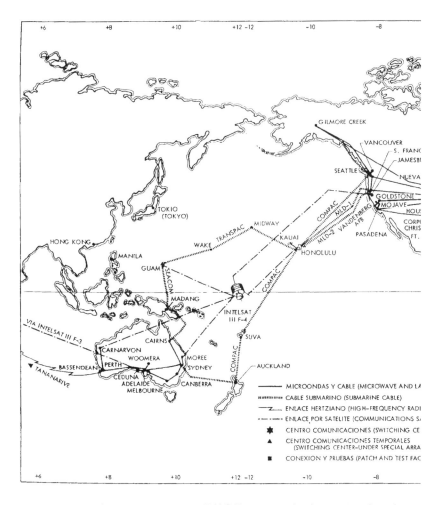

If tracking stations were the sensory organs of NASA's communications network and the Goddard Space Flight Center the brain, then NASCOM (the NASA Communications Network) could certainly be called its central nervous system, with over 2 million circuit

circuits between the spacecraft and the ground, therefore, had to be added to the system for the first time. (Television coverage inside the Mercury capsule was also proposed but would not be implemented until Apollo, partly because the communication technology was just not yet ready to handle the demands of such a bandwidth intensive item.) Added to all this was perhaps the most important difference between satellite and human spaceflight communications: reliability. If communications were lost during a pass for an observatory satellite, some data might be lost; and on a human flight, the consequences could be much more severe.[63]

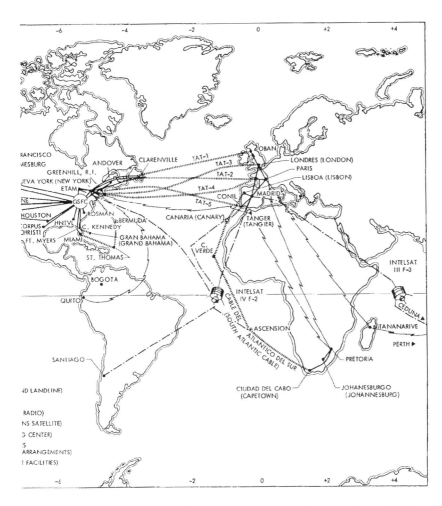

miles at its height of operations. Shown here is a layout of the circuits in 1972. Since the diagram was taken from a brochure describing the Madrid Station, some of the notes are in Spanish. (Madrid Space Station, INTA/NASA: Information Brochure, JPL P72-223)

These important differences were known to Langley engineers even in 1958 as they laid the foundation for a Mercury Network. It soon became apparent, however, that they could not just simply piece together a worldwide, real-time communication system from existing commercial or military circuits. Integration with Minitrack was also not possible; the requirements were just too different. Communication requirements for Mercury were, at the time, such a leap from anything else that had been done that Langley, and later Goddard, knew that an entirely different network had to be designed and built specifically to support the new man-in-space project.

This network, as it finally emerged, would consist of two communication circuits, one for voice and one for teletype. Teletype would use the then state-of-the-art Western Union 111 Torn-Tape Relay System, limited to a speed of 60 words per minute. The limitation was due to capabilities of the terminal equipment that was already being installed at the new MSFN sites. Initially, oversea switching centers were located at Adelaide, Australia (subsequently moved to Canberra), Honolulu, and London. Later, communication switching centers were added in Madrid, Guam, and the Kennedy Space Center. Goddard served as the hub and main switching center.[64]

"Data pipe" between Goddard and the Cape represented a substantial advance in wideband transmission capability over the previous generation with NASCOM. In 1961, for instance, four two-way, voice data circuits each capable of transmitting at one kilobit-per-second (1,000 bps) were installed between the Cape and Goddard. While this may not seem like something hardly even worth mentioning in today's world where data rates are measured in *gigabits*-per-second (that is, a billion bps), it was quite the improvement over the Minitrack days; one of these channels could now transmit 30 times as much data. This faster data requirement was needed because, prior to the establishment of the Mission Control Center in Houston, Goddard had to control the real-time displays at Mercury Control, then still located at Cape Canaveral. Soon, other wideband data lines were installed between GSFC and the STADAN stations, most notably at the new DAF of Rosman and Alaska. Finally, in preparation for sending astronauts into space, another important voice link was established in 1963 with a newly laid ocean floor cable connecting the new station on Bermuda with the Cape. This link could carry 2,000 bits-per-second of digital information and would continue to serve the Bermuda Station well into the Shuttle era.[65]

* * *

As NASA began sending men to the Moon, new stations were added and aircraft and ships were assigned to the network. Unlike Mercury or Gemini, live television was going to be used on Apollo and new equipment accommodating state-of-the-art transmission schemes at the S-band frequencies would be installed throughout the network. The role of communication satellites like *Syncom* would be called on to provide "switchboards in the sky." Developed by an industry team, their success would lead to the creation of the Cosmsat Corporation and an international consortium to manage and market worldwide satellite communication services. In time, this development would pay great dividends as the consortium's Intelsat ultimately ended up serving the very government agencies that helped to develop its technology.[66]

To see how sending humans into space remade the Goddard networks, one needs to turn the clock back to a time before there was a GSFC.

CHAPTER 3

THE MERCURY SPACE FLIGHT NETWORK

By the summer of 1961, the GSFC had fully established itself as the lead NASA center for directing science application satellite projects along with the communication, tracking, and data network that support them. In a little over two years since its founding as the first new field center established after the creation of NASA, GSFC's workforce had grown from 216 to 1,900, accounting for 11 percent of the space agency's total.[1] The STADAN was quite busy. It quickly matured during this time, supporting a wide array of satellite projects ranging from the original Vanguard and Explorer series to the newer TV and weather satellites such as TIROS (Television Infrared Observation Satellite) and even privately built communication satellites. The first satellite project which Goddard assumed full responsibility for was Explorer 6, launched on 7 August 1959. This began 16 years of GSFC association on this very successful series, which continued until 1981, with the orbital decay of Explorer 55. In April 1960, TIROS 1 downlinked the first ever global cloud-cover photographs taken from orbit. Solar science satellites were launched beginning with Solrad 1 in June 1960. AT&T's Telstar 1, the world's first commercial satellite, was launched on 10 July 1962.

The first Orbiting Solar Observatory (OSO 1) was launched on 7 March 1962. For years, John C. Lindsay, a former Associate Chief of the Space Sciences Division at Goddard, had investigated satellite designs using spin stabilization which enabled observations of the Sun in ways not possible from the surface of Earth. By 1960, the Ball Brothers Aerospace Corporation of Boulder, Colorado (at the time the leading developer and provider of stabilized pointing control devices for sounding rockets and balloon systems) had produced a successful engineering prototype of a spin-stabilized satellite. OSO1 produced many new findings. Among them was the discovery that the Sun's corona has openings (now called coronal holes) which were interpreted as huge, fast-moving bubbles rising through the corona.[2]

Even as application and science satellites were hitting full stride and making ever more exciting discoveries on a regular basis, events, though, were taking place which soon redefined NASA's priorities and led to the establishment of an all-together, different Goddard spaceflight network.

* * *

On 12 April 1961, the Soviet Union launched Yuri A. Gagarin, a 27-year-old Second Lieutenant in the Soviet Air Force, into orbit on Vostok (East) 1.[3] But unlike Sputnik, Gagarin's flight did not come as a total shock to the United States. What the Soviet Union accomplished four years earlier with Sputnik was the first sign to the international community of the existence of a full-fledged and very robust space program, one in which they proceeded with at an unrelenting pace. As early as the mid-1950s, a disturbing picture of Russia's growing capability in long-range, heavy payload rocketry emerged. The CIA, using the Lockheed U-2 high-altitude reconnaissance aircraft, had reported in 1957 the first Soviet ICBM on its launch pad at the Baikonur-Tyuratam space complex. Over the next two years, the world would know what the CIA had long suspected: the USSR was embarking on a series of space experiments designed to put a human into space.

In May 1960, the Soviet news agency Tass reported the launching of a "Spacecraft 1" into orbit. It weighed a massive 10,008-pounds and contained a "dummy cosmonaut." This was followed by more qualification flights carrying dogs which took place over the next several months. By the following April, the veil of secrecy surrounding Soviet space activities was rapidly breaking down.

It began with a rumor from Moscow on 7 April that a cosmonaut had been launched in secret. This was followed by a message wired to London on midnight of 11 April quoting unidentified sources as saying that a cosmonaut had returned from space and was undergoing physical examination for illness, and he was "suffering from post-flight effects of a nature more emotional than physical." Variations of this story immediately surfaced, including one by a French correspondent who suggested that the mystery cosmonaut was none other than Vladimir Ilyushin, son of the famous Soviet aircraft designer. The White House immediately denied the authenticity of these reports saying that the United States had no evidence whatsoever that such a flight had taken place. However, the next day, Moscow released the following official announcement:

> The world's first spaceship, Vostok, with a man on board was launched into orbit from the Soviet Union on 12 April 1961. The pilot space-navigator of the satellite-spaceship Vostok is a citizen of the USSR, Flight Major Yuri Gagarin. The launching of the multi-stage space rocket was successful and, after attaining the first escape velocity and the separation of the last stage of the carrier rocket, the spaceship went into free flight on a round-the-Earth orbit. According to the preliminary data, the period of revolution of the satellite-spacecraft round the Earth is 89.1 minutes. The minimum distance from the Earth at perigee is 175 km [109 miles] and the maximum at apogee is 302 [188 miles], and the angle of inclination of the orbit plane to the Equator is 65° 4'. The spacecraft with the navigator weighs 4,725 kg [10,420 pounds] excluding the weight of the final stage of the carrier rocket.[4]

Although the actual implementation of a U.S. man-in-space program had to await the creation of NASA, concept proposals for "manned-satellites" and how to track them had been in circulation as far back as the early 1950s. These took on a more serious form in 1956 when the Air Force embarked on a study known as "Project 7969" entitled, *A Manned Ballistic Rocket Research System*, designed specifically to investigate the requirements for a human orbiting laboratory. The formulation of ideas in that project did not directly translate into an actual program, but it did push the National Advisory Committee for Aeronautics to begin its own studies of human spaceflight.

The Secretary of Defense created, in early 1958, an Advanced Research Projects Agency (ARPA) to coordinate civilian and military research for piloted and unpiloted space projects; the idea being that ARPA would act as an interim space agency until Congress passed the appropriate legislation establishing NASA. In this capacity, ARPA articulated the first objectives of the American human space program. The goal was clear: "Achieve at the earliest practicable date orbital flight and successful recovery of a manned satellite, and to investigate the capabilities of man in this environment."[5] These goals were to be embodied in Project Mercury, America's first human spaceflight program.

Within the newly created space agency, the Langley Research Center located in Hampton, Virginia near the mouth of the Chesapeake Bay, was the home to two key groups entrusted with starting Project Mercury. The

STG was one of these, responsible for the overall implementation of the man-in-space program, formally designated as Project Mercury on 26 November 1958. The STG directed the development of both the launch vehicle and the spacecraft as well as the recruiting of flight crews and flight controllers. Another cadre of personnel formed the other group, informally called the Tracking System Study Group (TSSG). The TSSG helped the STG establish network support requirements for Project Mercury. Heading it was Edmond C. Buckley, who was later to be in charge of all NASA tracking activities as the Agency's first Associate Administrator for Tracking and Data Acquisition (from 1962 to 1968).

The STG was at the time under the jurisdiction of the newly established Space Projects Center. In 1961, this center moved to Greenbelt to form the nucleus of the new GSFC. By fall of that year, however, the STG had moved on to Houston, Texas to establish the MSC. With this move, MSC immediately became the lead NASA center for all human spaceflight activities and assumed overall responsibility for executing Project Mercury. The job of implementing an effective worldwide tracking and data acquisition network remained with GSFC. These initial shifts in center responsibilities, as it turned out, determined how each of these Field Centers eventually ended up supporting NASA's human spaceflight activities for the next 40 years, roles that continue today.[6]

One of the principals on the TSSG was H. William Wood. A member of Ed Buckley's team at Langley, Bill Wood was investigating X-ray and gamma ray sensing instruments for possible use aboard spacecraft when he got the call to help plan a network for Project Mercury. Said Wood:

> At Langley, we worked under the supervision of Hartley Soule, an assistant laboratory director who helped us cut through the red tape and made us realize early on that we were engaged in a very unique and challenging undertaking. However, the Langley Center did not want to become distracted from its primary role as a research organization. . . . In my opinion, it did not wish to evolve into an operational hub for space activities. Thus tracking, data acquisition and computer activities for Project Mercury were assigned to the Goddard Space Flight Center prior to the first manned orbital flight.[7]

Wood and others who agreed to help were told by Soule that they could return to Langley whenever they wished. Those who agreed to consider a transfer soon met at the new Space Projects Center with Director Harry J. Goett. Some ended up staying at Langley, and others transferred with the STG to the MSC in Houston.

Chapter 3 \ The Mercury Space Flight Network   69

America's first human spacecraft was Mercury. It flew six astronauts into space (two sub-orbital, four orbital) from May 1961 to May 1963. Weighing some 1,360 kilograms (3,000 pounds), the vehicle was launched into space atop the U.S. Army Redstone rocket and later, the Air Force Atlas-D intercontinental ballistic missile. Pictured is Wally Schirra's Mercury 8 capsule *Sigma 7* being readied in Hanger S at Cape Canaveral in September 1962. (NASA Image Number GPN-2000-001441)

The TSSG began addressing the network issue in earnest in the fall of 1958. The job soon turned out to be bigger than anyone had anticipated. Network requirements for human spaceflight were rather different than those needed to support unmanned satellites. Unlike the Minitrack Network—which passively received spacecraft signals and did not make an accurate orbit determination until several orbits had been complete—human spacecraft tracking had a much more *instantaneous contact requirement*. Towards that end, radar tracking rather than interferometry, was essential. Telemetry now had to be augmented, for example, to monitor the health of the astronaut. The complexity of command functions also greatly increased over what was needed for science satellites in order to ensure the safe orbiting and, more importantly, return of the crew. Finally, a manned spaceflight network was obviously needed to establish two-way voice communication between the ground and the vehicle.[8]

By the end of the year, the TSSG had identified three major areas of concern associated with a Mercury tracking network, concerns which had up until then been widely underestimated. First, and somewhat to their surprise, the group found that there was no such thing in existence as either a commercial or a military, real-time, worldwide communication system. Here was a vital ingredient of mission control that was completely missing! Second, there was no existence of reliable, high-capacity data links that could carry the large amounts of data between the computer facilities and the mission control center. Third, good radars were available, but they were designed primarily to track DOD ballistic missiles and reentry vehicles, their beams too narrow to expeditiously locate and keep track of a fast-moving spacecraft in orbit.[9]

The Space Task Group began to feel that the weight of the network job was diverting too much attention from its primary job of designing the Mercury spacecraft. In January 1959, Charles W. Mathews, a member of the STG, recommended to Abe Silverstein, Director of Space Flight Development at NASA Headquarters, that the STG be formally relieved of the responsibility for building the network and that the responsibility be given to the TSSG. This was a seemingly reasonable request since the TSSG had already been studying the network problem for nearly a year. On 16 February, Silverstein formally directed this change in a memorandum to J. W. Crowley, Director of Aeronautical and Space Research at NASA Headquarters.

Silverstein's memorandum asked the TSSG to "complete and refine" network plans so as to satisfy requirements generated by the STG and to "place and supervise" contracts for generating procurement specifications and a final network deployment. Langley officials were to make use, wherever practical, of DOD personnel and facilities in the Pacific Missile Range (PMR) in Point Mugu, California; the Atlantic Missile Range (AMR) in Cape Canaveral, Florida; the White Sands Missile Range (WSMR) in Alamogordo, New Mexico; and the Eglin Gulf Test Range (EGTR), Eglin

Air Force Base, Florida. In this manner, the practice of NASA leveraging DOD assets wherever practical began early on in the construction of the Mercury Space Flight Network (MSFN).

On 16 February 1959, the Langley network study group was officially renamed the Tracking And Ground Instrumentation Unit, or TAGIU. Two years later, TAGIU would be relocated to GSFC when the new Center opened. Leadership of TAGIU was assigned to Barry Graves who had been part of the TSSG since the beginning. Several key participants from TSSG also continued their assignments on the new team. Leading their respective disciplines were:

> George B. Graves, Jr., *Electronics and Head of TAGIU*
> James J. Donegan, *Computers*
> Ray W. Hooker, *Site Selection and A&E (Architecture and Engineering)*
> Paul Vavra, *Assistant to Graves*
> H. William Wood, *Logistics*

TAGIU quickly grew to 35 people. This nucleus was in essence heir to Langley's radar and high-speed flight experience, plus six months' worth of network studies. Although TAGIU engineers knew roughly what a Mercury Network should entail, they did not yet have the detailed specifications needed to procure hardware and to begin building up stations. Getting industry help under contract and enlisting the aid of the Air Force was among the first items of business. (The USAF was at the time building up their own network: the Air Force Satellite Control Network.) TAGIU's effort went from turning studies into requirements, requirements into specifications, specifications into hardware, and finally, hardware into stations. Even though the major functions of the Mercury Network had already been decided on: tracking, telemetry reception, voice communication, and capsule command; the team fully realized that these functions could be performed in a number of different ways.

By early spring 1959, TAGIU was ready to generate detailed network specifications. To this end, four contracts to industry were awarded:

1. Ford Aeronutronics, *Study radar coverage and trajectory computation requirements*

2. MIT Lincoln Laboratories, *General consultation and proposal evaluation*

3. RCA Service Corporation, *Write the network specifications*

4. Space Electronics, *Design the Mission Control Center*

Things moved quickly. Said Wood:

> We were told that we had about two years to complete the job; then we would return to basic research. Of course, that is not the way it turned out. . . . I was told one morning that I would be managing the RCA contract and that same afternoon we had our first meeting with the contractor. We moved pretty fast! At the time, a great deal of attention was being given to the Mercury spacecraft itself: how to get a man into orbit and how to return him safely to Earth. But very little attention had been given to communications and tracking which had to support those flights. So we were asked to get busy and develop a set of specifications. This we did by 1959. They were no thicker than an issue of the *Aviation Week* magazine.[10]

The network developed around 12 ground rules:

1. The launch azimuth of approximately 32.5° was fixed early on. It enabled maximum use of DOD range facilities and kept the ground track of the spacecraft over the continental United States much of the time to preclude over-flight of countries that might not cooperate. It also resulted in acceptable recovery areas close to U.S. Naval facilities.

2. The Atlantic Missile Range based at Cape Canaveral was to be employed for launch and recovery operations. This was an easy decision because of AMR's already superb facilities and experience.

3. The network was to be worldwide, using stations in foreign countries where necessary, and operate on a real-time basis to monitor spacecraft status and astronaut health.

4. The space medicine community strongly advocated continuous voice contact with the astronaut, but this requirement proved impractical. Despite the controversy, STG and TAGIU moved ahead using the goal of a maximum 10 minute, loss-of-signal and voice contact in between stations (the so-called 10-minute "dead time" rule).[11]

5. A centralized control center was to be built, but controllers would be located at each ground station in case of communications difficulties.

6   Proven tracking and communications equipment was to be employed to the extent possible with a minimum of research and nonrecurring engineering development.

7   In order to enable a go/no-go decision to be made, and to effect emergency reentry prior to orbit insertion before the spacecraft approached the African land mass, continuous tracking from network radars downrange of the launch site was needed to give the computers enough real-time data for accurate orbit prediction before the spacecraft passed beyond the Bermuda tracking station.

8   After orbit injection, intermittent tracking was sufficient. Continuous tracking was again to be required during deorbit and reentry so as to accurately pinpoint the splashdown point.

9   These existing radars would need acquisition aids because of their narrow beamwidths.

10 Only computers could cope with the flood of tracking and telemetry data arriving from all the network stations. A centralized computer facility with data link to the Mission Control Center would therefore be needed.

11 Frequent network simulations and exercises would be needed to train the operators.

12 Redundancy would be required to provide the reliability needed for manned flight.

The team started with the assumption that the network would support a three-orbit Mercury mission at an inclination of about 33° to the Equator. Next, it determined just how long a communication gap could be tolerated between stations. The first design iteration had 20 stations scattered around the globe; eventually, the number was reduced to 18. The procedure was definitely quite straight forward. As Wood put it, the team literally just "took [out] a world atlas and tentatively located these sites."[12]

By early 1960, site surveys for the entire network had been conducted and the building of the MSFN began. During Mercury, Mission Control was located at Cape Canaveral, Florida, adjacent to the launch site, while the network computer center was located at Goddard. All the computer activities were handled on a remote basis with Goddard computers, for example, to drive the displays and plot-boards at the Cape. (This would be the case until 1965, when starting with Gemini 4, the Manned Spacecraft Center in Houston acquired the necessary computer capability for human flight operations when the new Mission Control Center opened.)

Satellite programs and STADAN had already given GSFC quite a head start, providing the expertise and experience needed for network communications, simulations and data processing. Furthermore, Goddard management under Center Director Harry Goett desperately wanted the work. GSFC could claim, with justification, that to expand the staff and facilities at Greenbelt rather than build new ones at Langley or Houston was more economical and technically feasible. Some thought was given by Headquarters to locating the Mercury ground computers at Cape Canaveral. But the Cape was already heavily involved in Air Force and Navy launches. In addition, NASA wanted to keep its computing facility separate from the DOD. These factors all made Goddard the most logical choice to lead this work.

On a Mercury flight, the network computer facility would receive reams of data from its field stations. Jack Mengel, who four years earlier led the development of the Minitrack Network and who, along with most of his original Naval Research Laboratory staff now at Goddard, was asked by Graves to develop the MSFN computing capability. This made sense since Mengel's group was still in the throes of expanding their computing facilities and had just taken over the Vanguard Computation Center formerly under ownership of the Naval Research Laboratory. Before long, some 60 Goddard and IBM contractors were assigned to support the MSFN under Niles R. "Buck" Heller who became Chief of the Manned Space Flight Support Division at Goddard.[13]

The network relied on proven C-band and S-band radars for tracking; for redundancy, some stations used both. Radar skin-tracking (which uses direct reflections of radio waves transmitted from the ground to the target) while fine for tracking ballistic missiles, was not considered reliable enough at orbital altitudes to be used in the Mercury program. Instead, two radar-triggered onboard beacons (or transponders) at these wavelengths were located on the spacecraft to provide the ground with more robust radar returns.

The C-band FPS-16 and the S-band VERLORT (Very Long Range Tracking) radar built by the Reeves Instrument Corporation were specifically stipulated by the TAGIU, as both were considered proven systems, having been used for years by the DOD. Manufactured by RCA and used by the Army at White Sands since 1957, the FPS-16 was a 3.6-meter (12-foot) dish, high-precision radar capable of tracking to an accuracy of seven meters at a range of 925 kilometers (500 nautical miles). The VERLORT, a 3-meter (10-foot) transportable dish operated by trailer-housed equipment, had a slightly longer range of 1,300 kilometers (700 nautical miles). It was a radar developed in World War II for tracking aircraft. Limited in range and accuracy, it did have a reputation for being able to acquire an object with great reliability and begin tracking in a very short amount of time.

The only new piece of major tracking equipment specified was an Active Acquisition Aid (AAA) required for spacecraft acquisition by the narrow-beam radars. The acquisition antenna was actually an array of 18 small antennas mounted on a screen. By locking onto a radio signal transmitted from the spacecraft, this arrangement allowed it to begin coarse tracking to an accuracy of about half a degree. This pointing information was then made immediately available via a cable and switching system called the "Acquisition Bus" to aim the other antennas. The AAA, made by the Cubic Corporation, had a very wide beam of 20°, effective out to 1,500 kilometers (800 nautical miles).[14]

Proven equipment was also stipulated for other elements of the ground-to-space link, which included telemetry, command, and voice. TAGIU selected systems that were already available or, as a minimum, not too difficult to adapt from existing hardware. Telemetry from the Mercury spacecraft was downlinked at 226.2 MHz using the analog modulation technique of Pulse Amplitude Modulation (PAM)/Frequency Modulation (FM), a scheme commonly used for ballistic missile and aircraft testing work. Two separate telemetry links between network sites and the spacecraft were utilized for redundancy. Each station had dual-redundant FRW-2 transmitters for command uplink. These were commonly used by the military for high performance aircraft testing. The FRWs were connected to a 10 KW high power amplifier to give them the needed range to send commands to a spacecraft in low-Earth orbit. This capability enabled ground controllers at the MSFN sites to issue commands to a spacecraft in the event the astronauts were to somehow become incapacitated. Common VHF radio was used for two-way, air-to-ground voice communication; a backup was operated on the international military aircraft emergency frequency.[15]

A seminal ground rule that TAGIU stipulated was that nonrecurring engineering costs for equipment research and development could be minimized through reliance on off-the-shelf hardware. Radars, radios and computers existed, but no one had ever tried to mold them together into a cohesive, real-time network on a global scale. Not unlike a jigsaw puzzle, when TAGIU tried to put the electronic puzzle together, it found that most of the available pieces did not fit well together. And most did not have the requisite reliability to support human spaceflight. Modification of available equipment also became a frustrating problem. Modifications were often made after equipment had been fielded just to bring them up to MSFN standards. Equipment that could not be modified had to be backed up with 100 percent spares to ensure operating redundancy. Network equipment reassigned from the DOD was considerably more reliable than common off-the-shelf equipment, not surprising as military specifications were more rigorous and based on operating environments similar to those of the MSFN. However, obtaining priority military equipment was, for the most part, difficult due to availability.[16]

Before the MSFN could be used to support a mission, it of course had to be checked out, first on a station by station basis and then as an integrated network. One of the most valuable testing methods during network shakedown was simulated spacecraft tracking using aircraft flyovers. Although ground stations had a collimation tower at the edge of the antenna field, it was used only to get a first order calibration. Calibration airplanes, or "cal planes," flying at 7,600 meters (25,000 feet) were needed for final system calibration and verification of an antenna's ability to actually track a moving object. Graves had asked Bill Wood to develop a "real world" method to accept the stations after they were constructed. Wood's people came up with a variety of ideas, ranging from using balloons to devices mounted on towers. Each had its technical advantage as well as disadvantage, and there was really no clear-cut choice. During initial discussions at Goddard, there were, in fact, as many votes against using calibration planes as there were for it. When Barry Graves finally pressed for a recommendation, Wood would later say, "I suppose in self defense, I said airplanes."[17] To do this, NASA procured a Douglas DC-3, two DC-4s and later a Lockheed L-1649A Super Constellation (known both affectionately, and sometimes not so affectionately by those who had to track them, as "Connie") and outfitted them with Mercury TT&C flight-qualified electronics.[18]

Calibration aircraft were used in different ways to test the status of network stations. For STADAN, they were used to calibrate the Minitrack system and to conduct acceptance testing for new antennas. The DC-3 and DC-4 primarily supported the STADAN while the L-1649A supported the MSFN. Calibration of the Minitrack sites for instance, was performed by flying at 7,600 meters and making East-West crossings of the main Minitrack beam in 10 percent steps north and south of the station zenith. The RF (at 108MHz, and later, 136MHz) crossings of the main beam were electronically compared with the positions obtained from the flashing lamp on the belly of the aircraft and measured against the stellar background. Flyovers were done for a period of up to six weeks prior to a scheduled mission and usually took a couple of hours. The aircraft made multiple passes at predetermined altitudes, speeds and directions to simulate a spacecraft passing overhead. The antenna would attempt to autotrack the calplane during each pass. These station simulations ("sims") were designed to uncover any hidden anomalies and to help prepare station operators for their upcoming mission. Unlike a spacecraft orbiting Earth, however, the calibration airplanes did not fly in well-defined trajectories. As such, they actually turned out to be more difficult to track than the spacecraft. After a few exercises, the prevailing thought was that "if you could track the cal plane, you could track the real thing."[19]

Since the DC-3 was not pressurized, the crew had to go on oxygen bottles when flying over 3,600 meters (12,000 feet). Later, Hal Hoff at

Goddard required all NASA members of the calibration team to undergo high altitude chamber orientation at Andrews Air Force Base. This turned out to be smart. Former Goddard engineer Dave Harris recalled an episode where a crew member monitoring the transmission equipment suffered from a classic case of oxygen deprivation and the response of the others "was quite rapid."[20] There were also simulated ditching sessions held to prepare for any contingency. Besides serving as a flying target for tracking tests, another benefit of these calibration planes was to ferry necessities to the foreign sites. Toilet paper was a frequent cargo when heading to South America.

In addition to these site-specific exercises, computer-based simulations were used to exercise the entire network. The principal tool of computerized simulation was the Computation and Data Flow Integrated Subsystem, or CADFISS, developed by James Donegan and Goddard's IBM contractors. During an "integrated-sim," CADFISS interrogated the various stations via the Mercury ground communications network (and later the NASCOM) while GSFC computers analyzed the response and determined the status of each element in the dataflow. In this way, stations could be checked simultaneously, faults isolated and debriefings held with Goddard engineers directly "over the loop" immediately after a simulation session. As a mission date approached, these integrated-simulations were conducted with increasing frequency.[21]

✯ ✯ ✯

When the STG was officially formed on 7 October 1958, it fell under the jurisdiction of GSFC rather than LRC because NASA Headquarters had every intention of moving the group to the new Greenbelt Center. As the human spaceflight program expanded, however, it became apparent that Goddard would no longer be able to absorb STG responsibilities while still carrying out its prime responsibility of directing unmanned satellite programs. It was also evident that human spaceflight was going to continue well beyond just Project Mercury and that it required a new center of its own. Due to its very charter and the work it was already doing on Mercury, the STG was identified as the nucleus of the proposed MSC and in the fall 1961, the group literally picked up and moved to Houston (the location having recently been selected by Capital Hill as the site for the new Field Center).

The official word on delegating MSFN network responsibility to GSFC was given in a 3 April 1961 letter from Abe Silverstein, Head of the Office of Space Flight Programs at Headquarters, to Harry J. Goett, the first Director of Goddard, entitled "Mercury Network Operating Responsibilities." On 12 May 1961, Langley center director Floyd L. Thompson and Harry Goett met at Greenbelt to arrange the transfer of network operating responsibility. They established a committee (with a Langley chairman) to oversee the transfer which was scheduled to take place during the third quarter of calendar year 1961. To help meet this new directive, Goett brought in Ozro M. Covington to Goddard from the WSMR where he had been the Technical Director of the Army Signal Missile Support Agency. His experience in organizing the White Sands range instrumentation was exactly what Goett was looking for, experience that soon proved invaluable in the buildup of the MSFN.[22] This was a strategic move on Goett's part as Covington brought with him a very-abled manager named Henry Thompson from White Sands to serve as his deputy.[23] Thompson, in turn, recruited George Q. Clark from White Sands. (Clark later became the first Tracking and Data Relay Satellite System Project Manager at GSFC, responsible for both the pioneering technical solution and the innovative procurement approach. See Chapter 7.)

In addition to his tangible technical abilities, Covington had unique intangibles which were greatly needed to soothe rifts that immediately began developing between Greenbelt and Houston. It was in the midst of these organizational changes that "turf wars" between the two NASA Field Centers appeared regarding ownership and the role that the network was going to play in the upcoming human space program. Commenting on this touchy issue, Covington recalled:

> During those early days there were many debates and conflicts as to the assignment of the manned spaceflight program. There was a time when the Goddard Center was to be its base of operation. The center had already been charged with major unmanned flight programs, all of which also required sophisticated tracking and communications support. But soon it became evident that a separate NASA organization had to be created to plan and direct America's man-in-space efforts. It was, after all, a mammoth undertaking, particularly after President John F. Kennedy had told the world that the United States would send a man to the Moon and return him safely to Earth in the 1960–1970 decade. Indeed, this was not the only conflict as to who would be responsible for the manned missions. There were similar conflicts as to who would be charged with the difficult and challenging tracking and communications tasks supporting these flights. There was the U.S. Air Force which was launching and tracking its missiles from Cape Canaveral. It certainly wanted the job, and always there was the question of the wisdom of separating the mission control center and the network. As Goddard's representative for manned space flight, I attended many of those early planning sessions, and it was during one of those meetings that I had occasion to spend some two hours with Dr. Christopher Kraft who was then planning some of the early Mercury missions. We met in a hotel room

near Cape Canaveral and discussed at length the difficulties arising from the separation of mission and network responsibilities. We developed a very productive relationship which continued throughout the years, ranging from the Mercury and Gemini through the Apollo programs.[24]

The "Not Invented Here" syndrome was by no means unique to GSFC and MSC. In developing the Mercury Network and its Control Center, it was clear early on that NASA was conducting a very different kind of launch than what the Cape was used to. Besides the obvious presence of having a man on board, the payloads that NASA was launching were generally active in all phases of a launch, from prelaunch countdown through insertion into orbit. This was generally not the case, though, for DOD payloads which were mostly dormant and were powered up only after orbit insertion. NASA thus had to install its own network equipment at the Cape and at key downrange stations. The DOD and its contractors were naturally not very happy when TAGIU insisted that all equipment used to interface directly with the manned payload were to be designed and controlled by MSFN engineers and that they be operated in accordance with NASA procedures. Happy or not, there was really no other way to reliably implement the network.[25]

There was also not always complete agreement as to the GSFC role when it came to manned tracking. Goett questioned the wisdom of committing his people and resources to a major program not entirely under his control. But Covington would explain years later, Goett came to see it the other way:

> When he saw the impressive team we had assembled for the tracking job, he agreed and relented. 'Let's go after it', he agreed, and the Center did indeed, becom[ing] the nerve center of NASA's worldwide manned spaceflight tracking network, just as it was also supporting major unmanned missions in the near-Earth regions. We assembled a team of some 350 top flight people, supported by a large number of contractor employees. Goddard's administrative staff under Dr. Michael J. Vaccaro gave us the critical support services we required to tackle the job, from personnel recruiting to budget planning and contract administration. We needed this support very badly.[26]

Covington, for one, realized the importance of open communications between the two Centers and treated MSC as a customer rather than a rival:

> I considered it to be my primary function to maintain the closest possible liaison with our 'customer', the Manned [Spacecraft] Center in Houston. We created a reliable system of tracking sta-

tions around the globe along with communications links and computer facilities all designed to provide the mission controllers with the information they needed to get our men into space.[27]

At the conclusion of Project Gemini in 1967, Covington was named Goddard's Director for Manned Flight Support, overseeing all network activities in preparation for the Apollo Moon program. The management structure at NASA Headquarters also shifted during this time so as to meet the changing requirements of human spaceflight. Since leaving Langley in early 1958, Edmond C. Buckley, as Director of Space Flight Operations, had headed NASA's network operations, reporting directly to Abe Silverstein. To streamline operations, one of the first things Buckley did was to consolidate all of NASA's tracking and data networks (STADAN, MSFN and DSN) under him to form the Office of Tracking and Data Acquisition (OTDA) on 1 November 1961.[28]

On 21 May 1959, Graves and his team sent out requests for proposals formally soliciting industry bids to build a manned spaceflight tracking network for NASA. On 30 June, a team led by Western Electric Company, Inc. was awarded the prime contract. Of the roughly $80 million spent in constructing the MSFN, about 85 percent ($68 million) went to the Western Electric team. The company named Rod Goetchius as Program Manager. In addition to Western Electric, four major subcontractors were on the team. Their roles were:

> Western Electric: Overall program management, procurement, production, transportation, installation and testing of equipment. Design and implementation of the ground communication subsystem. Training of maintenance and operating personnel.
>
> Bell Telephone System: Analysis and development of operation plans and tests. Design of command and control displays at Cape Canaveral and Bermuda. (Bell selected Stromberg-Carlson to build and install the flight control displays.) Provide a simulation system for flight controllers and astronauts.
>
> Bendix Corporation: Design and fabricate of telemetry and tracking display equipment. Systems design, fabrication and integration of radars not already furnished by the government. (Bendix obtained new radars from RCA and Reeves Instrument Corporation.) Design and fabrication of all Mercury spacecraft communication equipment.
>
> IBM: Computer programming and operations at GSFC and Bermuda. Maintenance and operation of the launch and display subsystem at Cape Canaveral.[29]

Construction of the Mercury Space Flight Network ground stations proceeded swiftly in 1960 and 1961. Site surveys had established a set of minimum requirements for geographical locations:

> No physical obstructions to the transmission and reception of signals greater than 1° elevation angle
>
> Existence of adequate separation distance between receiving and transmitting antenna to prevent electromagnetic and physical interference
>
> Minimum outside radio interference (a quiet RF location)
>
> Existence of housing and utilities
>
> Availability of good roads[30]

One of the first stations to be established on foreign soil was also one of the most critical. With the exception of Cape Canaveral, Bermuda, some 1,450 kilometers (900 miles) due east of the Carolina coast, was the most complex and important of the 15 MSFN ground stations. This was because the flight path of the Mercury Atlas took it almost directly over the island, providing a short but crucial 25 second window to track and make decisions on its status during ascent into orbit.

As Bill Wood explained, the station was important at the time for these and several other reasons:

> The first station to be definitely decided upon was Bermuda. It was needed because of the high failure rate of the Atlas booster which in those early days was about 50 percent, and since an abort situation was a highly probable reality, we needed Bermuda to keep an eye on the Cape Canaveral launches and the first critical phases of the flight downrange. We planned a control center there as well as at the Cape to give us reliable communications and controls should we be forced to make abort decisions. We were, after all, dealing with manned missions and unproven, even unreliable communications.[31]

Lynwood C. "Lynn" Dunseith, one of the original MSFN engineers who in 1982 became the Director of the Data Systems and Analysis Directorate at the Johnson Space Center, also recalled that in those days:

> We had great difficulties with our communications from Bermuda, a key station during the launch phase of any mission. So, a control center was established there in the event Bermuda had to enter the picture for launch and abort decisions.[32]

The schedule was tight. On 15 March 1961, less than two months before Alan Shepard's scheduled suborbital flight, Washington finally reached a formal agreement with the United Kingdom of Great Britain and Northern Ireland to operate a tracking station on the island.[33] The main site was on Cooper's Island, a small 77-acre rock-coral shelf just off of Saint David's Island on the northern shores of Bermuda. The station was located on an eastward extension of Kindley Air Force Base, on land already managed by the United States Air Force. Its use dated back to a World War II agreement between President Roosevelt and Prime Minister Churchill. Another site was in Town Hill on the main island. The station cost $5 million to build in 1961.[34]

NASA employed a large workforce to operate this station—60 contractors along with 20 Bermudians. Since the tracking network (as well as the space agency itself) was still young, everyone was put to work immediately. Robert E. Spearing, who today is the Director of Space Communications at NASA Headquarters, was 22-years-old when he began working at the GSFC to help get Bermuda up and running. He recalled:

> When I started, I was placed with another engineer who was fairly experienced working on what we call UHF command systems. . . . These are high power transmitters that are used to actually send information to the spacecraft and are also used for what we call command destruct range safety options. About three or four months after I started, the fellow I was working with informed me that he was leaving the Agency for another job. I, of course, went to see his supervisor and asked him what I should do since this fellow was leaving. He looked up at me and he said 'You're it!' . . . That is the way NASA operated back in those days. We were not heavily populated. We were definitely not over-staffed and everyone was given a lot of responsibility to get things done in the best way you could.[35]

The criticality of the station to make the key go/no-go call was not lost on the Bermudians. When construction began in April 1961, the local island newspaper headlined "Bermuda Tracking Station Has Vital Part to Play in First Manned U.S. Spaceflight," going on to boast that "an order from here could bring it down."[36] For fiscal reasons (along with perhaps more importantly, diplomatic expediency and promotion of international goodwill), maximum use of the local workforce was exercised in Bermuda. This worked out well, for the most part, not only in Bermuda but in other places as well. Some problems did arise, particularly in nonindustrialized countries where work was mostly done by hand. Even in more advanced places such as Australia, contractors often ran into "specifications and standards" discrepancies where, for instance, electromechanical devices made in the USA did not interface correctly with local, European-based standards like voltages and the imperial versus metric systems—something that any traveler can relate to even today.[37]

Even as construction was underway on Bermuda, an agreement was reached to build another critical station, this one in the South Pacific. On 6 April 1961, Secretary of State for Foreign Affairs, H. C. Hainworth, exchanged signatures with Walworth Barbour, Charge d'Affaires ad interim

Designated as Site No. 11, the Canton Island tracking station was located four and one-half miles southeast of the airport terminal in an area used as a fighter strip during World War II. Designed only for telemetry and voice communications, the site was rather sparse with no radar or spacecraft command capability. Communication was maintained with the Mission Control Center by the Goddard teletype and voice loop networks while communication with the Hawaiian area was maintained over a teletype circuit and by a limited traffic single sideband circuit. The island was staffed by 47 U.S. contractors.[39] (Unnumbered photograph. Folder # 8810, NASA Historical Reference Collection, NASA History Division, NASA Headquarters, Washington DC.)

of the British government, formally establishing a tracking station on the small island of Canton. This was the culmination of planning by NASA which first started in 1959. When LRC was first delegated by Headquarters with the responsibility for establishing the Mercury network, a chief aim was to identify locations that maximized communication coverage over three orbits, this being the then planned duration of the Mercury orbital missions.

As the TAGIU examined the ground track map, it was found that a large gap existed from the time when the spacecraft left the Australian station at Woomera to when signals could be acquired by the Hawaii Station on Kauai. During the first orbit, the ground track of the Mercury capsule took it well south of the Hawaiian Islands such that acquisition-of-signal would not have been possible until the capsule was over the North American continent—a communication gap of over 25 minutes! A loss-of-signal of this duration was not considered at all acceptable when Mercury was first being planned (even though longer quiet periods were to become common place on later missions). A site had to be found to close this gap. It turned out that the ground track of the Mercury spacecraft took it almost directly over Canton Island in the Kiribati Republic.

Canton (also known by its Kiribati name of Kanton or Abariringa) is the largest and most northern of the Phoenix Islands, located in the middle of the Pacific Ocean just south of the Equator. It is a volcanic atoll, made of a low, narrow rim of land surrounding a large shallow lagoon. As with most atolls in this region, it is relatively small, only seven kilometers (four and one-half miles) wide on the west, from which it narrows to a southeast point some 14.5 kilometers (nine miles) away. Since its discovery by independent sailors in the early 1800s, mostly American whalers, the island served mainly as a stopping point for American and British ships traversing the Pacific shipping lanes. Canton broke into the news in 1937 when American and New Zealand astronomers chose it as a spot from which to view the total solar eclipse of 8 July; enough publicity was generated to at least put the tiny spot on the map. Prior to this, about the only news coming out of the island was the continual British efforts to reassert their jurisdiction over the Phoenix Islands and the speculated role Canton may have played in the disappearance of Amelia Earhart and her navigator Fred Noonan. In 1938 and 1939, Pan American Airways developed an extensive airport on the island, deepening and clearing the lagoon to initiate air travel to New Zealand using Canton as one of their ports of call.

On 9 August 1959, NASA Administrator Keith Glennan wrote a letter to the Department of the Interior proposing to establish a tracking station on the island. (At the time, the United States had codominion status along with the United Kingdom for the island.) In a reply to Mr. Glennan, Interior Secretary Fredrick A. Seaton granted permission, saying "We wish to assure you of any further cooperation your Administration may require of this Department in furtherance of this most important project."[38] In December, Langley sent a delegation to the island for a site survey and to begin negotiations with the local authorities. Construction began the next year, and the station was well prepared by the time John Glenn orbited Earth on 20 February 1962.

Located 3,200 kilometers (2,000 miles) northeast of Canton are the Hawaiian Islands. Situated in the middle of the Pacific Ocean, Hawaii provided an ideal setting for a network ground station, picking up spacecraft as they emerged from the South Pacific. Construction of the Kokee Park Tracking Station on the southwest hills of Kauai began in May 1960. Eleven months later, the station was completed, coming online in time to support the first unmanned orbital Mercury Atlas test flight (MA-4) in September 1961. Five months later, Glenn completed the first U.S. human orbital mission on MA-6.[39]

Even though Hawaii was designed to support the human spaceflight program, it was also frequently tasked to support Goddard's unmanned science satellites because of its good location and full compliment of equipment. In fact, multitasking of this station went beyond NASA. Hawaii was a shared effort between the space agency and the Navy's Pacific Missile Range (PMR). Beginning with its construction, the Kokee Station was a joint venture between the two departments, with the Navy operating a tracking and data acquisition facility on the grounds of what had been Bonham Air Force Base on the western side of the island. When construction was finished in August of 1961, it was integrated into the PMR command at Kaneohe, Oahu. In this arrangement, the Barking Sands (U.S. Navy) and Kokee (NASA) facilities operated as a single, integrated station even though they were separated by some 3.5 kilometers (2.2 miles) and 1,200 meters (4,000 feet) in elevation.[40]

In the spring of 1965, the DOD transferred control of the facility to the Air Force Western Test Range, headquartered at Vandenberg Air Force Base in Lompoc, California. According to Virg True, former Station Director, there was little change as result of this reorganization other than adding one more scheduling office into the mix. True himself, though, transferred to the Air Force in order to remain in what he felt was the more exciting space program as opposed to the Navy's missile programs. Such was the appeal of human spaceflight in the 1960s.[41]

It should come as little surprise that squabbles existed between NASA and the Air Force. Disagreements often existed on things ranging from operations control and information flow to daily responsibilities, particularly during human spaceflight operations. At Hawaii, these disagreements quickly magnified in a short amount of time soon after the station became operational. By late summer of 1965, something had to be done. The problem was eventually settled in sweeping fashion and in NASA's favor: the station was transferred to NASA. Again, daily operations at the station level changed very little. All range users received support on a priority basis, allotting first priority to NASA human flight activities, second priority to Air Force ICBM launches, followed by naval fleet missile evaluation and training exercises, and lastly, support of Nuclear Test

Cultural dichotomy was very evident in this 1962 photograph of the Kano tracking station in Nigeria. Security was not an issue as the station was an object of curiosity for the local Hausa villagers. (Unnumbered photograph, Folder Number 8819, NASA Historical Reference Collection, NASA History Division, NASA Headquarters, Washington DC.)

programs. All scheduling was coordinated through the Station Director who did his best to accommodate all users. In this arrangement, all agencies eventually became very cooperative and service was virtually never denied to any legitimate user.[42]

Contractors were relied upon throughout the network. At Hawaii, the Chance Vaught Aircraft Company and its Hawaiian subsidiary, Kentron Hawaii Limited, initially played the major role. But in 1971, the operations and maintenance effort of all network stations was combined under a single contract. Bendix, already the major NASA contractor for tracking and data acquisition, was awarded the contract, one which it would keep until the station closed in 1989.[43]

Hawaii was, by all accounts, the busiest station in the worldwide network, due primarily to its location. First, launches from the Cape at a 28° inclination (that is, launches directly due east from the Cape) yielded more visible passes at Bermuda, Carnarvon, and Hawaii than at any other station. Second, its location on a path from Vandenberg Air Force Base to the Kwajalein atoll provided an ideal location for midcourse tracking of Air Force ICBM tests. Third, PMR operations in the surface-to-air and air-to-air missile test programs required a large and rather continuous demand for radar and telemetry services. Fourth, periodic nuclear weapons testing in the South Pacific brought significant workload in support of Sandia Corporation and its subcontractors. Finally, the demands of the Pacific Fleet for support of weapons testing as an adjunct to the Vietnam War at the time called for significant resources from the station. All this yielded the heaviest workload experienced by any of the NASA stations.[44]

As the Mercury network took shape, ground station locations were determined by a number of factors. These ranged from geography to the willingness and political alliance of the local government, and to language. Perhaps no other station in the MSFN better illustrated a dichotomy in cultures than the one that sat among the shrubs and parched red clay on the outskirts of the ancient city of Kano in central Nigeria. On 19 October 1960, an agreement was reached in the capital city of Lagos, allowing the United States to construct "NASA Tracking Station 5" in the Federation of Nigeria. This news was announced with great fanfare to the Nigerians. The local Hausa villagers living around Kano described the site as "the place the Sardauna built to get the message from the stars."[45]

A NASA ground station often provided the most visible and sometimes the *only* tangible look into America's space program on foreign soil. Because of this, the station staff was often looked upon as unofficial ambassadors of international goodwill to the native populace, representing American goodwill and know-how in these countries. Kano exemplified this more than any other station. It was quite the "hi-tech" tourist attraction of its time, where open-door was the norm, drawing the curious from all parts of Nigeria. An unguarded gate leading to the station welcomed visitors. Albert E. Smith, the first Station Director said at the time, "There's nothing here that's classified; we are at home to all visitors."[46]

Life in Kano was simple, unchanged for centuries. Touareg warriors still lived in pressed-earth dwellings; naked children played along the roadside; camel caravans presided over by nomads shared the dusty washboard roads with donkey-drawn wagons and automobiles. Such was the setting for NASA's first tracking station in central Africa. Although the station would be short lived—officially closed down on 18 November 1966 just one week after Gemini 12 splashed down, a victim of technology evolution as NASA revamped the MSFN in preparation for Apollo—it met a crucial need at the time. Depending on the orbit, Kano provided anywhere from a three to six and a half minute communication window with the Mercury (and later Gemini) spacecraft as it passed over the continent after leaving the Grand Canary Island coverage area.

The station was also important because it could monitor (but not remotely trigger) a retrofire over Africa for an Indian Ocean recovery in the event of an emergency. It was operated and maintained by 30 Bendix and General Electric workers along with 10 technicians from the British Cable and Wireless Company who maintained the cross-continent and trans-Atlantic ocean floor cable between Kano and Cape Canaveral. The Air Force also had on-call a 30-person rescue team there to aid in ocean or land recovery operations if needed during Mercury and Gemini.

As the infrastructure of the MSFN was being established, international cooperation was just as crucial as it had been back in the Minitrack days. This was best illustrated at Guaymas, Mexico where diplomacy played the major role in getting the station built. The establishment of the Guaymas Station could best be described as a labored process, one that required great patience and perseverance. NASA considered a station in Mexico very important since it would be the first North American land station to establish contact, and if needed, enable the ground to command a first orbit retrofire for an Atlantic Ocean recovery in case of an emergency with the Mercury spacecraft.

In the spring of 1959, NASA presented to the State Department a list of foreign locations for tracking stations. The Department told NASA that it was reasonably optimistic about the chances to obtain entry to all areas but Mexico where, it felt, the space agency would have extreme difficulty due to internal political strife and anti-American sentiments. The Department agreed to provide assistance in every way possible in getting approval in Mexico. Initial overtures by the U.S. embassy in Mexico, however, received no response.

In the summer of 1959, Milton S. Eisenhower, advisor to his older brother President Dwight D. Eisenhower, on a trip to Mexico, made a personal appeal to President Adolfo Lopez Mateos to open negotiations. This resulted in at least an expression of interest. It was followed by discussions at a White House dinner attended by NASA Administrator Glennan during the visit of President Lopez Mateo and Mexican Ambassador Antonio Carrillo Flores later that summer. Again, interest was at least verbally expressed by the Mexican government to, at a minimum, consider the proposal. There then ensued a lengthy period of inactivity in which nothing constructive happened. NASA could not get an answer from anyone on where Mexico stood. It was emphasized by the U.S. Embassy throughout this process that the political climate in Mexico at that time was questionable at best. By then, planning of the Mercury network was well underway and the Mexican situation was beginning to impact the schedule. Great pressure came from the Langley team assigned to lead in the planning and implementation of the Mercury ground network to get an answer, either a yes or a no.

Edmond Buckley and E. J. Kerrigan made a visit to Mexico in January 1960 in a final effort to determine whether the Mexican government was really interested at all. With the very active cooperation of U.S. Embassy officials, discussions were finally held with the Acting Minister of Foreign Affairs and other top Mexican officials. The talks were deemed favorable, enough so to gain approval from President Lopez Mateo for further negotiations. Arrangements were made by Mexico City for meetings to be held between the University of Mexico and NASA on the basis of mutual scientific cooperation. This was the breakthrough that NASA had been looking for. Doing so provided the scientific impetus that Mexico deemed essential to provide the public support needed for any cooperative effort between the two countries. It was personally obvious to Edmond Buckley, though, that the actual possibility for mutually beneficial scientific cooperation of the sort desired by the Mexicans would be for projects *other than* Mercury. But he also felt that collaboration was necessary considering the importance of a Mercury station in Mexico.

Operating in this not entirely ideal but at least workable framework, negotiations proceeded over the next four months of 1960. The Mexican government presented several guidelines around which they felt the talks should be centered. These were designed primarily to assuage the prevailing negative American public sentiment:

1. Great care must be taken in all actions and public announcements so that the Mexican people not misunderstand the scientific and peaceful nature of the activity.

2. It was important that the military not participate in the operation of the station.

3. It was desirable that the activity be described as an international cooperative activity of many nations and not as a bilateral agreement between the United States and Mexico.

4. To further emphasize wide international participation, it was desirable that Australians or other nationals be present at the Mexican station; that Mexicans be present at other stations, for example, at Cape Canaveral.

5. It was preferable to have the agreement negotiated between the president of an American university and the president of the University of Mexico.[47]

A formal agreement between Mexico and the United States establishing the tracking station at Guaymas was finally signed on 12 April 1960.

With great fanfare, an inauguration ceremony officially opening the station was held 14 months later on 26 June 1961, presided by NASA Deputy Administrator Hugh Dryden, the Mexican Foreign Minister and others in a large Mexican delegation. The cost to establish the station was $2.25 million in 1960, one of the biggest projects in that area at the time.[48] The significance of an agreement establishing a spaceflight tracking station in Mexico was no small matter. Scientific and cost reasons aside, it was the first real cooperative project between the two neighboring states since before World War I and, as such, was quite momentous. Both the State Department and the U.S. Embassy made it clear that this represented a big step to bettering relations with its neighbor to the south, one that went beyond merely space exploration and Project Mercury.

One of the first actions taken during these negotiations was the establishment of the Mexico-U.S. Commission for Space Observations. The decision to establish a commission was not arrived at in a cavalier way. The U.S. State Department and General Counsel at the time required a good deal of thought before agreeing to such an arrangement, since intergovernmental commissions generally tend to establish rules which are binding on the countries involved, and thus require congressional approval. It was recognized that in this case, though, the commission would be established primarily to provide assurance to the Mexican people that this was truly a cooperative, civilian, scientific project in which their government was fully informed at the highest levels. Without the appointment of this commission, the tracking station in Mexico likely would not have materialized on schedule, if at all.

Mexico City's emphasis that, in their view, this was primarily a scientific and not political cooperative was reflected in the makeup of the commission. The Mexican Section was led by Ing. Ricardo Mongas Lopez, former Dean of the Institute of Geophysics at the University of Mexico. Mongas Lopez was a well renowned scientist with diverse experience in multiple fields who had served on many government panels. He was joined by Dr. Eugenio Mendez Docurro, Director of the National Polytechnic Institute and Ing. Jorge Suarez Dias, a former Dean of the Polytechnic Institute.

Their carefully selected counterpart, the U.S. Section, was composed of Edmond C. Buckley, Assistant Director for Space Flight Operations, NASA Headquarters; Ralph E. Cushman, Chief of Field Installations, NASA Headquarters Procurement Office; and G. Barry Graves, Head of the Langley TAGIU team responsible for planning and implementing the Mercury network. The U.S. also appointed Raymond Leddy, Counselor of the U.S. Embassy in Mexico City, as the permanent liaison officer for the U.S. Section of the commission. This last appointment was done so as to comply with Mexican wishes to have someone in their country who could handle day-to-day affairs.[49]

But the appointment of a commission did not translate into immediate progress. It was actually quite the contrary. Originally, meetings were to be held every two months. In reality, however, the commission only convened when pressure from the U.S. Embassy forced its scheduling. The Mexican government still tended to proceed cautiously and kept a close watch over the actions of the commission. This caution was evident when, in the first official meeting, Ambassador Oscar Rabasa, Director in Chief for American Affairs, kept a firm control of the proceedings until he was satisfied that the direction of the meeting was headed where it was supposed to: science; only at that point did he relinquish control to Chairman Mongas Lopez. This was somewhat of a sobering process for the United States in general, and NASA in particular.[50]

Mexico has always had (and still has) a deep government bureaucracy based on Latin American heritage. Thus, NASA needed real assistance to properly work with the Mexicans so as to avoid becoming snared in bureaucratic and jurisdictional entanglement. Take for example that as many as five to seven separate Mexican bureaus ordinarily had to approve imports into the country. The use of this commission, backed by the President of Mexico and his Foreign Affairs Office, proved crucial to help cut through bureaucratic redtape so that real work could get done. Although interbureau difficulties still arose, the United States did not become involved in them, but rather, utilized the Mexican Section of the commission to work out the often thorny issues.

In addition to dealing with governmental bureaucracy, an intangible and even more volatile issue was also at work, an issue that really called for careful diplomatic attention. Anti-American sentiment in Latin America was high at the time and strong feelings about the United States were rampant among the Mexican populace. In June 1960, Dr. Glennan invited members of the Mexican Section to Washington, at NASA's expense, to visit the Goddard, Wallops, and Langley facilities. Even though only two members came, they returned to Mexico with invaluable publicity from the traveling Mexican press. Over time, the Mexican Section of the commission turned out to be extremely helpful by giving wholehearted support to the station and keeping unfavorable press and perhaps possible demonstrations about the "United States Missile Station" at a minimum.[51]

The commission produced many favorable television and radio spots promoting the American tracking station and provided newspaper interviews in support of the project. Patience and perseverance paid off for NASA in Mexico. It was very apparent as the station finally opened that the commission (on both sides) had served its purpose well. Even Ing. Mongas Lopez and Raymond Leddy doubted that the Mexican government would ever allow it to be dissolved. "The assurance desired by both governments that the procedures be carried out in a manner that would promote good will between Mexico and the U.S. seems to have been observed," Edmond Buckley was to reflect years later on the effort it took to build up the Guaymas station. Simply put, "Different approaches were needed in different countries."[52]

The Mercury Control Center was located near the launch area at Cape Canaveral, apart from the computers which in those days were still centralized at the Goddard Space Flight Center in Maryland. Unlike the high resolution, computer graphical displays of today, the ground track of the capsule on the 'big board' was moved mechanically by wires. This picture was taken on 20 February 1962 during America's first human orbital flight, Mercury Atlas 6. (NASA Image Number 62-MA6-161)

While it was quite true that different approaches were indeed needed in different places, one common denominator generally permeated all negotiations regardless of the country involved. In setting up its network, NASA took great care to always emphasize the civilian nature of the project to the host country. The Agency's stance was that

> The National Aeronautics and Space Administration is a civilian scientific organization and that Project Mercury is a scientific experiment being conducted for the purpose of ascertaining the problems of man's existence in space. Furthermore, the results from this experiment will be made available to all the world.[53]

The promotion of the peaceful nature of the work and the sharing of results had worked well back in 1957 during the IGY when the NRL was first setting up Minitrack. Fueled by increasing East-West tensions, this approach was needed all the more as suspicion tended to exist among many countries as to the true nature of the American space program. This was especially true of African states. Ray W. Hooker, Assistant Chief of Langley's Engineering Service Division, reported following a trip to Africa in rather blunt and sobering terms that

> In the case of the Kano and Zanzibar sites, the British have sold the local government on the fact that this is an American experiment, harmless in nature and would contribute to the scientific knowledge of the world. [But] In both the Nigerian and Zanzibar governments, there is the general native population which is capable of believing almost anything and getting quite excited about it. A rumor was circulated in Kano at the time of the site team's visit there to the effect that the team was tied in with the French atomic bomb experiment in some manner.[54]

Such gross misconceptions on the part of the local government had to be resolved diplomatically. In the Nigerian situation, personal assurance to the Emir of Kano by the Langley advance delegation was done. To do this, NASA had to enlist someone the Emir trusted, in this case Arnold W. Frutkin from the National Academy of Sciences who had established considerable positive reputation with foreign countries during the days of the IGY.[55] The American Embassy or Consulate usually served as a good starting point for such negotiations, enabling dialogue with the right officials who were in a sufficiently high position to initiate action. The case of Nigeria was especially sensitive as the British were at the time trying to end their rule in that country on good terms and most of their actions had that particular goal in mind. This turned out to benefit NASA as the British helped to clarify the peaceful intent of the Americans.

While it is entirely true that NASA's charter has always been civilian in nature, the Agency nevertheless, and from the onset, has had to work closely with the DOD to accomplish certain missions. Still true today, this was inescapable in the early 1960s: the DOD owned all the launch facilities and had a wealth of experience and knowledge in the areas of missiles and rockets. It was also in the midst of setting up its own satellite network (the Air Force Satellite Control Network, or AFSCN). Securing the "high ground" was (and is) a national security objective for the DOD.

No where else did NASA rely on the DOD more than at Cape Canaveral, Florida, where launches took place. To manage this interface, the Air Force created the position of a Department of Defense Manager for Manned Spaceflight (DDMS). DDMS was given the job of interfacing with the civilian space agency's newly created Project Mercury representatives to work through common problems. When Mercury came onboard at the Cape, the DDMS supported NASA in such areas as launch operations, range safety, contractor support services and the construction of a new control center at the launch facility.

It was in this capacity that Henry H. Clements (Associate Director of the Johnson Space Center from 1981 to 1984) began his association with NASA. As a Captain in the Air Force, he was reassigned to the DDMS, becoming the first Network Controller (NC) during the suborbital Mercury Redstone missions. "This was really a one-man operation, keeping tab on some 15 remote sites, verifying that they were ready to support the mission and that the capsule communicators stationed at various tracking sites were in position and linked to the control center at the Cape via Goddard," Clements recalled in 1982 the rudimentary setup that was Mercury Control.

> Much of the equipment and technology used in those early days was rather rudimentary. We had a display map on which the capsule was simply moved by wires and the status of our tracking stations was merely indicated by red or green lights. Our tracking stations often found it difficult to stay on the radar beacon—a beeper signal generated by the orbiting space capsule. Then, there was the problem of a smooth handover from station to station as the spacecraft circled the Earth. There were also problems when sunspots caused microwave dropouts in ground communications and pipe layers accidentally cut vital telephone cables. Yet, despite it all, we had extremely high reliability due to the outstanding support from NASA's civil service team and our contractors. They all were very anxious for this program to succeed and for our astronauts to return safely. In this effort, reliable tracking, computer and communication support was vital.[56]

The State Department was instrumental in helping to secure international cooperation not only with third-world countries but also with established United States allies. An agreement with Spain was needed, for example, for a Grand Canary Island station. A good telemetry and radar capability in the Canaries was considered critical by mission managers for orbit establishment, particularly in the case of an abort.

The Canary Islands are an archipelago of seven volcanic islands located 180 kilometers (110 miles) off the Moroccan coast of northwest Africa. These islands were known from antiquity. Prior to their conquest in 1402 by Spain, they were inhabited by the Guanches, native peoples related to the early Berbers of North Africa. The conquest of the Canaries took almost 100 years and set a notorious precedent for the conquest of the New World. Due to the terrain and staunch resistance of the native Guanches, this conquest was not completed until 1496 when the Canaries were incorporated into the Castilian kingdom. The Spanish imposed a new economic model based on single-crop

Table 3-1: The Mercury Space Flight Network (Three Orbit Configuration)[57]

Station	Abbreviation	Location		S-Band Radar	C-Band Radar	Acquisition*	Telemetry	Command	CAPCOM**
North America									
Cape Canaveral, Florida	CNV	28°28'N	80°34'W	•	•	FA	•	•	•
Corpus Christi, Texas	TEX	27°39'N	97°23'W	•		FA	•		•
Eglin AFB, Florida	EGL	30°46'N	86°53'W	•		FA			
Guaymas, Mexico	GYM	27°57'N	110°43'W	•		FA	•	•	•
Point Arguello, California	CAL	34°39'N	120°36'W	•	•	FA	•	•	•
White Sands, New Mexico	WHS	32°21'N	106°22'W		•	FA			
Australia									
Muchea, Western Australia	MUC	31°35'S	115°56'E	•		FA	•	•	•
Woomera, South Australia	WOM	31°23'S	136°53'E		•	FA	•		•
Africa									
Kano, Nigeria	KNO	12°03'N	08°31'E			SA	•		•
Zanzibar	ZZB	06°13'S	39°13'E			SA	•		•
Atlantic									
Bermuda, United Kingdom	BDA	32°15'N	64°50'W	•	•	FA	•	•	•
Grand Bahamas, British West Indies	GBI	26°38'N	78°16'W		•	M	•		•
Grand Canary, Spain	CYI	27°44'N	15°36'W	•		FA	•		•
Grand Turk, British West Indies	GTK	21°28'N	71°08'W			M	•		
Pacific									
Canton Island, Kiribati Republic	CTN	02°50'S	171°40'W			SA	•		•
Kauai, Hawaii	HAW	22°07'N	157°40'W	•	•	FA	•	•	•
SHIPS									
Atlantic Ship (Rose Knot Victor)	ATS					FA	•		•
Indian Ocean Ship (Coastal Sentry Quebec)	CSQ					FA	•		•

* FA: fully automatic   SA: semiautomatic   M: manual   ** CAPCOM: capsule communicator

The staff at Canary Island about 1967. Standing: John Adams, Chuck Rouillier (Station Director), Tom White, Percy Montoya, Matt Harris, Ed Bender, Ed Crough. Sitting: Clay Krugman, Dick Kelly, Roger Lee, Glenn Smith, Russ Lutz. (Photograph courtesy of Gary Schulz)

cultivation, first the sugar cane followed by wine, an all-important trade item with England. The islands eventually became an important stopping point in the trade routes with the Americas, Africa and India, and the port of La Palma turned into one of the most important ports in the Spanish Empire.

At the beginning of the twentieth century, the English introduced a new cash-crop, the banana, and the islands prospered. In 1936, Francisco Franco traveled to the Canaries as General Commandant from where he launched the military uprising of 17 July. He was able to quickly take control of the archipelago. Despite the fact that there was never actually a war in the islands during the 1940s, it was one of the places where post-war repression was most severe. Organized opposition to Franco's regime did not begin to materialize until the late 1950s, when groups such as the Spanish Communist Party and various nationalist, leftist and independence factions such as the Free Canaries Movement came onto the political scene.

When NASA approached Madrid about Grand Canary Island on 10 September 1959, Mr. William Fraleigh, First Secretary of the Consul Political Office, was told that the situation was "rather delicate." Evidently, a South American leftist newspaper that was run by exiles of the Spanish government had published a misleading article disclosing the fact that Madrid was in the midst of negotiating with the United States to *give* a portion of the Canary Islands in order to establish an airbase. Madrid was rather sensitive about the matter since the Spanish Moroccans were already beginning to talk about the independence of the Canaries following the example set by the French Moroccans in 1956. The Spanish government quickly advised the State Department that care must be taken to clearly delineate that the MSFN station was not going to be related in any way to the U.S. airbase already there. Otherwise, Madrid could not guarantee permission to use the station. This bit of warning was passed down through Headquarters such that by the time formal negotiations were held, the civilian versus military nature of the station project did not become the showstopper that it easily could have been.[58]

NASA always tried, to the extent possible, not to disturb the everyday lives of the local populace when building its foreign stations, but relocating the local tenants was sometimes unavoidable. This was the case on Grand Canary Island. Here, three groups of transient dwellings housing migrant farmers were located within the site of the station boundary. NASA was at first inclined to leave these dwellings alone, intending that they continue to be used after the station was up and running. But as construction progressed, it became clear that this arrangement was probably not going to be in the best interest of the Agency; that leaving the structures in their present locations and allowing people to occupy them were going to inevitably interfere with station operations. The houses had to come down.

Whenever it was necessary to displace or disrupt local property, it was NASA's standard policy to determine the removal and replacement cost prior to starting any actual work. Alternatives were pursued if the cost was deemed unacceptable. In the case of Canary Island, an agreement was reached—with the assistance of the U.S. Naval Facilities Engineering Command—to replace the structures in kind but just outside the boundary of the tracking station.[59]

* * *

Construction of the MSFN began in April 1960; by mid-summer, all stations were under construction. The last one, Kano, Nigeria, was completed in March 1961. On 1 July 1961, 24 months after awarding contracts, NASA officially accepted the new MSFN. It was quite an impressive achievement, considering that the first American human orbital flight was not scheduled for another six months. A significant amount of research and development and nonrecurring engineering went into the effort. That also included all the land and ocean-floor cable links between the GSFC core, the Cape Canaveral launch site and the individual field stations. Table 3-1 presents a summary of the MSFN as it appeared during Project Mercury, and Appendix 1 shows the network on a world map.

The standard or baseline Mercury network was set up initially to track three orbits. This was augmented during the program so that by Mercury Atlas 8 (the fifth piloted flight), missions had been extended to six orbits. In order to accommodate the additional three orbits, the Atlantic Ship was equipped with command uplink transmitters, redesignated the Pacific Command Ship (PCS) and repositioned south of Japan. Three Navy ships (*American Mariner, Huntsville* and *Watertown*) also supported the mission from the vicinity of Midway Island in the north-central Pacific.

These were not the only augmentations. To support the final Mercury mission—L. Gordon Cooper, Jr.'s MA-9 launched on 15 May 1962 which was 22 orbits, the standard, three-orbit Mercury network had to be modified significantly as follows:

> Indian Ocean Ship *Coastal Sentry Quebec* was moved south of Japan.
>
> Atlantic Ship *Rose Knot Victor* was moved to the South Pacific near Easter Island.
>
> DOD ship *Range Tracker* was stationed northeast of Midway Island.
>
> DOD ship *Twin Falls Victory* was stationed between Bermuda and the U.S. East Coast.
>
> The DOD Eastern Test Range station on Ascension Island and Puerto Rico provided FPS-16 radar and telemetry recording.
>
> The new MSFN Gemini/Apollo station at Antigua was turned on to record telemetry and voice.
>
> Temporary voice communication sites were erected on Kwajalein Island, San Nicholas Island and Wake Island.
>
> DOD aircraft were assigned to provide voice and telemetry relays

Also starting with this mission, the radio link between Bermuda and Cape Canaveral was supplanted by submarine cable and the GSFC computers were upgraded to IBM 709s.[60]

Stations had different roles and different equipment depending on their roles. Not all stations used flight controllers for instance. Sites like White Sands and Eglin were used only for tracking. The primary stations with full TT&C capabilities were the Mercury Control Center (MCC) at Cape Canaveral, Florida; Bermuda; Muchea, Western Australia; Kauai, Hawaii; Point Arguello,

California; and Guaymas, Mexico. At these stations, commands—still few in number—could be transmitted up to the spacecraft as combination of tone pulses. Telemetry was received, demodulated and displayed in the form of strip charts and analog meter readings for flight controller evaluation. They read the health and status of the spacecraft, for example, marking readouts with different color grease pencils, each denoting the different times at which readings were taken. Between passes, summaries and reports were sent to the MCC via teletype; few telephone voice circuits were available at the time.[61]

The relative importance of a given site varied depending on what was happening during a flight. Cape Canaveral and Grand Bahamas, for example, provided vital telemetry, tracking, command, and voice coverage during launch and ascent. Conversely, Grand Turk Island provided radar tracking during Atlantic Ocean reentries. Bermuda bore the responsibility of deter-

Panoramic view of Launch Complex 14 as Mercury Atlas 9 is readied for flight in this picture taken eight days before its launch as the last Mercury mission. (NASA Image Number GPN-2000-000609)

mining whether a mission should continue or be aborted. Should Bermuda order a "no-go", Grand Canary Island then provided the crucial tracking needed for an emergency reentry. Finally, Kauai could transmit command tones up to the spacecraft to set off the timers used to fire deorbit retrorockets in the event of crew incapacitation.[62]

One capability that was common to just about all the stations was, as one would expect, air-to-ground voice communications. Not all stations used astronaut Capcoms, though (which by tradition is always an astronaut).[63] For stations where a NASA Capcom could not be assigned, a Communication Technician (or Comm Tech) handled voice calls to the astronaut. Before September 1962, the "Original 7" were all the astronauts that NASA had, so being able to speak with an astronaut on a space mission in those pioneering days was beyond privilege and quite an honor. The communications console at the Muchea station, for instance, was memorialized for years with a plaque which read:

> This plaque is to mark the spot where an Australian first spoke to a space traveler. The Australian was the communication technician at Muchea, Mr. Gerry O'Connor, and the space traveler was astronaut John Glenn.[64]

A key to maintaining network reliability was the utilization of built-in operational redundancy using spares and backup equipment (plus frequent exercise of that equipment). During the first piloted orbital flight on Mercury Atlas 6, controllers worst fears came true when the prime computer used for orbit determination at GSFC "crashed." But the built-in redundancy of the system had another computer available which was immediately brought online. Station personnel and equipment were kept sharp through exercises, simulations and frequent system checkouts. Although not part of the active network of stations when it first started, Wallops Island, just off the Virginia coast, served as a personnel training and network equipment test facility. This network training center was established by GSFC in July 1961 shortly after its networks became operational. There, equipment was prepared, tested and checked-out prior to shipment and field installation. At Wallops, all NASA and most of the contractor supervisory, maintenance and operation personnel were trained before their deployment overseas. Students at the Center were, for the most part, new engineering college graduates or electronics technicians in their 20s or early 30s. By 1963, the final year of Project Mercury, enrollment at the Center had reached 255 students.[65]

★ ★ ★

A hallmark of the NASA spaceflight tracking networks has been the simple but distinguishing fact that a mission has never been compromised because of network problems, an enduring value that Covington espoused to the men and women who worked under him. The MSFN, nevertheless, did have its share of drama during Mercury:

> Mercury Atlas 5 (MA-5): On the flight of Enos the chimpanzee, a tractor accidentally plowed up cable outside Tucson, Arizona, cutting off communications with the Hawaii and California stations. Unfortunately, the alternate DOD circuit that had been provided in case something like this happened was also severed at the same time. To compound the situation, telemetry revealed that fuel was literally boiling away from the capsule, thus vapors—not fuel—was getting to the spacecraft thrusters needed to control the orientation of the spacecraft in the weightlessness of space. An inordinate amount of fuel was being consumed based on the data obtained by the MSFN. Mission Control decided to cut the flight short by one orbit and commanded retrorocket fire from Guaymas (exactly what the station was there for). AT&T, who owned the ground cables, worked frantically to come up with a solution. Just as the spacecraft approached North America, NASA communications traffic was rerouted around the cable break. Mission Control had just 12 seconds left in the window to initiate reentry. Enos was saved.
>
> MA-6: As mentioned before, during astronaut John H. Glenn, Jr.'s pioneering flight, the backup computer at GSFC came in handy, performing computations for three minutes while the prime computer was being rebooted.
>
> MA-7: Using radar data from Point Arguello, California immediately following retrofire, the Goddard computers correctly determined that astronaut M. Scott Carpenter had overshot the intended splashdown point by 386 kilometers (240 miles). Mercury Control was at first skeptical when informed of this and the news was received with disbelief. Subsequent network tracks at Whites Sand and Corpus Christi confirmed Goddard's initial prediction of the overshoot.
>
> MA-8: Solar activity was high and serious communication problems were experienced by the MSFN. Navy instrumentation ships were also used to augment the ground stations. On the next to last orbit, a power failure occurred at Point Arguello and a backup DOD site on San Nicholas Island was used to provide radar tracking. As an experiment, astronaut Walter M. Schirra, Jr. turned off the tracking beacon on his spacecraft during the fourth and part of the fifth orbit to see if ground radars could skin-track the Mercury capsule reliably. They could not.

A problem also developed on the flight with the onboard thermal control system that was due to, of all things, wax buildup. The problem was serious enough that after the first orbit, MCC had to decide whether to bring Schirra back or give the problem a chance to correct itself on the next orbit, which it did. During this critical period, flight controllers had a constant flow of data on which mission-critical decisions were made for example, temperature measurements from the spacecraft to assure them that Schirra would not become dehydrated by excessive heat.

MA-9: On the last flight of Project Mercury, 15 incidents of radio interference were reported by the network. Bermuda, for example, reported hearing the Voice of America and that of a Greenville, North Carolina, amateur radio operator. DOD frequency controllers were able to contact the offending operator and clear the NASA channels. On several occasions, however, amateur radio operators actually succeeded in contacting astronaut Gordon Cooper! (He was directed by Mercury Control Center not to respond.) More serious were instances where industrial equipment generated excessive radio interference around ground stations. NASA, for the most part, was able to convince local equipment owners to shut off the equipment on station passes during the 34 hour flight.[66]

Mercury Atlas 9 splashed down on 16 May 1963, bringing America's first human space adventure to a successful end. Particular praise for the unsung heroes of the invisible network came from Cooper himself at the Project Mercury summary conference in October 1963. When asked about his experience and the people who made the program a success, Cooper paused and then replied:

> It would be difficult to single out any one group or organization for special praise because they were all a wonderful team and a smooth blending of extraordinarily competent technical skills. However, I do think that if one could be mentioned organizationally, certainly I would have mentioned the worldwide network. It is certainly comforting to know when you are out there, that the world's finest communications network and the finest electronic facilities that man can devise are functioning with a fantastic computer complex that will allow the onboard systems specialists to break out their diagrams and tell you immediately what your situation is in the event of trouble, and this is indeed what happened on several occasions. Without this marvelous organization, it might have been a little more difficult to get back home.[67]

By the end of Project Mercury, there was little doubt that the MSFN had more than met the expectations of its designers. In a span of less than five years, NASA's manned spaceflight network went from a concept on the drawing board of the Tracking System Study Group to a $125 million global network that safely brought four Americans back from orbit. This price tag—of which $53 million went into network facilities and $72 million to operating cost—came out to just under one-third of the $400 million total that NASA spent on Project Mercury. The yearly budget just to operate NASA's tracking stations stood at $225 million. To illustrate the changing times, just six years earlier, the entire Minitrack Network was constructed and operated for a mere $13 million.[68]

As 1963 turned into 1964, Goddard's two networks, the STADAN and the Manned Space Flight Network (note the name change from "Mercury" to "Manned") were a key and proven part of the five year old space agency. Workforce at the GSFC now accounted for 11 percent of the NASA total. Stations had been established spanning Alaska to Australia. Conceived as a "manned-satellite" program, Project Mercury was just the beginning. It allowed America to not only send astronauts into space for the first time, but also to man-rate its worldwide tracking and data acquisition network. This network matured quickly to become the dependable safeguard that NASA was counting on. Changes would soon be coming as the United States picked up its pace, sending more astronauts into Earth orbit and beyond in a determined race that would culminate with humans on the Moon.

CHAPTER 4

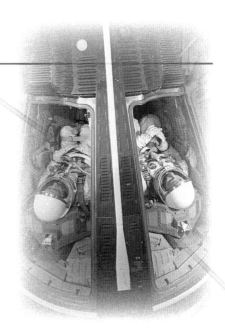

PREPARING FOR THE MOON

Despite the significant progress made during Project Mercury, in 1963 the United States still trailed (and trailed badly) the Soviet Union in terms of flight hours spent in space. The six Mercury missions flown between May 1961 and May 1963 had only accumulated a total of 53 hours in space. Thirty-four came on Mercury Atlas 9, Gordon Cooper's 22-orbit program finale. Of the six flights, two were suborbital. In contrast, Soviet Vostok cosmonauts had accumulated a total of 382 hours in space on six missions. Valentina Tereshkova, a 25-year-old textile worker from Yaroslavl who became the first woman in space in June 1963 on Vostok 6, was in orbit 17 hours longer than all the American astronauts put together.

It was clear by now that space had become the new global high ground for ideology and Cold War international prestige. "Now let the other countries try to catch us. Let the capitalist countries catch up with our country which has blazed the trail into outer space," was the unabashed challenge from Soviet Premier Nikita S. Khrushchev upon Gagarin's triumphant return from space.[1]

A week after the Gagarin flight, in a White House correspondence dated 20 April 1961, President John F. Kennedy gave Vice President Lyndon B. Johnson a directive. It had a definite sense of urgency. The President wrote:

> I would like, for you as Chairman of the Space Council, to be in charge of making an overall survey of where we stand in space. Do we have a chance of beating the Soviets by putting a laboratory in space, or by a trip around the Moon, or by a rocket to land on the Moon, or by a rocket to go to the Moon and back with a man? Is there any other space program which promises dramatic results in which we could win?[2]

Kennedy wanted results. Even more so, he wanted something dramatic, something that would capture the imagination of Americans everywhere to allow the U.S. to regain, in no uncertain terms, the upper hand in space.

To answer the President's directive, Johnson and Dr. Jerome Wiesner, Kennedy's science advisor, turned to NASA. Anticipating this, the Agency's top management triad of James E. Webb, the new Kennedy-appointed Administrator, Deputy Administrator Hugh L. Dryden and Associate Administrator Robert C. Seamans, Jr. had been working to prioritize a list of Agency objectives since the previous fall. To this end, they commissioned a study on 6 January 1961, chaired by George M. Low of the Manned Lunar Landing Task Group, to determine the technical, schedule, and cost requirements of a human lunar program. Table 4-1 lists the conclusions reached by the Low study.

Spurred on by these generally encouraging findings, Kennedy went forth with the commitment before a joint session of Congress on 25 May 1961, of "achieving the goal before the decade is out, of landing a man on the Moon and returning him safely to the Earth." This was a bold move by a young President who had been in office for just five months. While the

Table 4-1: Results of the Low Study on a Manned Lunar Program[3]

Mission	Spacecraft	Launch Vehicle	Date
Earth Orbiting 1 Man, Short Duration	Mercury	Atlas	1961
Earth Orbiting 3 Men, Long Duration	Apollo "A"	Saturn C-1	1965
Circumlunar, Lunar Orbit 3 Men	Apollo "B"	Saturn C-2	1967
Manned Lunar Landing Orbital Operations Direct Approach	Apollo "B" Apollo "B"	Saturn C-2 Nova	1968–1969 1970–1971

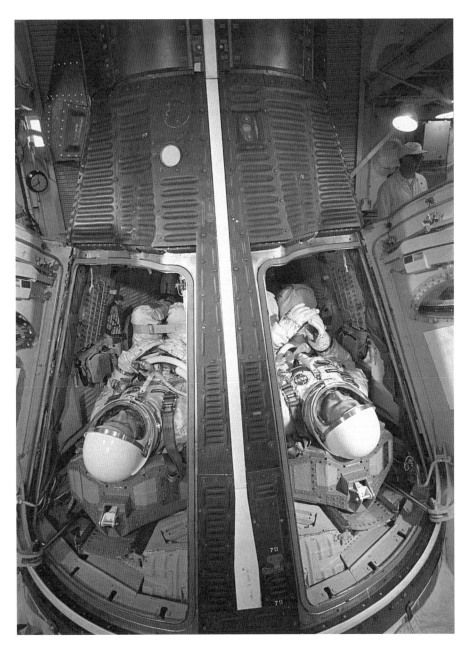

The Gemini spacecraft was basically a two-seat version of the Mercury capsule. It did, however, have an equipment section which enabled it to stay in space for up to two weeks. Gemini allowed NASA to gain the necessary experiences and man-hours in space needed before an attempt to the Moon was possible. Here, astronauts James McDivitt and Ed White train for their Gemini 4 mission in May of 1965.
(NASA Image Number GPN-2000-001018)

technical basis for his decision came from the NASA study, Kennedy felt that this gamble was one in which the United States had a chance to win and that it was sufficiently bold and dramatic enough to invigorate the nation and place America once again on the world center stage.

Before astronauts could fly to the Moon, many questions still had to be answered. For instance, what features of the Mercury spacecraft needed to be improved? Can a spacecraft be made with greater endurance so it can orbit Earth longer to find out the physiological affects of long-duration missions required to travel to the Moon and back? Can two spacecraft rendezvous and dock in space? Can astronauts work effectively outside the protection of his spacecraft? Even though America had decided to go to the Moon, NASA was not yet ready. To bridge the rather significant technology gap between Mercury and the emerging Apollo program, the Agency endorsed plans for a two-person spacecraft program called Mercury Mark II in December of 1961. The following spring, the name was changed and the program was officially christened Project Gemini—after the twin gods of Greek mythology—befitting of NASA's new two-person spacecraft.[4]

GSFC engineers made their first presentation to the newly formed Manned Spacecraft Center (MSC) on the outskirts of Houston in the first week of June 1962. The topic was technical requirements they would like to see implemented in a Gemini network. These included:

> Unification of all command, telemetry, and radio signals onto a single carrier frequency.
>
> Conversion from analog to the newer and much more bandwidth efficient Pulse Code Modulation (PCM) digital telemetry.
>
> Use of two acquisition aids at each tracking station (one for the Gemini spacecraft and one for the unmanned Agena docking target) and the ability to slave the radar to either vehicle.
>
> Modification of network station computers to accommodate processing both command uplink and telemetry downlink.[5]

Consumed with their primary job of developing the new two-seat Gemini spacecraft, MSC was lukewarm to the proposed changes. In their mind, they were just too much of a departure from what had just been done successfully on Mercury. Houston's thinking was correct. The Goddard suggestions, taken collectively, did in fact represent a major change in the way tracking and data acquisition would be done. The proposed technique was a harbinger of the (Unified S-Band) system that would later be used on the Apollo spacecraft. USB *was* revolutionary in its time, enabling spacecraft command, telemetry, voice, and television to all be transmitted using a single, combined data link. The technique was not entirely new, however, to NASA as the DSN had used USB since 1958.

The proposed changes broke ground with the conservative reliance on time-tested technologies such as analog telemetry, which had been in use since the 1940s. By the mid-1960s, digital systems had been under research at the White Sands Missile Range for some time. NASA had even tried it experimentally at the Bermuda Station on the final Mercury mission. After initial discussions, Houston agreed to make the switch to PCM telemetry but objected to the others on the grounds that complete dependence upon a single TT&C link could lead to total mission failure if just part of the system failed.

Most of the Goddard proposals were in effect rejected. Despite this initial disagreement, GSFC knew what they had and was convinced it would work. The two NASA Centers held a series of technical interchange meetings and working groups to discuss the changes over the next 12 months, with Greenbelt making its case for the new tracking and communication technique. By June 1963, Houston was persuaded for the most part, agreeing to the proposed changes but with one important stipulation: that computers at network stations be employed only for telemetry processing but not for commanding. The idea was to preclude inadvertent or erroneous commands from being uplinked to the spacecraft in the event of a computer anomaly.[6]

One Gemini guideline that had a significant effect upon the MSFN was the relaxation of the 10-minute "dead-time", which was now relaxed to one primary ground contact per orbit. Astronaut performance and the Mercury spacecraft had shown that having the ability to remotely send commands to the spacecraft from every network outpost, while nice, did not turn out to be the necessary requirement that it was thought to be. With this decision, the MSFN no longer had to spread its valuable resources equally over the globe. It could now concentrate on a limited number of primary sites supplemented with a number of secondary stations. In this arrangement, primary stations were those that had command uplink capability in addition to voice, radar and telemetry while secondary stations did not have command capability.[7]

Another change in network philosophy was network centralization in terms of mission control and mission computing. Back before John Glenn's first orbital flight, many had simply presumed, even at Goddard, that some of the communication links between Mission Control and the tracking stations would be lost, at least intermittently. But this did not turn out to be the case at all as Mercury proved that reliable network communications were the rule, not the exception. NASA then had the confidence to remove flight controllers from the network stations and centralize all control activities at the new MCC in Houston. As a precaution, Capcoms remained at a few of the primary ground stations where there was still lingering skepticism on the part of MSC about the reliability of communications.

At least this was the official position coming out of Houston. It was well known within NASA circles that such assignments were a way for Donald K. "Deke" Slayton, head of Flight Crew Operations at MSC, to give his astronauts some much needed rest and relaxation at attractive places. As former Flight Director Eugene F. Kranz put it, "Slayton would send astronauts out at the very last moment to all of the sites that were generally good locations to go to—Bermuda, Hawaii, California, Australia."[8] This was generally not a problem for those working at the station, except when the astronaut crossed the line and began "throwing his weight around" as happened when Pete Conrad showed up in Australia on Gemini 3 saying that Slayton wanted him to be in charge during the mission.

The other network centralization implemented by GSFC involved the computer system. On Project Mercury, computing was performed in Greenbelt, Maryland. The only other network computers were at the Florida launch site itself and at Bermuda. This architecture—identical to what was used on Mercury—continued through the first Gemini mission (Gemini 3) in March 1965. As preliminary telemetry processing plans were first being laid for Gemini, this was the computing baseline computer that engineers worked from. A rather limited architecture, it was capable of processing and sending only four groups of spacecraft health and status parameters back to the MCC for monitoring and evaluation. To meet the increased data requirements of the more complex Gemini spacecraft, the MSFN now had two UNIVAC 1218 computers installed at each primary outpost. Additional submarine (ocean-floor) cables were also laid to meet the increased data flow demands. These improvements had the aggregate effect of greatly improving real-time data decommutation and processing allowing much more spacecraft information to now be sent to Mission Control than was possible on Mercury. Former MSC network chief Lyn Dunseith captured it succinctly when he said, "Voice, telemetry, command, and tracking data acquired by the Goddard managed communications and tracking network represented some of the most critical information available to the flight controllers at their display consoles"[9]

As network changes continued and Gemini missions took place, Houston gained more and more confidence in the network. Take the role of computers. Two U1218 computers were originally set up in dual redundant mode, operating in parallel to process telemetry data. As they began demonstrating their reliability and as spacecraft TT&C burdens increased, MSC relented, finally agreeing to let computers handle both telemetry and command ("fire retro rockets," "turn on telemetry transmitters," "ring astro alarm," etc.). The digital processing capability of the U1218s made a dramatic jump during Gemini, increasing from 2 input/output lines to 32, with transmission rates reaching the then state-of-the-art 50,000 bits-per-second.

The now famous Mission Control Center at the Manned Spacecraft Center (MSC) in Houston, officially assumed mission network operations beginning with the second piloted Gemini flight (Gemini 5) in August of 1965. Two identical Mission Operations Control Rooms, or MOCR, were located on the second and third floors of Building 30 on the grounds of MSC. In 1996, the Department of Interior designated NASA's Mission Control Center as a National Historical Landmark. Pictured here is Mission Control during Gemini 5. (NASA Image Number GPN-2000-001405)

Eventually, one computer was tasked entirely to telemetry while the other to commands.[10]

Project Mercury had shown mission control and mission computing to be so inter-related that the Office of Tracking and Data Acquisition at NASA Headquarters decided that they should be best managed by the same Center. Since the MCC was going to be at the MSC, the MSFN computing system was reassigned to Houston. Gemini 4 in June of 1965 marked GSFC's finale as the primary computing center for NASA human spaceflight. On this flight, MSC computers were placed in a so-called "ghost mode" where they were checked out and accepted in preparation for its upcoming assumption of primary computing duties. When Gemini 5 left the launch pad on 21 August 1965, the MSC in Houston officially took over the mission computing function from Goddard. From that point on, the GSFC system was relegated to a backup role and employed mainly for network development, testing, and mission simulations, a role it performed until the end of Apollo.[11]

Mercury flights had been very basic, limited to circular, low-Earth orbits of less than 320 kilometers (200 miles) in altitude. Gemini, though, would fly many high apogee elliptical orbits, some as far as 1,600 kilometers

(1,000 miles). To improve tracking at these altitudes, RCA FPQ-6 skin-track C-band radars were added to the network. One of the most accurate and powerful tracking radars of the time, the FPQ-6 had an output power of 2.8 megawatts and was effective out to 60,000 kilometers (37,000 miles). All equipment was housed in a two story building. Its operation was fairly simple. It could be operated from a single console by two or three technicians depending on the tracking mode used. A team of at least seven people was required, however, for maintenance of the equipment. The reflector was an 8.8-meter (29-foot) dish and the combined weight of the moving parts and hydraulic drive was over 30 tons, controllable using a small joystick on the control console. For rigidity and stability, the antenna tower foundation extended nearly 10 meters (30 feet) underground.[12] The older VERLORT and FPS-16 radars used on Mercury were kept in service. This provided redundancy so that, in the event of spacecraft beacon failure, the MSFN could still skin-track. With these combined capabilities, the potential for any tracking losses or blackouts was greatly reduced, if not eliminated altogether.[13]

Lighter TELTRAC telemetry antennas and associated telemetry equipment were also installed across the MSFN to serve as acquisition aid for simultaneous tracking of both the Gemini and the unmanned Agena docking target during rendezvous missions. A major objective of Project Gemini was to demonstrate and test-out the rendezvous procedures being developed for the upcoming Apollo lunar missions. These missions required the Command Module (CM) and the Lunar Module (LM) to rendezvous with each other as the latter returned from the surface. On Gemini, the unmanned Agena spacecraft served as a surrogate rendezvous and docking target. For command uplink, the network continued to rely on FRW-2 UHF transmitters using 10 kilowatt high-power amplifiers.[14]

Communications between the MCC and the ground stations also became much more efficient during Gemini. Air-to-ground voice transmissions, in particular, garnered special attention. Former Project Gemini Director at NASA Headquarters, William C. Schneider, recalled that

> Early in Project Gemini . . . we found that voice communications from the spacecraft left much to be desired. A near-perfect mission received bad notices because the people on Earth couldn't hear what was happening. So we went to work to fine-tune the system to be ready for the more advanced Gemini and Apollo flights.[15]

This is true even today. Despite crystal clear digital videos from space, the quality of voice transmissions—which is limited by the microphones worn by the astronauts—still leaves room for improvement.

The first stations to transmit telemetry back to Houston were Bermuda and the early-ops sites downrange of the Kennedy Space Center (KSC) at Grand Bahama, Grand Turk, and Antigua. When Houston supported its first mission in

June 1965, the telemetry transmission rate from Bermuda to Houston was 2,400 bits-per-second (2.4-kbps). Commands could be sent from the MCC to remote ground stations in one of two ways. In one method, used for routine or so-called housekeeping commands, Mission Control teletyped the command sequences prior to a scheduled pass over a given site which were then stored at the station. Later on as the orbiting spacecraft passed over the station, an onsite technician would uplink them up to the craft. For more urgent matters, Houston could send command messages over the 2.4-kbps master circuit to KSC for immediate relay via dedicated government priority "T-1" landlines and submarine cables to the next MSFN station in the spacecraft's ground track.

Communications between Houston and NASA tracking ships was enhanced whenever possible by collocating NASA vessels with Navy communication ships. This provided a network of UHF daisy-chain, relay points from sea-to-land and vice versa. The *Coastal Sentry Quebec,* a converted Class 1 World War II freighter, was usually situated in the Western Pacific covering the South Pacific gap between Australia and Hawaii. The Air Force Eastern Test Range and Western Test range operated the *Rose Knot Victor* and *Range Tracker,* which were moved around in the South Pacific, Atlantic, or Indian Oceans depending on a specific mission's requirement.[16]

One final measure of the increasing capability of the ground communication network was at GSFC itself, where the SCAMA (Switching, Conferencing, and Monitoring Arrangement) was updated. SCAMA was the telephone switchboard at the Center that handled all voice communications from around the world. In the early days of Project Mercury, it could simultaneously conference only 10 worldwide voice circuits. This number jumped

Wives of Gemini 4 astronauts James A. McDivitt and Edward H. White talk with their husbands in orbit from the new Mission Control Center at the Manned Spacecraft Center on 3 June 1965. Patricia White is on the left, and Patricia McDivitt is on the right. (NASA Image Number S65-28922)

Table 4-2 The Manned Space Flight Network in the Mid-1960s[17]

Station (location)	Abbreviation	Network Role*	Ownership**
North America			
Canaveral (Kennedy Space Center, Florida)	CNV	Primary	NASA
Texas (Corpus Christi, Texas)	TEX	Primary	NASA
Eglin (Florida)	EGL	Secondary	DOD
Goddard Space Flight Center (Greenbelt, Maryland)	GSFC	Secondary	NASA
Guaymas (Mexico)	GYM	Primary	NASA
Houston (Texas)	HOU	Primary	NASA
California (Point Arguello, California)	CAL	Primary	DOD
Wallops (Wallops Island, Virginia)	WLP	Secondary	NASA
White Sands (New Mexico)	WHS	Secondary	NASA
Atlantic			
Antigua (British West Indies)	ANT	Secondary	DOD
Ascension (Ascension Island, United Kingdom)	ASC	Secondary	DOD
Bermuda (United Kingdom)	BDA	Primary	NASA
Grand Bahama (British West Indies)	GBI	Secondary	DOD
Grand Canary (Spain)	CYI	Primary	NASA
Grand Turk (British West Indies)	GTK	Secondary	DOD
Africa			
Kano (Nigeria)	KNO	Secondary	NASA
Pretoria (South Africa)	PRE	Secondary	DOD
Tananarive (Malagasy Republic)	TAN	Secondary	NASA
Australia			
Carnarvon (Western Australia)	CRO	Primary	WRE
Perth (Western Australia)	MUC	Secondary	WRE
Woomera (South Australia)	WOM	Secondary	WRE
Pacific			
Canton (Kiribati Republic)	CTN	Secondary	NASA
Hawaii (Kauai, Hawaii)	HAW	Primary	NASA
Ships			
Coastal Sentry Quebec	CSQ	Primary	NASA
Range Tracker	RTK	Secondary	DOD
Rose Knot Victor	RKV	Primary	DOD

Remarks

Launch Control Center

Located at the abandoned Rodd Naval Auxiliary Air Station; an original Mercury station

Located at the Air Force Eglin Gulf Test Range 50 miles northwest of Panama City, FL; an original Mercury station

Overall network responsibility; development and test facility

Located in northwest Mexico on the shores of the Gulf of California; an original Mercury station

Manned spaceflight Mission Control Center

Located some 40 miles north of Santa Barbara, part of the Navy Pacific Missile Range; an original Mercury station

Training and test facility just off the shores of Virginia

Located on the grounds of the Army's White Sands Missile Range near Alamogordo; an original Mercury station

Air Force ETR station

Air Force ETR station

Go/No-Go decision site; an original Mercury station

Air Force ETR radar site

Located 120 miles off the African coast; critical abort tracking site; an original Mercury station

Air Force ETR radar site

Original Mercury station in west-central Africa

Air Force ETR station

Replaced the Zanzibar Station; last land site before crossing the Indian Ocean to Australia

Collocated with the NASA STADAN site

The original Mercury site at Muchea was used until Perth became operational; call sign was retained

Original Mercury station; collocated with STADAN site

Original Mercury station

Original Mercury station

Usually stationed in the western Pacific near Japan

Usually stationed in the central Pacific near Midway Island

Usually stationed in the south Pacific off the South American coast

*Primary stations were those that could uplink system commands to the spacecraft. Secondary stations were those used primarily for radar and telemetry downlink. All had UHF air-to-ground voice capability.
**DOD: Department of Defense; NASA: National Aeronautics and Space Administration; WRE: Weapons Research Establishment, Australian Department of Supply

to 220 when the Mercury Space Flight Network became the Manned Space Flight Network for Project Gemini.[18]

From mid-1963 into the spring of 1964, a number of tracking stations were added. Overall, the MSFN expanded from 14 land stations to 23 (9 primary, 14 secondary) plus an additional Navy ship. As before, coordination with the DOD played a central role in this evolvement; DOD support was just as essential for Gemini as it had been for Mercury, the STADAN and Minitrack. By the time Project Gemini came around, coordination between the two departments at the working level was well established. The Air Force, in particular, remained a key player in the MSFN, providing support via the Eastern and Western Test Ranges. The network for manned spaceflight tracking was indeed a well-balanced, well-orchestrated effort between NASA and the DOD, with the latter even assuming primary station responsibilities at some places. Table 4-2 summarizes some key characteristics of NASA's Manned Space Flight Network as it appeared in 1965 and 1966 when America flew 20 astronauts into space. (Also see Appendix 1.)

✯ ✯ ✯

Network expansion in the mid-1960s was not designed merely to meet Project Gemini requirements. It prepared the MSFN for the soon to come, and the ultimate goal, of Apollo flights to the Moon. Since Apollo would be progressively more complex—first Earth orbit missions followed by circumlunar and finally lunar landing flights—network complexity also increased incrementally. Augmentation to many existing stations, along with new stations with totally new capabilities, was necessary. Several new sites around the world were founded during 1964. One of them was on Ascension Island, the network's most isolated location.

Located just south of the Equator in the Atlantic some halfway between South America and Africa, the desolate 88-square-kilometer (34-square mile) island was originally discovered by the Portuguese on Ascension Day in 1501. Due to its remote location, it remained unoccupied until 1815, when it was garrisoned by the British Navy in an effort to prevent any attempt to snatch Napoleon Bonaparte from St. Helena some 11,000 kilometers (6,850 miles) to the south. At the turn of the century, the British Cable and Wireless Company set up a relay station on the island for telegraph cables that ran between Britain to Cape Town and South America. Little activity took place on the island after that until the Second World War, when it took on more importance, becoming a key refueling base for cargo planes of the *Cannonball Express* which the militarized Pan Am crews flew, rushing high priority supplies between Miami, Florida and Karachi, India. "If you can't go to the Moon, the next best place is Ascension Island," was ironically the airline's advertising catch phrase in those days.[19]

Ascension Island emerged as a key network location during Apollo network planning in 1964. From August 1964 to July 1965, Ed Buckley initiated a series of technical notes and memorandums to Bob Seamans and GSFC and KSC Center Directors Harry J. Goett and Albert F. Siepert, pushing to establish Ascension and Antigua in time to support Apollo. A feasibility study was conducted. It concluded that the various program requirements in the planned Apollo (and Deep Space) missions confirmed the necessity of putting a station in the middle of the South Atlantic. Voicing their support, the JPL in Pasadena also independently concluded that flights of certain lunar exploratory probes would have to be delayed until this station came on line.[20]

In August of that year, the U.S. approached the British government with a proposal to add a spaceflight tracking station on the island to support both piloted and unpiloted missions. No difficulties were expected as Ascension already played host to a U.S. Air Force radar installation. The island was also the mid-Atlantic relay point for data coming from and going to Africa via cable. An agreement was reached three months later between the two governments paving the way for NASA to establish a MSFN station on the dormant volcanic island.

To minimize construction and operating costs, as well as potential interference to existing and future facilities, island assets supporting the DSN and those for Apollo were consolidated into a single complex at a desolate area on the southeast side of the island aptly named Devil's Ashpit. Engineers chose Devil's Ashpit as it was in a very RF quiet location, being separated from the Air Force Eastern Test Range radar site and two British ground stations by the 859-meter (2,819-foot) Green Mountain. All community support and common use facilities such as barracks, the mess hall and recreational facilities for the men stationed there were integrated with the existing Air Force station already on the island.

Under this arrangement, NASA operated and maintained all its technical facilities on Ascension while the Air Force provided logistical support to NASA. (A very similar agreement between the DOD and NASA was reached for operations on Antigua in the Eastern Caribbean.) Transportation of supplies was mostly provided by the Air Force Military Airlift Command. Potable drinking water was always a concern on the remote island. To alleviate this burden, a 144,000-liter-per-day (36,000-gallon-per-day) fresh water desalinization plant was one of the first facilities constructed on the island.[21]

Civil engineering upgrades (road work, ground preparation, power) at Devils Ashpit began in late 1964, first on the Deep Space side followed by the Apollo side. Construction followed in February 1965 on the Deep Space 9-meter (30-foot) antenna and its 55-square-meter (600-square-foot) air conditioned service building. It was operational six months later. This was soon followed by another 9-meter antenna, this one for Apollo, with its own 37-square-meter (400-square) foot air conditioned service building. This powerful system was to be used specifically for high gain USB communications with the spacecraft (2.1 GHz operating frequency with a +43 dB antenna gain) with a 10 kilowatt command transmitter—sufficient for sending commands to the Apollo spacecraft in the near-Earth portions of its journey.[22]

By January 1966, construction was finished for the most part, in time to support mission AS-201 on 26 February, the first test flight of the Saturn 1B launch vehicle. Not as large as its "big brother" the Saturn V, at 68.3 meters (224 feet) the Saturn 1B was still by far the most massive launch vehicle NASA had ever flown, capable of delivering 18,600 kilograms (41,000 pounds) into low-Earth orbit. In addition to the large tracking antennas, a 30-meter (100-foot) free-standing collimation tower with a 9.3-square meter (100-square foot) air conditioned service building was added to support the autotracking antennas. NASA did not skimp in establishing the Ascension Station, spending some $10.8 million in 1965. When it was all done, Ascension (ACN) proved to be a state-of-the-art, full service station, with operations conducted at a brand new 1,330-square meter (14,300-square foot) air conditioned operations building. Rounding out the facilities on MSFN side was a 185-square meter (2,000-square foot) storage building and a 2,500-kilowatt power plant.[23]

With the rapid buildup on the island came traffic problems, which NASA had anticipated. At the request of the representative of the local British government on Ascension, the Agency constructed access roads on a new southern route to the station from the airport. The route traversed the south facing slopes of Green Mountain allowing traffic to bypass the area around Two Boats Village in the more heavily populated central part of the island.

NASA began bringing the Ascension Station online in the spring of 1965, phasing in approximately 10 people each month. ACN was exercised as a secondary tracking station during Project Gemini in preparation for its fulltime role on Apollo. By the following March, some 110 station workers were on the island. Due to its remote location and sustainment cost, normally half of the contingent assigned to Ascension was transient personnel who was on the island only during actual missions. The station was unique as it was the only "singles-only" outpost in the network. The prime contractor Bendix apparently thought its remote location and harsh living conditions would pose a hardship, and so company employees were not allowed to bring their families.[24]

A particular concern on Apollo was the launch phase of its trajectory. Attenuation of communication signals by the Saturn V rocket plume placed some limitations on the spacecraft's S-band antenna. USB stations, therefore, had to be placed closer together than first planned. The problem was not only one of needing to be geographically positioned correctly to see the vehicle from the ground, but also one of being able to maintain a reliable, low bit-error rate and continuous telemetry link between the two. To meet this Atlantic Ocean Area support requirement, NASA had to have a string

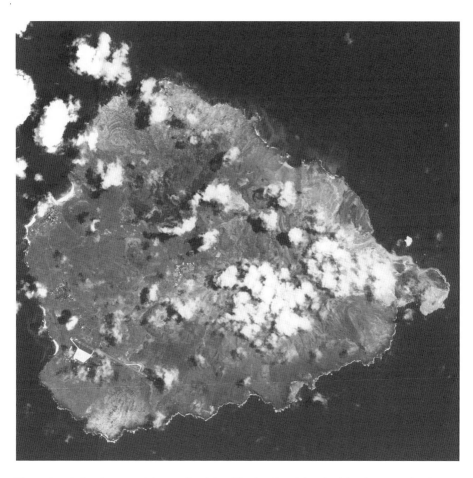

The volcanic landscape of Ascension Island is clearly evident in this photograph taken by the Ikonos satellite on 24 February 2003. The island is less than 14 kilometers (9 miles) wide. The MSFN Apollo station was located on the eastern side of the island, just to the right of the center, large cloud cover on this picture. (From the NASA Earth Observatory Data & Images archive)

of stations along the ground track at the Cape, Bermuda, Grand Bahama, Antigua, Grand Canary, and Ascension. This chain of stations was needed so as to provide communications coverage for the range of launch azimuths (the direction a rocket is launched with respect to true North) being planned to accommodate the various lunar landing sites. For instance, for launches of 72°, Cape Canaveral, Grand Bahama, Bermuda, and Grand Canary provided support. For the more southerly launch angle of 108°, Cape Canaveral, Grand

Bahama, Antigua, and Ascension provided support. For launches in between, a combination of these sites was used.[25]

In the summer of 1965, NASA approached the United Kingdom to discuss adding stations on Antigua and Grand Bahama Island. Diplomacy was once again the key. Paving the way for formal negotiations between the two governments, the senior British representative to the Air Force Eastern Test Range at Cape Canaveral, who had earlier arranged for and participated in the site surveys of Antigua and Grand Bahama Island with NASA officials, had earlier (informally) acquainted the British Colonial Office of the proposed NASA needs on these two territories. This preliminary work greatly expedited formal negotiations with the London Embassy when the time came. The selection of these two islands was by no means arbitrary and was the end result of surveys conducted on several South Caribbean islands including Barbados, Saint Incia, and Eleuthera by joint NASA and Air Force teams.[26]

Antigua, a 280-square kilometer (108-square mile) island in the British West Indies, already had an Air Force ground station which at the time was being used by NASA as a secondary station for voice communications with the Gemini spacecraft. The Antiguan government enthusiastically embraced the idea of establishing a "Moon Station" on their island. The rare opportunity to play host to one of the tracking stations for Apollo with its publicity and potential economic fallout were just too good to pass up. This enthusiasm was shown by the actions that quickly followed the initial discussions.

On 20 July 1966 (exactly three years to the day before the Apollo 11 lunar landing), Chief Minister Bird of Antigua signed an agreement with NASA making available a 168-acre plot of land near Dow Hill for NASA to construct a station. Since approximately one-third of this land was privately owned at the time, the Antiguans agreed to negotiate the purchase of this land from the island owners and finance it themselves. NASA would pay the Antiguan government a bargain sum of $336,000 ($2,000 per acre) plus interest over the next eight years under the agreement, as long as NASA guaranteed full payment even if the station were to be abandoned prior to 1974.[27]

NASA, at its own expense, widened and paved the roads needed to access the station from the airport, the existing Air Force base and from the local municipalities. In this mutually beneficial arrangement, rights-of-way and easements needed for widening the roads and for installing communication lines were furnished by the Antiguans at no cost to NASA. It was estimated by the Office of Tracking and Data Acquisition that about half a million dollars of road improvements were made on the island in 1966.

As soon as the roads were completed, a single 9-meter (30-foot) USB antenna system was constructed at Dow Hill near the Shirley Heights region on the southern tip of the island. Logistics and site support were provided by the U.S. Air Force. All the personnel support facilities such as barracks, mess hall and recreation for the new NASA station were integrated into the existing Air Force base, as was done on Ascension Island. As a further sign of interagency cooperation and cost savings, site construction was managed for NASA by the U.S. Navy Bureau of Yards and Docks under a continuing arrangement with the Air Force Eastern Test Range. It did not take long for Antigua to become operational in May of 1967. The station reached its peak of operations two years later with 92 people, mainly Bendix and its subcontractors, assigned to the island.[28]

The complexity of NASA's working relationship with other U.S. government agencies increased as the MSFN expanded. Take Canton Island in the Kiribatis. Prior to NASA assuming responsibility for Canton, three American agencies used the island under an agreement with the Kiribati government. The Federal Aviation Administration (FAA) had 50 people on the island to provide refueling and communication services. Meanwhile, the United States Navy had 26 people on the island assigned to the Pacific Missile Range. In addition, there were six people who worked for the Weather Bureau. After the completion of Mercury Atlas 9 in May of 1963, a two year period of relative inactivity in human spaceflight ensued. Still, NASA's expenditures on Canton was $1.2 million a year even though there were no missions to support.[29]

Original plans had called for NASA to operate Canton only through the first three Gemini flights, or about the middle of 1965. But during this hiatus in missions, the role of Canton was reevaluated by OTDA. Meanwhile, the FAA officially notified NASA in early 1964 of its intention to withdraw operations from the island. Up until that point, NASA was fully prepared to continue supporting Canton. With this sudden withdrawal, OTDA now felt like the FAA had suddenly left it "under the gun" to make a decision as to whether or not the Agency was going to continue supporting work on the island. OTDA needed additional time to consider the alternatives. In particular, it wanted to know whether the two other agencies, namely, DOD and the Weather Bureau, still had any requirements for the island.

In a meeting held at the Department of Commerce on 31 July 1964, the various stakeholders of Canton laid out each of their agency's position for the island. NASA had two requirements. One was for tracking and data acquisition on the first orbit after launch. Canton, as a secondary station, had only voice and telemetry. But it would be decisive in the event of a first orbit abort. Under the planned trajectories for Gemini and early Apollo, Canton was the last ground station that would be in contact with the spacecraft before retrofire sequence had to be carried out.

The second NASA requirement was one that was still several years away. Apollo reentry in the Pacific Ocean could be either in the Northern or Southern Hemisphere. Canton Island was ideally situated, being just to the south in one case and just north in the other. In other words, at this stage in Apollo planning, it appeared that Canton would be a key weather observation site for both Pacific reentry areas. But in 1964, this requirement was still several years in the future, and it was difficult to justify NASA expenditures to keep a site operational to meet a possible future requirement that was perhaps as far as five years away.

At this meeting, the DOD indicated they really no longer had any need for Canton that was directly related to defense. It had two positions, though, both based on an unwritten, good-faith commitment specifically to support NASA. The first was the Navy's original plan to support the Agency for one more year, until July 1965. The Navy was prepared to honor this commitment using remaining FAA funds available for base support through fiscal year 1965. The Air Force then "volunteered" to pick up the support after 30 June if NASA still had a requirement, but would do so only on a fully cost reimbursable basis from NASA since they no longer had a specific requirement for Canton. It was further made clear that NASA would have to let the Air Force know within the next 30 days whether Canton would be needed so that they could make appropriate budgetary plans.

The Weather Bureau's position was that Canton Island was the only equatorial site it had which possessed a full weather balloon observation capability to above 30 kilometers (100,000 feet). It was therefore extremely important for meteorological research purposes as well as providing data for storms moving northward towards Hawaii and southward towards American Samoa. Even though this presented a very important (almost mandatory) requirement to the Weather Bureau, they made it clear to the other stakeholders that the bureau would be forced to reconsider this requirement if it were required to fully support the island on its own. The bureau's conclusion was essentially that they would be willing to pay its pro rata share of cost, provided "it didn't cost too much."[30]

The final arrangement reached between the agencies was for one single agency to manage and fund both the technical and administrative support on Canton, coordinating the latter with other interested stakeholders. NASA, with the most at stake, ended up assuming the lead role. In a letter written on 21 January 1965 to James Webb, Ed Buckley recommended this action, specifically pointed out that Bendix is also the major support contractor to the other agencies on Canton. To this end, he suggested that Headquarters could easily arrange to amend the Bendix contract to include Canton without significant change in technical personnel.[31] Many FAA government workers were also receptive to on-the-job transfer to contractor employment status. Jurisdiction and logistical responsibility on Canton began to transition from the DOD to NASA with the launch of the first crewed Gemini flight in March of 1965. This transition was complete by the time the second Gemini mission took place in June. DOD operations on Canton, along with what was left of the FAA, were completely phased out.

But it turned out that NASA was not to keep Canton open very long either. In November 1966, after Gemini 12 splashed down bringing the program to a successful conclusion, the Office of Tracking and Data Acquisition and the Office of Manned Space Flight jointly conducted a thorough review of MSFN requirements. Their conclusion was that future requirements in the mid-Pacific could be met more flexibly and effectively by one of the Apollo reentry ships. One important reason was that a ship could provide S-band support, something that the Canton Island ground station could not.

With its fate sealed, a step-by-step phase out of the station followed. Canton Island ceased participation in all network activities by July 1967. An advance notice was given by the State Department to the British government followed by a final meeting held at the Department of the Interior on 10 August. It was verified then that no other U.S. government agency was interested in assuming responsibilities for the island. This was soon followed by a FAA Notice to Airmen that, except for emergencies, the Canton Island Airport would be closed to all traffic on 1 September.

Station staff was immediately reduced and preparations initiated for assuming a caretaker status until a complete evacuation of the station could take place. By September 1967, the approximately $3.2 million of NASA equipment invested in the station had been removed and reassigned to other stations. Contractors and their families left the island. The final inspection flight left Canton on 20 December 1967, ferrying out the remaining few facility support workers, along with a handful of Standard Oil engineers and geologists who had remained on the island to finish out their scientific research.

Such joint and sometimes convoluted decisions were not uncommon since NASA (a civilian organization) and the DOD (a military organization) both had—and continue to have—a stake in the frontier of space. They generally served each other's interests well. Issues relating to the sharing of cost and resources could be found simmering but were often easily settled with the stroke of a pen. They ranged from the trivial such as funding of recreational facilities to who would provide office equipment, to the more serious requirements of cooperative use of water production, transportation, station operations, and maintenance costs.

In the summer of 1965, for instance, a dispute over who should pay the cost of running the power generation equipment on Ascension Island had gotten quite bitter with neither the onsite DOD official nor the NASA Station Director budging an inch. To break the stalemate, Ed Buckley recommended that the KSC make available $58,500 to Patrick Air Force Base to run the power plant.[32] Some of these conflicts could have been perceived as perhaps a bit petty by those looking in from the outside. But for the people stationed at remote locations, these otherwise "petty" issues could directly affect everyday quality of life where access to resources could not always be taken for granted.

Senior management on both sides back in the States was often called on to keep such issues from escalating.

While interagency problems were one thing, strife between collocated STADAN, MSFN, and DSN personnel was a different matter all together. It had the potential to not only be ugly but also impact the ability of a station to perform its mission. A case in point was Ascension Island in 1967. As the MSFN started operations, a lack of cooperation began to develop between Bendix contractors working the manned spaceflight system and those working the unmanned planetary Deep Space program. This refusal to coordinate their efforts eventually led to the Goddard side of the house failing to adequately provide their assistance to help maintain and support the DSN antenna. The condition deteriorated to the point where Ascension was faced with the problem of having to refurbish the antenna which had become severely corroded. Most of the metal had to be refinished and the electrical wiring replaced.

In the spring of 1967, James Bavely, Chief of Network Operations at the Office of Tracking and Data Acquisition, directed E. J. Stockwell and a representative from Goddard to the island to get first hand information on the problem. The detrimental effect on station maintenance eventually got to the point where installation of a radome to protect the antenna was considered. However, with the need at the time to use MASERs to support Deep Space missions, the real possibility existed of burning a hole in the radome. A solution would have been to remove the radome during a mission and then replace it afterwards. The alternative (and the one eventually implemented) was much more attractive: proper preventative maintenance through better cooperation among those working the two networks on the island.[33]

While this was happening, one of the most volatile episodes to befall the network played out halfway across the world on the eastern shores of Africa. The MSFN station in Tananarive, Madagascar was one in which the beginning and end were tied to the political unrest and instability of not one but two governments. Unfavorable circumstances surrounding the government of the host country have, on just a handful of occasions, led to station closures. These included, for example, Cuba (see Chapter 2) and South Africa (see Chapter 6). None, though, were as severe and dangerous as what happened in East Africa in 1963 and 1964, where the disruption impacted operations for both the MSFN and STADAN.

Just off the eastern coast of Africa some 5° south of the Equator is the island of Zanzibar. Its written history dates back to the Persian empire of the sixteenth century. Occupying a prominent spot along the east African shipping lanes, the control of Zanzibar was the object of multiple conflicts that occurred amongst various ruling sultans in the 1800s. The British Empire with its powerful Navy was also gradually taking over offshore islands in the area during this time. In 1890, the island became a protectorate of the United

Kingdom, anchoring the Commonwealth's very important Bombay-Zanzibar shipping route to the Far East.

Zanzibar was needed to provide spacecraft coverage after loss-of-signal at Kano, Nigeria. It was also the last land station to see a spacecraft before it crossed the Indian Ocean. Despite some mistrust on the part of the local British client Sultan, on 14 October 1960, the United States signed an agreement with the United Kingdom to place a NASA station on the island. With the completion of stations in Kano and Zanzibar in 1961, the MSFN was essentially finished. The Zanzibar Station was located about 16 kilometers (10 miles) east of Stone Town near the village of Tunguu. Assigned in a totally foreign environment, the Americans, to their delight, were well received by the local villagers. Technicians and their families blended in as just another minority group in the ethnically diverse region. The staff usually lived in Stone Town with their families when they were not at the station supporting a mission.

Although the station successfully supported all four Mercury orbital flights without major disruption, it could not avoid operating under a continuous umbrella of scrutiny from the local authority. Zanzibar, in the early 1960s, was a highly unstable country of some 300,000 people, ripe for strife with factions like the proindependence Afro-Shirazi Party and the Ittihad ul'Umma, pro-Peking, pro-communist party that favored Chinese expansion into East Africa, all trying to seize power. Tensions had reached the point by 1963 where NASA was realistically concerned and keenly aware that hostilities could erupt with little or no warning. In July 1963, less than two months after the conclusion of Project Mercury, the State Department issued a memorandum warning of imminent potential riots in Zanzibar pending the outcome of national elections. In a Confidential letter (since declassified) to Goddard Center Director Harry Goett foreshadowing things to come, NASA Headquarters recommended that station personnel, as a precaution, formulate 1) an emergency escape plan, and 2) a plan to reduce staffing of the station to a caretaker status that could be implemented by no later than 3 July, when elections were slated to begin.[34]

A period of political unrest did follow the elections but station personnel did not have to implement their evacuation plan, at least not yet. But the situation deteriorated rapidly soon thereafter. By year's end, the British, who had dealt with over 70 years of factions and strife on the east African colony, finally granted Zanzibar its independence, establishing it as a constitutional monarchy on 19 December 1963. This state of independence was short lived, though, as the ruling Sultan was overthrown less than a month later in a bloody military coup instituted by the Afro-Shirazi Party. The new socialist regime then went on to merge with the neighboring mainland state of Tanganyika to form the country of Tanzania.

Meanwhile, the original tracking network agreement had expired in July 1963 after the last Mercury mission. NASA was literally in the midst

Bendix FIELD ENGINEERING CORPORATION

SUBSIDIARY OF THE BENDIX CORPORATION

OWINGS MILLS, MARYLAND

March 11, 1964

Mr. Edmond C. Buckley
Director, Office of Tracking & Data Acquisition
National Aeronautics & Space Administration
400 Maryland Avenue, S. W.
Washington 25, D. C.

Dear Mr. Buckley:

    Several weeks ago our employees and their dependents were involved in a very hectic but successful evacuation from the Island of Zanzibar.

    An emergency operation such as this requires strong, capable leadership to ensure a timely and successful completion. I would like to take this opportunity to pass on my appreciation for the outstanding leadership displayed by Mr. Frederick Picard. Our Mr. Burch, the Bendix M & O Supervisor, speaks highly of Mr. Picard's actions during this very trying period. He relates to us that Mr. Picard spent an uninterrupted twenty-four hour period in an incessant effort to arrange a safe, orderly evacuation for American personnel.

    He further relates an incident which I feel typifies and climaxes Mr. Picard's efforts in behalf of our people. All Americans and a number of British residents had gathered in the English Club as a central point of refuge. During the night the rebels were breaking down doors of neighboring buildings and were entering these buildings. When this group approached the English Club door, Mr. Picard went outside and bravely placed himself in front of the door. He spoke to the group in Swahili and succeeded in preventing their entry into the Club.. This was indeed an act of courage.

    The Field Engineering profession places our personnel in all corners of the world. It is gratifying to know that dedicated men like Mr. Picard are standing by for help when a situation as severe as the Zanzibar incident occurs.

    Once again, my deepest thanks to Mr. Picard for a job "well done."

                                              Sincerely yours,

                                              L. F. Graff
                                              President

LFG/es

Letter of gratitude from Bendix to NASA on successfully navigating the circumstances of what could have been a tragedy in Zanzibar. From a station safety standpoint, this was probably the most tense moment in the history of the Goddard networks.[35] (Folder Number 8824, NASA Historical Reference Collection, NASA History Division, NASA Headquarters, Washington DC)

of negotiating a renewal with the Sultan when the revolt happened. With safety of Americans now at stake, NASA Associate Administrator Robert C. Seamans ordered the immediate evacuation of the station on 14 January 1964. The three dependents of Station Director Tom Spencer (Spencer himself was away on business in Malagasy at the time), along with eight Bendix workers and their 18 dependents were evacuated out of the country by the U.S. Navy as a destroyer stood by on alert offshore.[36]

The drama of these events (coming on the heels of Gordon Cooper's triumphant 24-hours in space just six months earlier) was well publicized in the United States. The behind-the-scene (and, some would later say, heroic) diplomatic intervention of State Department officials in Zanzibar to buy more time making possible a rescue mission and to keep the events from escalating into an international incident was critical, the importance of which cannot be overstated. In letters of appreciation to Secretary of the Navy Paul H. Nitze and Secretary of State Dean Rusk, NASA Administrator James E. Webb officially recognized the cooperative efforts of their departments, and credited several individuals by name, especially mentioning on-the-spot actions of Charge d'Affaires Fredrick Picard, in resolving this incident as successfully as possible without tragedy.[37] After the station staff was safely home, L. F. Griffin, then President of Bendix Field Engineering Corporation, expressed his appreciation and gratitude in a letter to the space agency, personally thanking Buckley "in behalf of our people."

But after the staff were evacuated, there still remained the question of what to do with the approximately $3 million worth of communications equipment that was abandoned. The new Zanzibari President Abeid Amani Karume, who had originally given the United States a 60-day window to remove the assets, withdrew that offer and demanded in a meeting with Frank C. Carlucci, the new Charge d'Affaires, that the U.S. completely rid all station equipment from the country by the end of that April. (It was thought at the time that President Karume did this as a reprisal to statements made by William H. Attwood, U.S. Ambassador to Kenya, that communist China was turning Zanzibar into "a kind of non-African state" to be used as a staging area for their base of operations against other governments in Africa.) Also to be dismantled as an adjunct to the Zanzibar tracking station was a nearby communication facility that relayed data to Kano, Nigeria, for transatlantic communications to Florida and Houston via cable.[38]

While the State Department deemed it necessary to physically remove station assets in the interest of preserving national security, NASA's position was somewhat different. Norm Brockett, the Director of Network Operations and Facilities, thought that the actual reuse value of the equipment was fairly negligible when compared with the risk to Americans who would have to be flown back into Zanzibar, tear down, load the hardware, and then be flown back out again. (A team of 19 workers were, in fact, standing by in neighboring Nairobi for just this purpose.) With NASA making it quite clear that any effort to remove station equipment would be carried out only because it was the desire of the State Department to do so, no attempt was ever made. President Karume's deadline came and went. No retrieval team was sent and the station was eventually abandoned in place.[39]

NASA's official stance at the time was that the loss of Zanzibar would have no real effect on future plans for Gemini and Apollo. These programs would stay the course. But the Office of Manned Space Flight had in the meantime placed a requirement on Gemini for voice communication and telemetry in that geographical area: on at least 50 percent of the orbits, Zanzibar was the last station just before the astronauts fired retrorockets for deorbit and reentry.

To meet this strictly technical requirement, a couple of contingency plans were considered, both involving the use of tracking ships. One was to move the Indian Ocean Ship *Coastal Sentry Quebec* farther west. The OTDA quickly eliminated this option, though, since it would have left an unacceptably large void over the Indian Ocean prior to acquisition-of-signal at Carnarvon, Western Australia. OTDA also looked into what it would take to acquire and configure another ship off the coast of east Africa. It was also quickly determined, however, that this could not be done in time for the first crewed Gemini flight, at the time scheduled for October of 1964. (It had already been postponed from April. Gemini 3 eventually flew in March 1965.) Fiscal constraint was also a factor as the annual cost to operate a ship was expensive, over twice that of a land station. NASA had to find a more permanent solution.

A few locations were considered where a transportable system could be emplaced, such as in Southern Rhodesia (Zimbabwe) or in South Africa, where there was already a DSN and STADAN site. Yet another possibility was the Malagasy Republic on the island of Madagascar. Just three months earlier on 19 December 1963, the U.S. had entered into a 10-year agreement for the installation of a transportable STADAN station outside the port city of Majunga in northwest Madagascar. The two countries had reached the agreement in accordance with the spirit of a United Nations resolution calling for the application of results of space research to benefit all peoples. In addition to generating much needed weather forecasts, especially during hurricane season, the station would provide jobs for some 200 local residents in nontechnical positions for handling day-to-day maintenance work.[40]

Initial equipment consisting of five 30-foot trailers—one each housing a 136.2 KHz and 400 MHz telemetry receiver—were set up at Majunga. More equipment soon began arriving from the Australian sites of Muchea and Woomera, which were phased out at the conclusion of Project Mercury. A MPS-26 radar was temporarily deployed prior to the addition of a FPQ-6 radar. All together, NASA spent some $600,000 in additional funds to finish-out a transportable station in the east Africa region to replace the one lost at Zanzibar.[41]

In the summer of 1964, as it became apparent that the first mission of the new Gemini spacecraft was not going to occur until the following year, Goddard officials began giving thought to moving the station to a more permanent establishment. In September, construction began at Imerintsiatosika, 24 kilometers (15 miles) outside the capital city of Tananarive (now Antananarivo) in the central High Plateau region of the island. By the time the move was completed, American staff at Tananarive had increased from 21 to 58. As a transportable STADAN site, the station had been one of the simplest in the network, requiring only 18 Bendix and Motorola contractors along with three GSFC-assigned supervisors. These requirements were further reduced in between missions when it was routinely reduced to caretaker status, requiring only an American representative onsite to supervise the Malagasy nationals employed to take care of day-to-day maintenance. By the time Tananarive ramped up to support Gemini 3 on 23 March 1965, such down times were a thing of the past as 44 fulltime American contractor employees along with 13 trained Malagasy nationals were reporting to the NASA Station Director.[42]

The disruptive environment that plagued the station at Zanzibar was a sharp contrast to what NASA experienced on the island of Guam. The Agency's work on the island would turn out to be one of the most amicable and long lasting in the history of the NASA networks, one that continues to this day. It began in the spring of 1964 as OTDA began looking at new locations for the Apollo network. To support lunar flights, several new capabilities were required:

> Tracking and data acquisition for Apollo rendezvous tests in Earth orbit
>
> Establishing the spacecraft orbit in preparation for and to make the go/no-go decision for Trans-Lunar Injection (TLI)
>
> Continuous voice and telemetry contact during the critical lunar injection phase
>
> Continuation of coverage during premidcourse flight to confirm the "go" status of the lunar mission on the outbound trajectory

NASA needed a ground station to provide coverage in the broad ocean area between loss-of-signal at Australia (Honeysuckle Creek) and acquisition-of-signal in Hawaii. After looking at trajectory ground tracks, mission planners determined that the Mariana Islands afforded the best geographical location from which the Apollo requirements could be met in the Pacific. Site survey teams were sent to Saipan, Tinian, and Guam in April 1964. They found that although suitable geographic locations existed on each of these islands, Guam was the best for several reasons. First, an international ocean cable between the island and the U.S. had just recently been put into service. Second, radio noise in the southern part of the island was virtually nonexistent (a very RF quiet, -87.5 dB per square meter). Third, Guam already had an established and well used logistics pipeline to the United States. Finally, it did not hurt that the proposed site was on a private parcel of land owned by U.S. citizens that could be leased.

Located in a large, flat valley some 25 kilometers (16 miles) southeast of the capital city Agana (Hagatna), the Guam MSFN station occupied an area known as Dandan, which means "to knock at the door" in the native Chamorro language. The 550-square kilometer (212-square mile) island is some 6,500 kilometers (4,000 miles) west of Hawaii. It is today one of five well traveled insular areas of the United States (the other four being American Samoa, Northern Mariana Islands, Puerto Rico, and the Virgin Islands). The origin of its once primitive habitat is, surprisingly, completely obscure. The ancient inhabitants left no decipherable records. Latte stones found upon the arrival of European discoverers were so ancient that neither their origin nor their true purpose is known.

Spain first laid formal claim to Guam in 1565, 44 years after its discovery by Ferdinand Magellan, but actual occupation of the island did not begin to take place until 1668, when Padre Luis de Sanvitores led a group of missionaries onto the island. Spanish rule ended in 1898, following the Spanish-American War when Spain ceded Guam and the Philippines to the United States. President McKinley then placed the administration of the island in the hands of the Navy and for expediency, appointed the Naval Station Commander as the governor. The island fell to the Japanese in World War II and became the scene of some of the fiercest battles of the war. It was recaptured in the summer of 1944 when U.S. marines once again raised the Stars and Stripes over the island in its island-hopping campaign towards Tokyo. Five years after the war, Congress passed an act making Guamanians citizens of the United States, giving Guam self-government under a U.S.-appointed civilian governor.

One factor that made Guam very attractive besides its excellent geographical location was the unabashed enthusiasm of the host. Manuel Flores Leon Guerrero, the 50-year old American appointed Governor of Guam, made it no secret to the survey delegation that there was no better place in the Pacific to locate the first new Apollo tracking station than on his island. Affable and gregarious, Guerrero proactively campaigned for the proposed station, taking a very personal and active interest in the whole affair. He personally entertained the survey team and hosted a reception at the gubernatorial mansion so that NASA officials could meet face-to-face the leading citizens and merchants of Guam. He then volunteered the services of the government of Guam to aid the Americans in any way possible.

Two areas of vital importance to NASA were specifically addressed by Guerrero. First, he offered to secure or aid in securing the land necessary for the Apollo station. To this end, he offered to buy the necessary land in the name of the government of Guam and lease it to NASA. But if that did not work, he offered to negotiate for the direct lease to NASA by the owner, and if needed, to negotiate for the purchase of the land by NASA. As eventually implemented, land was leased by the owners to NASA.[43] The second issue pertained to the island's support of the NASA contractor employees who would be stationed on Guam. He felt certain that private enterprises would be up to the economic challenge of providing housing and community services that would be needed, offering his personal commitment to stimulate the private sector.

Two months after the survey team's return from Guam, Governor Guerrero personally visited Ed Buckley in Washington to again express his government's eagerness for a Guam station. The campaigning paid off. On 10 June 1964, the position of the OTDA was put forth in a letter to Hugh Dryden. In the letter, Buckley wrote, "The interest and support by the government of Guam will facilitate an early decision probably this week on the final site selection in southern Guam," and recommended the obligation of $170,000 in advanced funds to the Bureau of Yards and Boats to begin design of the Dandan site.[44]

Construction of the Guam Station began in January 1965. There was pressure to get the station operational, not only due to the pace of the Apollo schedule but also because there was fear on the part of the Agency that the DOD might, in some way, lay claim to the job first. In 1965, the Air Force was also building its own ground station on Guam (on the north part of the island, not the south where NASA was). Charles Force, Guam's first Station Director, said there was a feeling that "if NASA got its Guam station up first, then [we] would have the role supporting the manned missions, whereas if [we] didn't put one there, then the Air Force would have that role, and NASA didn't want the Air Force to have a key NASA station."[45] That fear may have driven the pace of construction as the station was completely done by September 1966. (As for the DOD station, it too became operational but was used strictly for its own purposes.)

Guam's capabilities were second to none, including its centerpiece, a USB 9-meter antenna system that provided telemetry, tracking, commanding, and voice communications to the Apollo spacecraft. Backup TT&C functions in the VHF range were accomplished using TELTRAC, SATAN, and Satellite Command Antenna on Medium Pedestal (SCAMP). The Dandan site had a large central operations building and a "diner".[46] A NASCOM Switching Center to handle Pacific circuits was later added on the south side of the building. Three other structures housed water, fire, power, flammable storage, and automotive equipment. The collimation tower and other support equipment buildings were located on a hill about three kilometers (two miles) from the main operations building (see figure on next page. As the station

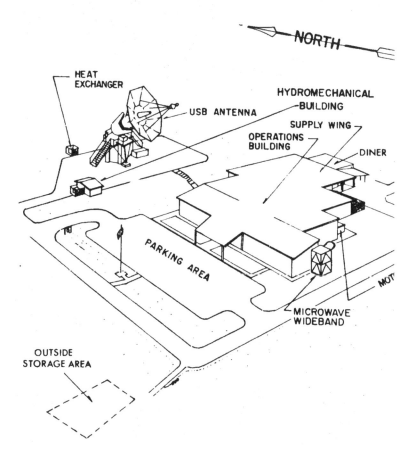

The Guam Apollo Station had as its centerpiece a 9-meter USB antenna used to communicate with the Apollo spacecraft during the near-Earth and trans-lunar portions of the mission. The station was one of the longest lived in the network, operating for over

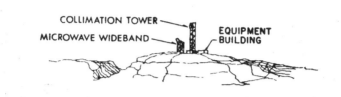

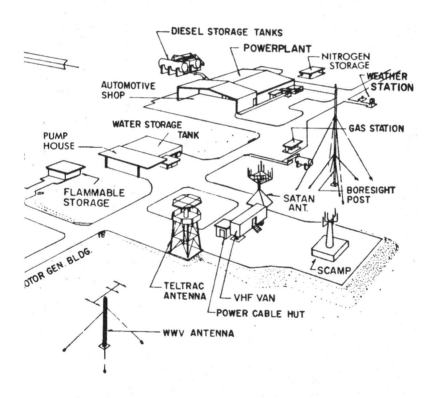

two decades, from 1967 to 1989. (Folder Number 8813, NASA Historical Reference Collection, NASA History Division, NASA Headquarters, Washington DC)

neared completion, it was integrated without delay into the manned network towards the end of Gemini, checked out and declared operational in March 1967, in time to support the historic (unmanned) first flight of the Saturn V on Apollo 4.

Like Bermuda five years earlier, the Guamanians were very proud of their station. NASA had the very active support of the community in that regard, from the Governor down. Force recalled a story as the station was about to open:

> When the station became operational, I decided it would be appropriate to have a dedication ceremony for the station, so I tried to get somebody from NASA Headquarters to participate. They declined.... [Apparently] Guam wasn't high enough up on their priority list for whatever reasons that they were going to participate. But when I called the Governor's office, they were out there with 'bells on' immediately and everything. We did have a very nice dedication ceremony. The Navy, who had physically constructed the station, heavily participated as did the government of Guam.

Station Director Charles Force welcomes dignitaries and guests to the Guam Station dedication ceremony on 21 January 1967. Seated to his right are Jose A. Leon Guerrero and A. W. Baumgartner, Bishop of Guam; to his left are Governor and Mrs. Guerrero; Marilyn Force, President of the Apollo Wives Club; and Cdr. Eugene Pickett, Officer In Charge of Construction, USN; who oversaw the station construction for NASA and presented a symbolic key to the facilities. (Photograph courtesy of Charles Force)

We had a lot of good local publicity. That was the first time I think it dawned on the local people that they had a future role in that station, and from there on out, we had very popular support.[47]

A bronze plaque at the station's main entrance proclaimed:

> This Apollo Tracking and Data Acquisition Facility, established by the Goddard Space Flight Center of the National Aeronautics and Space Administration, is hereby dedicated to providing exemplary support for the peaceful exploration of Space as mankind, using his God given powers, ventures forth to other celestial bodies in his continuing search for knowledge[48]

Guam was significant in that it was the first station built from the ground up specifically for Apollo. Though it had to endure its fair share of typhoons—being located in the middle of the Western Pacific typhoon alley—the station went on to support all six Moon landings as well as Apollo 13, Skylab, Apollo-Soyuz, and the Space Shuttle until 1989, plus numerous scientific satellite programs. Following a 10-year hiatus in the 1990s, NASA once again chose Guam, this time as host to the overseas ground terminal for the TDRSS (see Chapter 7). The establishment of the original tracking station is still considered one of Guam's crowning achievements and a source of pride for the Guamanians.

One of the existing stations overhauled during Gemini in preparation for the coming lunar landing program was Bermuda. As on Mercury and Gemini, Bermuda was a critical station immediately after launch and would now monitor the ascent of the Saturn V into orbit. First of the downrange stations to electronically see the rocket, Bermuda provided the critical go/no-go data to Mission Control for flight continuation or abort decision making. It was located in the right place, enabling one to observe a large portion of the S-II second stage burn and most of the S-IVB third stage burn at high elevation angles.

Apollo presented several first time technical challenges to the network. Saturn launches out of KSC with azimuths between 72 to 90° required the addition of a C-band radar capability on Bermuda to meet Houston's flight mission rules for acquisition of data needed to evaluate the spacecraft while it was in Earth "parking orbit" prior to the TLI burn. These evaluations served three primary purposes: guidance system analysis, propulsion system analysis, and overall malfunction analysis of the Apollo spacecraft prior to committing it on a trajectory to the Moon.

On 8 April 1965, Goddard awarded RCA a $4.6 million contract to provide an Apollo tracking and data acquisition system on Bermuda. The company was to install its most sophisticated long range radar, the FPQ-6, on

Cooper's Island. This C-Band system was state of the art for its time, accurate to within two meters (six feet) at 48,000 kilometers (30,000 miles). The previous system on Bermuda, the RCA FPS-16, tracked only to an accuracy of 5 meters (15 feet) at 800 kilometers (500 miles). It was kept as a backup. Bermuda was the second "Q-6" in the MSFN, the first having been installed earlier in Carnarvon, Australia to support Gemini and the early Saturn booster development tests.[49]

On 10 March 1965, Ed Buckley submitted a $1.6 million request to James Webb to consolidate and upgrade the existing MSFN facility on Bermuda to meet the combined requirements for projects Gemini and Apollo. This much needed upgrade was designed to put under one roof, the various telemetry facilities located in prefabricated metal structures and in trailers scattered about Town Hill and Cooper's Island. The corrosive effect of sea salt spray and moisture had over the years taken its toll, making a facility construction project imperative if NASA were to entertain any thought of continued operations on the island.

The upgrade was very thorough. It included an air conditioned, 1,100-square meter (12,000-square foot) Operations Building along with a 300-square meter (3,200-square foot) Generator Building to house the diesel generator. Adjacent to the USB antenna was a windowless 45-square meter (500-square foot) building housing the hydro-mechanical equipment to point the massive antenna. Concrete foundations were also dug for the 9-meter (30-foot) dish and the collimation tower. Extensive cabling between the existing Tracking and Communications Building and the new Operations Building were installed; an existing microwave terminal was relocated. Maintenance and administration staff increased by 30 percent. Twenty-six additional technicians were soon added as the site ramped up to support Gemini and Apollo. Once the Cooper's Island upgrade was complete, the old telemetry site at Town Hill was dismantled.[50]

By far the biggest change in gearing up for Apollo was the use of USB. It affected, rather extensively, network operations. Adding to the complexity was that some USB stations had dual capability and could support two spacecraft—the Apollo Command/Service Module (CSM) and the Lunar Module (LM), for example—simultaneously if they were in the antenna beam. Others were "single" and could handle only one spacecraft at a time. To illustrate the complexity of network planning during this time, one can look at how USB capability was added at Grand Bahama and Grand Turk.

The first thing that GSFC and MSC did was to correlate USB antenna patterns with trajectories to arrive at a preliminary set of ground station locations. This was done for Apollo even before the first Gemini mission took place. The result of this preliminary investigation, along with a later GSFC/KSC meeting held on 1 September 1964, was presented to the 11th Manned Spaceflight Instrumentation and Communications Panel in October 1964.

Chapter 4 \ Preparing For The Moon

These studies showed that because of a severe antenna pattern pull towards the rear of the launch vehicle, a serious gap in Apollo Saturn V command coverage would be encountered somewhere between the Merritt Island Launch Area (MILA, just downrange of the launch point) and at Bermuda or Antigua—which one depending on the actual launch azimuth. The immediate recommendation of the panel was that alternate locations at MILA and Cape Kennedy be considered for a USB site. Houston also suggested that additional stations at Grand Bahama, Grand Turk, and Vero Beach be considered. Their priorities were to be made mandatory, highly desirable, and desirable. But since abort requirements and antenna configurations used in the studies were new and still evolving at the time, the panel also recommended that more analysis be performed. To this end, a new sub panel was formed. The mission of this Subpanel on Launch Area Instrumentation was to make a comprehensive assessment of additional coverage requirements that were still needed. W. F. Varson from GSFC was appointed chairman of this subpanel.[51]

Its first meeting was held on 22 October 1964. At the end of the day, Varson's team had reached three conclusions: 1) The need to select a generally southern MILA location for the launch area USB station; 2) Continuous coverage from launch to Grand Bahama Island was probably not going to be feasible and that a station at Vero Beach would have to be considered if continuous coverage were to be made mandatory; and 3) Further analysis was again still necessary prior to committing to building a station at Vero Beach. The next meeting of the subpanel (now redesignated the "USB Implementation Subpanel") was held at Greenbelt on 10 November 1964. There, a more definitive plan of action began to materialize. The panel gave the go-ahead for a transportable USB system to be placed at MILA. It also made the very key decision that the three stages of the Saturn V launch vehicle *would not* require continuous coverage from launch to orbit, but that additional coverage for the Apollo spacecraft itself (the CSM) *would be* required to close a two to three minute gap between the Cape and Bermuda. It was concluded at this meeting that this additional requirement could be met by placing a transportable system on Grand Bahama supplemented with a planned Air Force USB station on Grand Turk. This action essentially took Vero Beach out of the picture.

An all-hands meeting took place 10 days later, this time with Varson's panel meeting with Major General Samuel C. Phillips, then the Director of the Apollo Program. Solutions for USB coverage were presented advocating the emplacement of a station on Grand a and possibly one on Grand Turk. The panel also recommended that any site selected between Cape Canaveral and Bermuda—to ensure link closure immediately down-range of the launch area—be transportable so as to accommodate various launch azimuths. Based on these recommendations, it appeared that Grand Bahama would definitely be needed but that the probability of a station on Grand Turk was still "50-50" at best. Despite this uncertainty regarding Grand Turk, launch area abort

A gathering of NASA Station Directors at GSFC in 1968. Front Row (left to right): Bill Wood (Head of the MSFN Operations Branch), Walt LaFleur (STADIR Bermuda), Bryan Lowe (STADIR Honeysuckle), Don Grey (Honeysuckle), Tecwyn Roberts (Chief of the Manned Flight Operations Division), Virgil True (STADIR Hawaii), Dale Call, unidentified; Second Row: unidentified, Jack Dowling (STADIR MILA), George Fariss (STADIR Goldstone), Fred Healey (Assistant STADIR Bermuda), Charles Force (STADIR Guam), Dan Hunter (Assistant STADIR Madrid), Chuck Jackson (Chief of the Logistics Management Office); Back Row: Larry Odenthal (STADIR Grand Bahama), Lewis Wainright (STADIR Carnarvon), Otto Womack (STADIR Guaymas), Hank Schultz, (STADIR Corpus Christi), Otto Thiele, (NASA Representative on the Vanguard), Bill Easter, Joe Garvey, (STADIR Antigua), Chuck Rouillier (STADIR Grand Canary). (NASA Image Number G-68-206)

coverage was considered sufficiently critical that steps had to be taken so as to prepare a location on the island should it be called on. The fiscal year 1965 budget process was already well underway by this time and the Air Force in the mean time decided not to put its own USB system on Grand Turk. The Agency thus decided that the best approach was to request FY 1966 funds be allocated for transportable systems on both Grand Bahama and Grand Turk.

But the Grand Turk issue was still up in the air as late as March 1965. Engineering analysis continued at GSFC and MSC, but no definite conclusions were reached. The analysis was not easy since uncertainties still existed in the Apollo spacecraft antenna patterns and in the predicted magnitude of the Saturn V booster plume attenuation. A progress report was submitted to the 12th Manned Spaceflight Instrumentation and Communications Panel

Gemini 5 is launched from Launch Complex 19 atop its Titan II booster for an 8 day mission, 21 August 1965. At launch, the vehicle stood 33 meters (109 feet) tall and weighed 154,200 kilograms (340,000 pounds). Although Gemini carried a crew of two, the entire vehicle was not greatly bigger than the single-seat Mercury Atlas, which stood 29 meters (94.3 feet) and weighed 117,930 kilograms (260,000 pounds) at launch. (NASA Image Number 65P-0160)

on 25 February. Varson felt that conclusions one way or the other regarding Grand Turk could be reached by the end of March and recommended that the Apollo Program Office be briefed as soon as his team was ready. A month later, the panel was ready with its decision.

The final conclusion of the Varson subpanel was presented to Phillips on 1 April. It recommended that a single USB transportable system be stationed at Grand Bahama with the capability to support a single spacecraft. The Grand Turk USB site, which throughout this process had consistently been deemed secondary and needed only for contingencies, was duly eliminated.[52]

In addition to augmenting early-ops operations in the Caribbean, the Guaymas Station in Mexico was also upgraded to accommodate a 9-meter (30-foot) USB single spacecraft system (one transmitter, two receivers). The

United States had renegotiated with Mexico City when Project Mercury ended in May 1963 to expand the station for tracking of unmanned science satellites. International goodwill between the two governments was further promoted as the United States and Mexico agreed upon other areas of scientific cooperation, in particular, meteorological sounding rocket programs. Just three weeks before Gemini 3 on 4 March 1965, an agreement was reached to extend operations at Guaymas to the year 1970. Over the next two years, upgrades were done to bring the station inline with the other primary sites to enable simultaneous tracking of both the Gemini spacecraft and the Agena rendezvous target.[53]

Construction began in the fall of 1965 and the upgraded Guaymas station was declared fully operational by GSFC in the spring of 1967. The $5 million expansion was a rather large project that necessitated the facility grounds to increase dramatically, from 30 to 114 acres. This was needed to ensure a noninterfering perimeter and to eliminate potential obstructions and personnel trespasses into the antenna beam—a real hazard when the antenna was transmitting. Strict perimeter control was required since the antenna would be, for the most part, operating at low elevation pointing angles from its location in northwest Mexico.[54]

As the first Apollo flight drew near and tracking stations were geared up, these foreign outposts began to take on more and more visibility on the international scene. The one person in charge of a station was the Station Director, or STADIR. As his title suggests, the STADIR was the person ultimately responsible for the everyday operations of a tracking station. But running the station turned out to be only one part of the job. The STADIR of a foreign station had another big responsibility: act as a spokesman for NASA. This "other duty as assigned" made publicly representing NASA a routine part of the job. In this regard, overseas NASA STADIRs were part of the Embassy staff, subject to direction from the Ambassador.

In the 1960s, the world was watching as America prepared to send men to the Moon. NASA was fully aware that the country's prestige (and Cold War standing in the international community) rested on the outcome. As Project Gemini continued to pioneer a series of American space firsts, international interest in the U.S. space program was intense. How a station was run could play a key role in influencing the public opinion in that country, being that it was often the most visible (and sometimes only) evidence of the space agency on foreign turf. Every local government, in addition, wanted reassurance that they were playing an important part in going to the Moon. This was especially important at locations where American sentiment may not have been at the best.

Sometimes a STADIR asked for guidance from GSFC management or Headquarters on handling of public affairs; sometimes they were just directed as to what to say. Other times, it was a little of both, as illustrated by the following letter from Ed Buckley to Morton Berndt, the Guaymas STADIR in 1965, on how he should convey the importance of the station to the local press (the word "Guaymas" appears four times in the statement; the word "important" six):

> If it should prove necessary during the coming missions to explain the importance of the Guaymas Station to the press, I suggest that you speak along the following lines:
> It is important to recognize that the data from the spacecraft systems and the astronaut's performance during every orbit passing Guaymas is a very important piece of information to the overall conduct of the flight. The Guaymas site is a very important part of the network from a standpoint of operational control, and the information to be gathered from this site during all periods of the operation is very important to the overall program. Guaymas is used in many ways such as the place where important retrofire information is obtained and initial contact with the North American continent after long periods of silence from the spacecraft while over the Pacific. We should never lose sight of the importance of Guaymas to the conduct of the manned space flight operation.[55]

By 1967, the MSFN had matured into a sprawling but centralized structure, an interconnected framework of over two dozen ground stations spanning three continents. It supported 10 very successful Gemini flights from March 1965 to November 1966. These missions produced a series of impressive firsts: NASA's first two-person spaceflight (Gemini 3); America's first extravehicular activity (EVA) or spacewalk (Gemini 4); the world's first spacecraft rendezvous (Gemini 6 and 7); the first docking (Gemini 8); and the highest apogee orbit to date of 1,370 kilometers (850 miles) above the surface of Earth (Gemini 11).[56]

The record was indeed impressive. By the end of the program, the United States had leapfrogged the Soviet Union in almost every aspect of human spaceflight. Americans had flown into space 16 times, accumulating over 1,000 hours in mission time (Gemini 7 alone completed a two-week marathon, 220 orbit flight). In sharp contrast, the Soviet pace slowed considerably after Tereshkova's Vostok 6. Only two Voskhod (USSR's two-person craft) flights took place during this time, bringing the Soviet time spent in space to 432 hours.[57]

The bridge to the Moon had been built. President Kennedy's goal of placing an American on the lunar surface by 1970 now seemed much more achievable.

NASA's tracking network was ready.

CHAPTER 5

THE APOLLO YEARS

As Apollo became the centerpiece of the national space program, major decisions had to be made about the proposed missions before tracking and data acquisition requirements could be fully defined. Tracking Apollo was obviously going to be much more than just an extension of tracking Mercury and Gemini, both of which remained in Earth's orbit. The complexity of Apollo trajectories and its flight phases were many:

1   The spacecraft was launched from the KSC into a parking orbit around Earth.

2   The vehicle was inserted from this parking orbit into lunar trajectory in a maneuver called Trans-Lunar Injection, or TLI.

3   The vehicle coasted on a ballistic trajectory for three days, from Earth to the vicinity of the Moon, making minor course corrections when needed.

4   The spacecraft performed a braking maneuver placing it in orbit around the Moon.

5   A Lunar Module (LM) separated from the Command/Service Module (CSM) to descend to the lunar surface.

6   After exploring the surface, the Ascent Stage of the LM lifted-off from the Moon and rendezvoused with the CSM in lunar orbit.

7   The LM was jettisoned after which the CSM performed a burn to insert it into an Earth-bound trajectory in a maneuver called Trans-Earth Injection, or TEI.

8   The vehicle coasted in a ballistic trajectory for two days back to Earth, making minor course corrections when needed.

9   The CM reentered Earth's atmosphere along a narrow corridor at 40,000 kilometers (25,000 miles) per hour.

10 The Command Module parachutes to a predetermined splash-down location in the Pacific Ocean.[1]

Many of the fundamental steps outlined above required capabilities well beyond the Mercury and Gemini configurations of the MSFN. Tracking and communicating with a spacecraft a quarter of a million miles away posed many new and different challenges for the network. For instance, ground stations required new equipment to expand into a USB system where tracking, telemetry, and command used a single carrier frequency. More powerful 26-meter (85-foot) dish antennas such as those used by the DSN to communicate with planetary space probes were added to meet the much more demanding range and data requirements. These were supplemented with 12-meter (40-foot) antennas to provide wider beamwidth coverage across this vast distance. The GRARR system was added to track the Apollo spacecraft while it was out of radar range. Rounding out the changes were new ground stations along with a contingent of ships and planes to fill coverage gaps and meet data relay requirements.[2]

Studies for the Apollo network began at Goddard in early 1962 in the TDSD. TDSD originally envisioned a network based on the emerging Mercury and Gemini MSFN stations, supplemented by STADAN sites. In this early plan, MSFN radars would be used for low-Earth orbit support of the Apollo spacecraft prior to the Trans-Lunar Injection (TLI) burn committing it on a trajectory to the Moon. The existing sites were prepared to handle this role, a role that was very similar to that of Projects Mercury and Gemini. This ostensibly made sense as technical and cost considerations both advocated that an Apollo network be built around the existing Gemini network of radar stations. In this way, the Apollo network would not have to be built from scratch.

This vintage 1964 drawing shows the relative sizes of the Mercury, Gemini and Apollo spacecraft, as well as the Atlas, Titan and Saturn V used to launch them. The combined weight of the Apollo Command Module, Service Module and Lunar Module at launch was 47,630 kilograms (105,000 pounds). By comparison, typical weight of the Mercury capsule was only 1,950 kilograms (4,300 pounds) and the Gemini 3,760 kilograms (8,300 pounds). (NASA Image Number S64-22331)

Augmentation of the Gemini network with range and range rate equipment along with the use of large S-band antennas for portions of the mission away from Earth were well understood early on in these Goddard trade studies. Table 5-1 shows the original Apollo network as envisioned in 1962.

This plan called for three block upgrades to bring the network up to its final form to support the original Apollo timetable. The so-called "1B Network" would have been used to support early test flights of the Apollo spacecraft in low-Earth orbit launched on the Saturn 1B rocket (missions AS-111 through AS-114). This first iteration would have essentially used the primary MSFN Gemini sites to provide radar tracking and TT&C support. The "V Network" would have been an interim block upgrade to support Earth orbit and high apogee missions of the Apollo spacecraft launched on the massive Saturn V launch vehicle (missions AS-201 through AS-205). Apollo Ships would have started joining the network along with an upgrade of the

Table 5-1: Apollo Network as First Proposed in 1962

Station	Earth Orbit Missions (Early) The "1B Network"	Earth Orbit Missions (Late) The "V Network"	Lunar Missions
Coastal Sentry Quebec	•		
Grand Canary Island	•		
Bermuda	•	•	•
Cape Canaveral	•	•	•
Carnarvon	•	•	•
Guaymas	•	•	
Hawaii	•	•	•
White Sands	•	•	
Madagascar		•	•
Apollo Ship 1 (Atlantic)		•	•
Antigua			•
Canberra			•
Houston			•
Palermo			•
Apollo Ship 2 (Indian)			•
Apollo Ship 3 (Indian)			•
Apollo Ship 4 (Pacific)			•
Apollo Ship 5 (Pacific)			•

Madagascar site for full global USB capability. A third and final block upgrade completing the Apollo network would have added four more ships, airplanes, a USB site on Antigua, plus three 26-meter USB facilities to be located in Houston, Texas; Canberra, Australia; and Palermo, Sicily.[3]

As it turned out, but for use of existing MSFN radar sites, the first incarnation of the actual Apollo network bore little resemblance to what was first proposed. By the fall of 1962, TDSD had decided against using STADAN stations for Apollo, opting instead to collocate with major DSN sites. This rather significant decision was based on a combination of factors: 1) The requirement to have a backup for the 26-meter USB antenna; 2) Similar requirements for long range spacecraft communications on Apollo and deep space missions; and 3) STADAN scheduling concerns. The STADAN was fully occupied with its mission of supporting unmanned application and science satellites, the number of which NASA continually added into Earth orbit.

Early planning had pinpointed fairly well the necessary primary ground stations for the near-Earth phases of Apollo missions. Secondary sites were added as planning progressed. Twelve-meter telemetry antennas at existing MSFN stations were replaced with a new generation of smaller, 9-meter (30-foot) USB antennas. NASA continued to pool its own MSFN equipment with DOD assets to fill needs where necessary. Five instrumentation ships and eight aircraft were also employed. By the time Gemini 12 splashed down in November 1966, the first of the MSFN stations to be reconfigured for Project Apollo had appeared. In March 1967, Guam came online as the first new site constructed specifically for Apollo. The process of assembling the remainder of the Apollo stations continued through the following year and was essentially completed by February of 1968.[4]

During this time, the early test missions actually began before the network was completed. Apollo 4, the first flight of the Saturn V, took place on 9 November 1967 with partial participation of the emerging network. This flight was an important milestone that demonstrated Saturn V performance and verified the CM heat shield ability to withstand the 2,750°C (5000°F) searing heat experienced on reentry. The following month witnessed the launch of the first Test and Training Satellite (TTS-1), designed specifically to exercise the capabilities of the Apollo MSFN. (TTS checkouts continued sporadically over the next several years with TTS-2 in December 1968 just prior to the first circumlunar flight of Apollo 8, and TTS-3 in September 1971.)[5] In January 1968, the network supported TT&C activities of the LM on its first unmanned test flight on Apollo 5.

A major difference between the earlier planned and the final configuration of the network was the location of the all important 26-meter (85-foot) USB sites for tracking and communications during the lunar phase of the mission when Apollo was in the vicinity of and on the Moon. The underlying geographical requirement was actually very simple: provide continuous coverage with three stations separated by approximately 120° in longitude. In North America, engineers liked the original plan calling for a Houston USB site since it would have eliminated the need for White Sands and Guaymas. But TDSD's decision to collocate the Apollo antenna with DSN made this impractical. Because Houston was only 20° east of Goldstone, California, where there was already a DSN station, there was really no justification to put a USB station near Houston—as the original plan had called for. (The Goldstone Communications Complex in the Mojave Desert would become the largest concentration of NASA tracking and data acquisition equipment in the world, encompassing sites for all three networks: DSN, MSFN, and STADAN.)[6]

Locating the 26-meter (85-foot) antenna near a backup was a written requirement for Apollo lunar operations. Redundancy using the DSN relied on a microwave relay connection between the MSFN primary antenna and the JPL-directed DSN antenna. The DSN system was referred to as an Apollo "wing-station" in this arrangement. With this link, the DSN antenna was slaved to and driven by the MSFN antenna, providing a full backup capability. At Goldstone, the original Pioneer site (DSS 11) served as the Apollo wing-station.

In the Southern Hemisphere, the proposed Canberra, Australia station (Honeysuckle Creek) was kept as in the original plan and collocated near the Tidbinbilla DSN wing-station some 30 kilometers (20 miles) away. That left a third site which had to be in the European area. Factors such as cost of operations, ease of accessibility, topology, and as always, cooperation of the foreign government involved, all went into the decision. On 28 January 1964, the United States and Spain reached an agreement to put the third 26-meter (85-foot) MSFN station at Fresnedillas some 50 kilometers (30 miles) west of Madrid, again located near a DSN site that was then being built (the Robledo DSN Station).

These, the three most powerful primary stations, were joined in the network by 11 other ground locations also classified as primary but featuring

Aerial view of the Apollo Station at Honeysuckle Creek, Australia with its 26-meter (85-foot) Unified S-Band antenna. At the upper left are the diesel fuel tanks for the power generators. Because of its remote location in a national forest, this crucial Apollo Station was run entirely off generator power. (Un-numbered photograph, Box 18, NASA Australian Operations Office, Yarralumla, ACT)

the smaller 9-meter antennas. In February 1965, Goddard awarded the Dallas Division of the Collins Radio Company $2.74 million to install the USB systems at the three sites. It was the follow-on to the $20 million contract that Collins received the previous year to install the 9-meter (30-foot) systems.[7]

The Apollo 26-meter diameter tracking antenna was quite large, the biggest of its kind in the Goddard networks—only the 70, 64, and 34-meter (230, 210, and 111-foot) dishes of the DSN were bigger. A novel sight is seeing these big dishes move, almost effortlessly, as they tracked an object across the sky. Much of this had to do with how well the weight of the antennas was balanced. Its ability to move smoothly and point accurately to within 1/100th of a degree directly affected how well it could stay tracked—or autotrack—on a spacecraft. These antennas were moved using gear-box mechanisms (gimbals) driven by hydraulic servos. With the large dish carefully balanced using counterweights, relatively low torque electric motors could be used to drive even the largest antennas. Most of today's modern tracking antennas allow for rotation in all three axes. In the 1960s, however, systems could move only in two axes. Many, like the MSFN 26-meter antenna, had a so-called 'X-Y mount' where an X-axis gear wheel drove the antenna in the north-south direction while the Y-axis gear wheel (mounted above the X-axis) drove the antenna in the east-west direction. This design allowed horizon-to-horizon tracking as the antenna could be pointed on the horizon in any direction to pick up a spacecraft ascending into view. These largest of the MSFN antennas could move at a good pace, tracking a spacecraft at rates of up to three degrees per second in both axes.[8]

In addition to DSN, several STADAN and DOD stations were also assigned to support Apollo in a backup or standby capacity. Three STADAN stations in the Southern Hemisphere—Lima, South Africa, and Tananarive—were tasked as needed. But it was the Air Force that furnished the majority of the supplemental stations, some of which were also located near MSFN sites. These were mainly radar sites in the Eastern Test Range; none were involved in USB operations. Across the network, different stations had different jobs. For example, the three 26 meters provided coverage for operations in the lunar vicinity and for EVA while the astronauts were on the lunar surface. The 9-meter antennas monitored the spacecraft during its transit to and from the Moon. Bermuda continued in its familiar role as the go/no-go decision site. Stations like Carnarvon and Hawaii were critical for near-Earth portions of a mission, both during outward bound (TLI) and when returning from the Moon and reentry. Grand Bahama, Antigua, and Ascension monitored, respectively, the early (S-IC first stage) and late (S-II and S-IVB second and third stages) phases of the Saturn V's powered flight into orbit.[9]

✯   ✯   ✯

The station on the desolate outskirts of the town of Carnarvon (CRO) was located 960 kilometers (600 miles) north of Perth, the largest city in Western Australia. The township derived its name from Lord Carnarvon, a former Secretary of State for Colonies in Britain. The NASA station was a popular tourist attraction along with Carnarvon's "Blows", natural hole formations in the rocky Australian coastline that, due to high pressure caused by pounding seas, caused water to shoot up like fountains. CRO was operational from 1964 to 1974. (Photograph courtesy of CSIRO)

Range instrumented ships had been an integral part of the manned network since Project Mercury. Ships have the distinct advantage over land stations because of their mobility; their big disadvantage is the higher operating cost (about twice that of land stations). Early network plans in 1962 had called for five Apollo Instrumentation Ships (AIS), two to be assigned to the Indian Ocean, two to the Pacific, and one to the Atlantic. By early 1966, however, Goddard had refined the plan so as to accommodate several Apollo mission profiles to where three TLI insertion ships were needed, one each for the Atlantic, Pacific, and Indian Oceans. In addition, two reentry ships were to be stationed in the Pacific. The five ships assigned to the Apollo network replaced the three that had been in service from the Mercury years, including the aging *Coastal Sentry Quebec* and *Rose Knot Victor*. In October 1968, just prior to Apollo 7 (the first human flight of the new Block II CSM), NASA returned the *Watertown*—one of its two reentry ships—back to the U.S. Navy.

TDSD evidently felt confident that it had adequate coverage in the Pacific with Guam, Hawaii, plus the *Huntsville*, to the point that a second ship was really not necessary.

This contingent of ships was the AIS fleet as deployed through Apollo 11. They had the obvious advantage over their land counterparts in that they were able to change their area of coverage from mission to mission depending on what was needed. On Apollo 8—the historic first human circumlunar flight—for instance, one insertion ship (*Vanguard*) was stationed in the Atlantic and one was in the Indian Ocean (*Mercury*). The third insertion ship (*Redstone*) along with the reentry ship (*Huntsville*) took up positions in the Pacific.[10]

Apollo was launched from the Kennedy Space Center at azimuths between 72 and 108°, depending on the particular mission (90° is a launch due east). Culminating the boost phase was the first burn of the S-IVB third stage of the Saturn V launch vehicle to provide the necessary impulse to insert the spacecraft into Earth orbit. As early as 1964, the OTDA had imposed the requirement for continuous two-way voice communications, reception of telemetry, command capability, and tracking during ascent into Earth orbit. The primary use of the tracking data was to verify that a proper parking orbit had been achieved, while command uplink and telemetry downlink were requirements for flight control operations to evaluate the health and status of the spacecraft and astronauts.

Since the third stage burn occurred about 2,250 kilometers (1,400 miles) downrange of the Cape, it was outside the coverage area of the Bermuda Station, and for most launch azimuths, also outside that of the Antigua Station. It was thus necessary to have a station farther downrange in the mid-Atlantic that was east of both of these islands. The ideal spot for such a station was at 24° North by 48° West. Unfortunately, no island or suitable land mass exists in the immediate vicinity of that location. Therefore, a ship was needed.[11]

While the first burn of the S-IVB got Apollo into Earth orbit, it could not yet begin the trek to the Moon. That was done with the TLI, a second burn of the S-IVB, raising the velocity of the spacecraft by some 3,550 meters per second (11,700 feet-per-second) to attain escape velocity. TLI was one of the most critical events of a flight, one that had to be monitored reliably. Once the burn was completed, the spacecraft was committed on a trajectory to the Moon and the three astronauts would not be able to return to Earth for at least four days—even on a so-called "free-return trajectory" where the spacecraft made a giant "figure 8" around the Moon and coasted back to Earth without making any additional engine burns.

Apollo mission requirements at the time called for tracking to begin no later than seven minutes after the end of the TLI burn to provide Mission Control with the necessary attitude data to make the important go/no-go decision on "transposition and docking"—a tricky maneuver in which the CSM travels a short distance away from the LM, turns around, docks with it and then pulls the LM out of the adaptor housing and away from the spent S-IVB third stage. In a 1964 memorandum from the OTDA to Donald Crabill of the Bureau of the Budget, Gerald Truszynski pointed out that while the South Africa and Madagascar stations could provide post injection coverage in that area, it would only be partial and would not be as complete compared to a ship stationed in the Indian Ocean.[12] Truszynski also pointed out two other factors favoring a sea-based solution. First, a ship was already being planned to alleviate coverage gaps on non-Apollo missions. Second, the State Department did a study on the long term political stability of South Africa which "did not assure retention of a critical major Apollo support station in the time period required."[13]

The return phase of Apollo also required some special coverage planning. As the CM reentered the atmosphere at the end of a mission, it could, by rolling the craft, control its lift-to-drag ratio making it possible for landing to occur in a fairly long corridor 2,200 to 9,250 kilometers (1,200 to 5,000 nautical miles) downrange from the point where it first entered the atmosphere. To pinpoint the expected splashdown location, network engineers had determined that a tracking contact of approximately three minutes in duration had to be made starting at the end of the initial telemetry blackout period. With the blackout window spanning 370 to 1,850 kilometers (200 to 1,000 nautical miles) downrange of the initial entry point, coverage had to be available out to 3,330 kilometers (1,800 nautical miles) from the point where the CM first entered the atmosphere in order to meet the three minute requirement.[14]

That was not the only factor. Depending on the mission, Apollo splashdown could occur either in the northern recovery area in the vicinity of Hawaii or in a southern area near Samoa. This left a lot of ocean to be covered. While there were islands in the western Pacific which could have been used as land stations, a total of seven sites would have been needed just to meet this three minute requirement, a requirement that could be met by using just three ships.

From an *overall* cost standpoint, though, it turned out that there was actually very little difference between using ships versus using land stations to cover post-injection, insertion, and reentry tracking. Here's why: Of the proposed five ships, OTDA had determined that all but one could have been replaced by land stations given the proper political environment. These four ships could have been substituted with eight new ground stations. In 1964, each new station cost about $12 million to build. Thus, the initial investment for land stations would have been in the neighborhood of $96 million. From NASA's experience with the Navy, the cost of obtaining and refurbishing four ships would have amounted to $98 million. Hence, there was only a two percent difference between the two solutions in terms of initial cost investment. As for annual operating cost, the rule of thumb was that a ship cost twice as much to operate as a land station. There was thus little difference in operating four ships versus eight land stations.[15]

With one of those rare occasions when cost was not a major player, OTDA went ahead with the ship-based solution based on the technical advantages:

> Position of the ships may be changed to meet requirements of individual missions whereas land locations were fixed.
>
> It was only necessary to maintain four communication links back to the United States instead eight, thereby reducing mission complexity.

To implement this solution, NASA acquired three "19-class" T-2 tankers and converted them into highly instrumented vessels equivalent in many respects to a primary ground station. Each ship possessed the same C-band radar and the same 9-meter USB antenna common to the Apollo prime stations.[16] The three ships, the *Mercury, Redstone,* and *Vanguard,* provided the network with the required flexibility to support various launch azimuths, Earth orbit insertion points and differing TLI points—all mission dependent parameters. In this way, all critical flight phases were covered and tracking gaps reduced.

These ships were large—a necessity, serving as stable platforms under severe sea states. The *Vanguard*, for instance, measured 181 meters (595 feet) in length with a 23-meter (75-foot) beam. It had a cruising speed of 26 kilometers per hour (14 knots) and a dash speed of 31.5 kilometers per hour (17 knots). These were tracking stations in every respect, capable of remaining at sea for two months, supporting a full Military Sea Transport Service crew and more than 200 field technicians. With enough electricity to supply a town of 5,000 people, they were equipped with facilities such as a store, barbershop, weight room, and a movie lounge. There was a hospital on board as well.

Serving as reentry ships in the Pacific were the *Huntsville* and *Watertown.* Being converted World War II "Victory" ships, these were somewhat smaller than the three insertion ships, measuring 139 meters (455 feet) long by 19 meters (62 feet) wide. They could accommodate 130 technicians and carried the same range of TT&C hardware as their larger counterparts, with the exception of a smaller, 3.6-meter (12-foot) diameter USB antenna.[17]

Taking these old World War II ships out of mothballs and retrofitting them into the space age was, as one can imagine, no simple job. Such an undertaking presented many technical challenges which NASA was not at liberty, in this case, to work out by itself. This was because as a part of the FY 1964 congressional action on NASA funding, Congress had instructed the space agency and the Department of Defense to work together and pool resources for the expressed purpose of acquiring range instrumentation ships.

A converted Navy tanker, the Vanguard was one of the so-called "insertion ships" that tracked and communicated with the Apollo spacecraft as it performed the Trans-Lunar Injection burn, sending astronauts on their way to the Moon. (Folder 8788, NASA Historical Reference Collection, NASA History Division, NASA Headquarters, Washington DC)

Congress knew this was not going to be an easy task, as both organizations—one civilian, the other military—needed to determine what was the best method of meeting joint ship requirements and to establish rules of operations. One thing was clear. Since these were going to be sea faring vessels, the DOD would have the lead responsibility for them.

To execute this agreement, the DOD established an Instrumentation Ships Project Office responsible for procuring and modifying the ships. The office was run by the Navy with representatives from both the Air Force and NASA. It quickly drew up specifications and bidding plans for the ships such that by September 1964, a competitive contract had been let. A $77.5 million fixed-price contract was awarded to the General Dynamics Corporation to convert and instrument three ships taken out of storage. Part of the work included installing, checking out, and integrating some $35 million worth of government furnished range instrumentation equipment onboard the vessels.[18]

From day one, the delivery timetable for the ships was inextricably tied to the development schedule of the Saturn IB and Saturn V launch vehicles. In order to support the flight test schedule, the original delivery dates planned for the three ships were for April, July and October of 1966. In late 1965, a requirement was added to install satellite communication terminals on the ships to ensure that communications between the ships and the Mission Control Center in Houston would not be at risk. To accommodate this new requirement, General Dynamics slipped the delivery by several months, to July and December of 1966 and January 1967.

As the first ship (*Vanguard*) approached completion and sea trials were being conducted in June of 1966, a number of technical problems began to surface. Many of these were of the type that could not have been detected until the complete system was tested at sea when the full dynamic effects of rough seas and high winds were combined. But however formidable, these problems were within the scope of the contract and were therefore, General Dynamic's responsibility to correct. Fixes proved inadequate and the delivery schedule suffered, slipping on a month-by-month basis. Listed is a sampling of the technical problems that arose, and their solution:

> The 9-meter diameter telemetry antenna did not operate satisfactorily over the entire required frequency range (from VHF to S-band). The antenna feed had to be redesigned and reinstalled.
>
> The same telemetry antenna also had a serious vibration problem. This was corrected after much engineering analysis by structurally stiffening the dish and by installing an electrical filter that eliminated spurious signals (RF noise) from the antenna drive mechanism.
>
> The command uplink antenna was simply too dynamically unstable in high wind conditions. It was completely redesigned to improve its aerodynamics and to make it smaller and lighter so as to improve servo drive response.
>
> The high frequency radio transmitting antennas—three on each ship—could not operate at full power because of electrical insulation problems. These had to be redesigned and replaced.
>
> The servo drive system for the satellite communications antenna, along with the antenna feed itself, did not perform according to specifications. The sensitivity was too low because of the poor quality of the antenna sub-reflector surface. These problems were rectified through redesign and remanufacturing of the hardware.[19]

As it became evident that these problems were impacting ship delivery, NASA sharply increased its day-to-day workings with the Navy and with General Dynamics. Managers from Goddard and Headquarters even went directly to top General Dynamics management, requesting them to bring the strongest possible management effort to bear on these problems so as to ensure adequate and timely solutions. Subcontractor problems were also uncovered. General Dynamics in turn, as part of their increased effort, brought in special consultants from outside the company and from academia. An MIT professor, for instance, was brought in to tackle the difficulties with excessive antenna vibrations.

NASA had to walk a fine line. Since the contract for the AIS was actually a Navy contract executed by its Instrumentation Ships Project Office, its actions with General Dynamics had to always be taken in full coordination with that office. The Navy cooperated and responded to NASA, passing its own rather strong terms down to its contractors. The strong management tactics worked. General Dynamics responded to the government pressure by instituting more frequent and detailed top-level management reviews of the project. They also assigned a senior company official at the vice president level to work full-time overseeing the project, this in addition to the Program Manager already assigned. The company also tightened up scheduling control over Bendix, their main subcontractor, and instituted bi-weekly senior management reviews attended by the President and Vice President of General Dynamics Electronics as well as the Executive Group Vice President from their Headquarters.[20]

Results were slow at first. For a few months, there seemed to be little progress. The pace eventually picked up, though, and much time was made up in the last few months of the delivery schedule. A limited ship capability was finally fielded in the fall of 1967 just in time to support the November launch of the first uncrewed Saturn V on Apollo 4. By the time the first crewed flight of the huge launch vehicle took place in December of the following year (Apollo 8), the AIS fleet was ready and at full strength.

In 1964, the OTDA had estimated the initial investment for the five ship AIS fleet at $98 million. The actual price tag, however, turned out to be $186.6 million, almost twice as much as predicted. On top of that, the annual cost of operating the ships had, by 1969, reached $5 million for each of the three insertion ships and $3.5 million apiece for the smaller *Huntsville* and *Watertown*.[21] In a cost saving move, NASA returned the USNS *Watertown* back to the Navy after the launch of the ATS-D satellite in August of that year. This raised concern within Congress, some thinking that the space agency was putting cost ahead of safety.

In reality, this move was based on changes in mission requirements that had been taking place. In the early stages of Apollo planning, reentry in either the Northern or the Southern Hemisphere was simultaneously considered to accommodate maximum flexibility in lunar mission planning, particularly for variable times of stay and departure from the Moon. As NASA progressed through the early Apollo/Saturn V development flights, it became evident to mission planners that a preselection of the return flight trajectory had to be made well in advance of launch. This change in requirement reduced the reentry zones that needed to be covered from two to one. As a result, only one reentry ship would be needed; hence, the release of the *Watertown* before the first crewed mission was even flown.

On top of this, as more Apollo/Saturn test flights took place, more and more information was gained across the board reducing the amount of uncertainty in the performance of the CM in such areas as reentry aerodynamics, heat-shield performance, and the capability of the onboard guidance system to achieve a controlled and accurate reentry. All these served to reduce the landing footprint, to the point where recovery aircraft could now handle nearly all the reentry communication and tracking functions. This development eventually led to the release of the USNS *Huntsville* back to the Navy at the conclusion of Apollo 11.

Similar significant reduction in coverage requirements for the outward bound (specifically, TLI) portion of a lunar mission was also taking place. This could be attributed to three things all having to do with raised confidence that mission planners now had in the performance of the Apollo spacecraft and its Saturn V launch vehicle. The first was a reduction in the launch window. To the Agency's delight, Apollo/Saturn V test launches to date had all occurred on time and at the beginning of a launch window. As a result of this demonstrated launch-on-time capability, the probability of missing a launch window on a given day was considered an acceptable risk, one which in no way compromised crew safety. A shorter window, in turn, engendered a reduction in the needed TLI coverage area.[22]

The second reason was also related to launch-on-time confidence. From orbit mechanics, the location over Earth at which trajectory injection for lunar flight must take place was determined by the time and date of launch

Smaller than the insertion ship, the *Redstone* was one of the World War II liberty ships that was converted into a reentry ship used to track and communicate with the Apollo Command Module as it reentered Earth's atmosphere towards a splash down in the Pacific. (Folder 8788, NASA Historical Reference Collection, NASA History Division, NASA Headquarters, Washington DC)

and the relative positions of Earth and its Moon. In NASA's planning for missions through the first lunar landing—to ensure the maximum number of chances for a launch—mission planners planned for a wide spread of launch azimuths, which meant that the TLI burn could take place anywhere over a wide geographical area. With the now reduced spread of launch azimuths, this coverage area could also be reduced.[23]

Finally, battery lifetime of the tracking beacon aboard the S-IVB third stage had been extended by nearly 50 percent since the first Saturn V launch on Apollo 4, 9 November 1967. This had ramifications to network requirements because after the TLI burn, the Apollo spacecraft, still attached to the burnt-out S-IVB, must be precisely tracked in preparation for transposition and docking. With the increased third stage beacon life, considerably more time was now available for the ground to perform this track, to the point where engineers could afford to wait until a land station came into view. Ship requirement for TLI tracking could thus be alleviated.[24]

All these factors allowed injection tracking to now be done by the Apollo land stations supplemented with a small number of instrumented aircraft. The net effect of these developments enabled the network to eventually relinquish two of the three injection tracking ships—the *Redstone* and the *Mercury*—starting with Apollo 12. It thus left the *Vanguard* as the only remaining Apollo ship operating after Apollo 11. At $6 million a year, it was the most expensive to operate, but was well used, supporting not only human space missions but also NASA projects such as the Pioneer deep space probes. TT&C equipment from the ships was returned to the MSFN equipment pool and redistributed for use at ground stations and on aircraft.[25]

☆ ☆ ☆

In addition to the instrumentation ships, eight Apollo Range Instrumentation Aircraft, or ARIA, served the network as airborne communication points relaying voice transmissions between the spacecraft and Houston. These aircraft were deployed—either in the Pacific or in the Atlantic depending on the relative positions of the Moon with respect to Earth—during each mission launch window. Without the vantage point of these airborne platforms flying some 10,500 meters (35,000 feet) above the ocean, as many as 20 to 30 relay ships would have been required just to relay communications between the spacecraft and ground stations.[26]

But even from their birds-eye vantage point, eight ARIAs were still needed to provide coverage in the Pacific and four in the Atlantic. NASA had originally planned on a fleet of 12 aircraft. In 1964, an Office of Manned Space Flight study concluded that a reduction in the area coverage per Apollo mission could be tolerated within the so-called "delta-V budget" of the spacecraft. What this meant was that, based on the propulsion capability (limitation) of the Saturn V launch vehicle then under development, the location where injection into lunar trajectory could take place on a given mission had to be either over the Pacific or the Atlantic, but not both. The TLI area thus had to be designated well in advance of a particular mission, as well as the reentry area. Mission coverage requirements, therefore, changed from two-ocean support to single-ocean support and the number of ARIAs reduced from 12 to 8 (6 for primary mission support, 2 for backup) for Pacific operations and down to 4 for Atlantic support. This amounted to a savings of $32.4 million.[27]

As Apollo preparations matured over the next three years, GSFC and MSC began to see that this reduction in the ARIA fleet was going to present coverage limitations. On some flights, it was inevitable that lunar trajectory injection was going to shift to a different location as the launch window progressed. If a launch were delayed and it became necessary to move to the other ocean, the entire mission timeline would then have to be adjusted since it took approximately 60 hours to reposition the fleet of aircraft from one ocean theater to the other. This was yet another seemingly simple but important reason why NASA always wanted Apollo Moon missions to take place as early as possible in a given launch window.[28]

The ARIAs were converted C-135A cargo airframes that NASA acquired on long term loan from the Air Force. They were heavily instrumented. Externally, the most obvious difference in the aircraft from regular C-135s was a large bulbous nose—a 3-meter (10-foot) radome that housed the world's largest airborne steerable antenna at the time. The antenna itself was a 2-meter (7-foot) S-band parabolic dish used for telemetry and voice. In addition to the "droop snoot" nose as it soon came to be known, the ARIA—designated the EC-135N—had a probe antenna on each wing tip that was used to enhance high frequency radio transmission and reception. A high frequency trailing wire antenna was added to the bottom of the fuselage. The aircraft was also heavily modified inside the fuselage to accommodate the suite of core electronics and facilities were added for eight more crew members.[29]

ARIA capabilities normally consisted of the following:

For telemetry reception and recording: single USB link, an S-band Pulse Code Modulation link, 6 VHF links.

Telemetry was usually recorded live and then "dumped" over the first available Apollo site (ship or ground station) for transmission to Houston.

USB and VHF voice reception and recording for real-time spacecraft/MCC voice relay.

Two-way, 100 words-per-minute teletype.[30]

The Apollo Range Instrumentation Aircraft (ARIA) served as airborne relay points between the Apollo spacecraft and the rest of the network. About the size of a 707 jetliner, NASA borrowed these converted Air Force C-135A cargo airplanes to support launch and reentry communications during the Apollo years. (Folder 8788, NASA Historical Reference Collection, NASA History Division, NASA Headquarters, Washington DC)

ARIAs had a nominal crew of 16. They were based at Patrick Air Force Base and flew out several days prior to a launch to their forward station in the mission operations area: Hickam Air Force Base in Hawaii or Ascension Island in the Atlantic.[31] Then on the day of the mission, the plane would fly to its assigned airspace to support launch or recovery.

Just as NASA had an agreement with the Air Force for launch support at Cape Canaveral, it had a similar agreement for the ARIA. Under a 10 November 1965 NASA-DOD cost sharing memorandum of agreement, the National Range Division (NRD) of the Air Force Systems Command (AFSC) had overall responsibility for the ARIA project. NASA provided the specifications and labor for its equipment and instrumentation needs while the Air Force provided structural modifications and the general onboard range equipment. There was a further breakdown of labor since, another division within AFSC (the Electronic Systems Division, or ESD), was responsible for the detailed

Definition and Acquisition Phases of the project. Management and engineering change control was thus maintained by NASA and NRD through representation in the ESD project office and an ARIA Project Configuration Control Board.

In this somewhat convoluted arrangement, the Air Force NRD operated, maintained and provided logistical support for the aircraft for NASA. Scheduling and aircraft availability was maintained through a senior-level joint NASA/DOD panel.[32] GSFC was the executing agent in administering and managing the NASA portion of the ARIA program. To this end, it was responsible for three things: 1) generate the necessary specifications for the communications equipment needed for Apollo; 2) ensure that the ARIA met overall Apollo requirements; and 3) integrate these aircraft into the MSFN.[33]

To modify the aircraft, the Air Force contracted Douglas Aircraft Company of Tulsa, Oklahoma, to serve as prime contractor with BFEC as their subcontractor. In this arrangement, Douglas was responsible for modifying the airframe while Bendix was responsible for supplying the generic and Apollo-specific suite of range instrumentation equipment to be installed on the aircraft. Contractor work during Apollo was driven by a tight schedule and ARIA was no exception. To meet delivery milestones, ESD issued the Douglas team with a fixed price contract heavy on delivery and performance incentives. While the target cost was $27.2 million, the contract could be worth well over $30 million if all the incentives were awarded. The first ARIA—scheduled for delivery in the first-quarter of 1966—was delayed and finally delivered to the Air Force near the end of the year, just in time to pass its first live test on Gemini 12 in November 1966.[34]

The remaining seven ARIAs trickled in throughout 1967 and into the following January. At its peak, Douglas (later McDonnell-Douglas after its 1967 merger with the McDonnell Aircraft Corporation) had over 300 people working on the ARIA program at its Tulsa plant.[35] After Apollo concluded in 1975, the word "Apollo" was changed to "Advanced" and the Air Force fleet of aircraft continued serving under the ARIA name, successfully supporting a host of NASA satellite launches, Skylab and planetary probes such as Viking and Voyager. Over the next 30 years, the DOD has maintained ownership of the aircraft which have been used primarily to support military ballistic missile testing activities.

In all its years of near flawless service, there was only one major accident. But it was tragic. On the morning of 6 May 1981, one of the planes—ARIA 328—took off from Wright-Patterson Air Force Base in Fairborn, Ohio, on a training mission. All 21 onboard perished just an hour later in a horrific crash. Among those killed were three civilians, two of whom were wives of crew members who were on the flight as part of a program for them to become more familiar with their husbands' work. Today, a living memorial dedicated to those who perished resides near the place where ARIA 328 took off that ill-fated morning. A bronze plaque, along with 21 flowering

Table 5-2: The Manned Space Flight Network as Implemented for Apollo*[36]

Station	Abbreviation	USB Antenna 12'	USB Antenna 30'	USB Antenna 85'	C-Band Radar	VHF TM Downlink	UHF CMD Uplink	Other
Primary Stations								
Antigua	ANT	•				•	•	
Ascension	ACN	•				•		
Bermuda	BDA	•			•	•	•	
Canberra	HSK			•				
Carnarvon	CRO	•			•	•	•	
Corpus Christi	TEX	•				•	•	
Goldstone	GDS		•					
Grand Bahamas	GBM	•				•	•	
Grand Canary	CYI	•			•	•	•	
Guam	GWM	•				•	•	
Guaymas	GYM	•				•	•	
Hawaii	HAW	•			•	•	•	
Madrid	MAD		•					
Merritt Island	MIL	•				•	•	
Ships and Aircraft*								
Huntsville	HTV	•			•	•	•	
Mercury	MER	•			•	•	•	
Redstone	RED	•			•	•	•	
Vanguard	VAN	•			•	•	•	
Watertown	WTN	•			•	•	•	
Aircraft (8)***	ARIA							•

crab apple trees, each symbolizing a lost soul, rests in the memorial garden at the United States Air Force Museum in Dayton, Ohio.[37] On 24 August 2001, the last ARIA flight landed at Edwards Air Force Base to bring the airborne tracking program to an end.

Besides ground stations, the AIS and the ARIA, NASA added a fourth tracking element during Apollo. This one was in space. To further cut down on potential communication gaps, the Agency called on the services of two communication satellites operated by the International Telecommunications Satellite Consortium, or Intelsat. One was the Intelsat Atlantic satellite, located in geosynchronous orbit at 6° west longitude off the coast of Africa. From this vantage point, it could provide communication relay for the Indian Ocean ship (usually the *Mercury*), the Ascension Island Station, the Atlantic Ocean ship (usually the *Vanguard*), and the Canary Island

Station	Abbreviation	USB Antenna 12'	USB Antenna 30'	USB Antenna 85'	C-Band Radar	VHF TM Downlink	UHF CMD Uplink	Other
NASA Support Stations								
Canberra (DSN)	CNBX			•				
Goldstone (DSN)	GDSX			•				
Lima (STADAN)	LIMA							•
Madrid (DSN)	MAD			•				
Pretoria (STADAN)	PRE				•			
Tananarive (STADAN)	TAN							•
White Sands (MSFN/DOD)	WHS				•			
Woomera (MFSN)	WOM							•
DOD Support Stations								
Antigua	ANT				•			
Ascension	ASC				•			
Cape Canaveral	CNV				•			
Grand Bahama	GBI				•			
Merritt Island	MLA				•			
Patrick AFB	PAT				•			
Vandenberg AFB	CAL				•			

*Nominal configuration 1968-1972
**The *Huntsville* and *Mercury* were usually stationed in the Pacific to monitor orbit injection, reentry and recovery. The *Redstone* was usually on station in the Indian Ocean with the *Vanguard* in the Atlantic. The *Watertown* was deployed only for the early developmental flights and was removed from service in October 1968 prior to the first human flight (Apollo 7).
***Eight Apollo Range Instrumentation Aircraft were used as communication relays to support operations in areas where there were no grounds stations, especially during reentry and landing.

Station. The other was the Intelsat Pacific satellite, located at approximately 5° west of the international dateline over the Kiribatis in the mid-Pacific. It served the Australian stations, Guam and Hawaii as well as the Pacific ships *Huntsville* and *Redstone*. Eventually, reliance on Intelsat for communications relay would free up the ARIA to focus on real time USB support.[38]

Table 5-2 summarizes the Apollo Network as it was eventually established in 1968 (also see map in Appendix 1). This was essentially the configuration used throughout the lunar landing program, with ships, aircraft, and satellites being augmented on a mission by mission basis.

✯   ✯   ✯

As the MSFN was being modified for Apollo, centralization, network communications and the ability to make decisions at remote ground stations were again topics that came to the forefront. In the Mercury days, the network had flight controllers and a Capcom at all of the prime stations mainly because worldwide, real-time communications were still in its infancy. Thus each primary site had the means to make critical decisions and to execute command instructions in the case of a communications failure. In addition, Bermuda had a computer of its own, to help make the vital go/no-go decision should communication with the main computing center at Goddard be severed.

On Apollo, GSFC, MSC, and Headquarters jointly agreed to use computers at outlying network stations. This was a radical move and a fundamental change away from what had been done up until then. The decision was not reached without controversy. Network philosophy had always been that ground stations would transmit raw, unprocessed (or slightly processed) data back to a central computing center—first located at Goddard and later moved to Houston. The expansion of NASCOM had made this possible. With Apollo, however, data rates increased several-fold over Gemini, and live television (a high bandwidth item) was added.[39]

To handle these faster processing requirements, 14 land stations were each fitted with two Univac 642B data processing computers to support both telemetry and command.[40] (Two units were needed to allow for simultaneous tasking of telemetry and command.) The old Gemini sites that had been equipped with the aging Univac 1218s were upgraded. But the increased real-time flow of information led to a buffering problem: the outlying sites could receive far more data from the Apollo spacecraft than could be transmitted in real time to Houston over NASCOM, which could still only handle a maximum traffic rate of 2.4 kilobits-per-second (kbps). This had potentially crippling consequences. Network engineers at Goddard devised a solution which was to essentially compress each station's aggregate data link into discrete 2.4 kbps frames or packets grouped into specific data types. By doing so, flight controllers in Houston could remotely select and query telemetry information of their choosing for review. Even though flight control consoles were installed on AISs and at some ground stations, they became unnecessary once the buffering problem was solved. These remote flight consoles were never used on a mission since controllers could review data and issue commands to the spacecraft from the MCC in Houston.[41]

To fully appreciate the complexity of network operations during Apollo, it is important to look at the level of teamwork that went into running the MSFN. Even though the MCC had control of the spacecraft from the ground during a mission, smooth network operations required coordination (and cooperation) between the two primary NASA centers involved: GSFC and the MSC. Before the actual mission, Greenbelt acted as the manager of network activities, preparing equipment and personnel across the network for readiness. It had the final say in pronouncing the network ready (green) or not (red). Once a mission began, though, Houston assumed control. Goddard continued in a support role to ensure overall network viability, monitoring the activities of the ground stations, ships, and aircraft for the duration of the flight. At his console in the MCC Mission Operations Control Room (the famous "Front Room" familiar to the world), the Network Controller (NC)—along with his contingent of support staff in the "unseen" Back Room—monitored network operations, maintaining contact with two key Goddard figures, namely the Network Operations Manager and his boss, the Network Director.[42]

Perhaps the most elegant achievement in terms of communications technology on Apollo was the use of USB. The use of a unified carrier yielded immediate benefits for the spacecraft, saving the space, weight, and power needed to accommodate other subsystems. Furthermore, communications at S-band (1550 to 5200 MHz) was much more powerful, accommodating more data than possible at the lower frequencies. But most importantly, it had the range to reach the Moon. Two competing USB systems were actually available for Project Apollo. One, under development by GSFC, was essentially an extension of the GRARR, used to support STADAN for NASA's science satellites. The other was a JPL product originally intended for deep space use. NASA would select the JPL system but modified it for Apollo. Even with this new capability, the MSFN still kept most of the pre-USB equipment operational, the majority of stations—both old and new—still fielding VHF hardware for backup.

NASA *was* sending men to the Moon, but like any other television or radio station, it still had to ask for permission in order to transmit at certain frequencies. Throughout Apollo, the International Communication Union granted the MSFN transmission only on a secondary basis. What this meant was that NASA was legally required to shut down if its transmissions interfered with other authorized users. But the Agency could not complain of interference from these primary users. Sure enough, the frequency range of the Apollo USB system did overlap with the band then assigned to commercial television broadcasting. GSFC and Headquarters identified and addressed this issue early on in the planning of the Apollo network and all conflicts were successfully resolved before flights took place. No significant frequency interference problems ever developed during the 15 times that Apollo flew.[43]

✳ ✳ ✳

By the spring of 1967, with the final USB upgrade at Guaymas completed, the MSFN was ready to support the first human flight of the new

Apollo spacecraft. In preparation for sending men to the Moon, the GSFC had in three years established seven new ground stations and extensively modified seven others. It had now been over a year since Americans last flew in space and President Kennedy's commitment of a Moon landing before the end of the decade was fast approaching; "Go Fever" was in full swing.

Then tragedy struck.

On the evening of 27 January 1967 during a "plugs out" launch pad countdown test at the KSC, a fire erupted in the pure oxygen atmosphere inside the Apollo Saturn 204 CM, killing astronauts Virgil I. Grissom, Edward H. White, II and Roger B. Chaffee. Super-heated flames consumed the spacecraft within 20 seconds. Grissom, White, and Chaffee didn't stand a chance. Countdown tests involved the network stations, and Apollo 1 was no different. Shock and grief quickly spread to the stations. Overseas, where the local populace took pride in hosting "their" ground station as part of the American space program felt the sadness. Official statements of condolences poured in from around the world. On Guam, its 9th legislature passed Resolution Number 118, officially expressing the grief of the Guamanians. The Honorable E. S. Terlaje, Acting Legislative Secretary, requested NASA send copies of the Resolution to the families of the three astronauts, which James Webb did.[44]

The Fire, as it simply came to be known, severely impacted all aspects of Apollo. Foremost were program timeline and flight schedule. Instead of the first flight taking placing in early 1967, it was delayed for over 18 months, eventually to October of the following year. Despite this deadly setback, uncrewed launches testing the CSM, Lunar Module, and the Saturn V launch vehicle continued as NASA endeavored to recover and rebuild the program.

In the revised timetable, NASA defined seven flights (missions A through G) designed to incrementally lead to a Moon landing (the G mission) by 1970. In the fall of 1967, the giant Saturn V launch vehicle developed by the MSFC was ready for its first all-up test. On the morning of 9 November, the Apollo network was put to the test for the first time, tracking the Saturn V stack on its maiden flight. Apollo 4 was sent into a high apogee (18,079 kilometer, 11,234 mile) elliptical orbit around Earth. At the conclusion of the flight, the Service Module pointed towards Earth and fired its 20,500 pound (91,200 newton) thrust engine to accelerate the spacecraft to a velocity of 40,200 kilometers (25,000 miles) per hour, replicating return and reentry from a lunar mission.

Telemetry received onboard the *Huntsville* and at the Hawaii Station showed no degradation in the cabin environment, verifying the design of the CM heat shield to withstand the 2,760°C (5,000°F) temperature of reentry. This was a significant step in the program since the ability of the ablative shield to protect a returning spacecraft at such velocities was largely unknown at the time. Guided by terminal tracking data from the network, the USS *Bennington* successfully recovered the Command Module

Chapter 5 \ The Apollo Years   167

Astronauts (left to right) Gus Grissom, Ed White and Roger Chaffee stand for photographers in front of Launch Complex 34 housing their Apollo 1 Saturn 1B vehicle. Ten days after this photograph was taken, the crew perished in a pad fire. (NASA Image Number GRN-2000-000618)

west of the Hawaiian Islands some nine hours after launch. Technically, programmatically and—perhaps most importantly—psychologically, Apollo 4 (the "A" mission) was an important and successful event, especially in light of the number of firsts it tackled. For the tracking network, it was the first shake-down of the MSFN for Apollo. The fact that everything worked so well with so little trouble gave NASA much needed confidence and a giant psychological boost. As Apollo Program Director Samuel Phillips phrased it, "Apollo [was] on the way to the Moon."[45]

Apollo 4 was followed in January 1968 by Apollo 5, which flew for the first time the LM made by the Grumman Aircraft Engineering Corporation. The spacecraft was put through its paces using command uplinks from the ground, successfully demonstrating system performance including critical restarts of the LM ascent and descent stage engines. At one point, Houston sent a "switch-off" signal to the guidance computer and flew the LM

A view of the Apollo 16 Lunar Module Orion shows the location of the 0.6 meter (2 foot) S-band antenna near the top of the Ascent Module. During their post mission press conference, the crew called attention to the steerable antenna which was frozen along a yaw axis during much of the flight. Also visible to the left of the S-band dish antenna are the VHF and EVA antennas. This photograph was taken by lunar module pilot Charles M. Duke, Jr. during the mission's first extravehicular activity on 21 April 1972. (NASA Image SAS16-113-18334)

in real time from the ground through a series of simulated landing maneuvers using only command uplinks.[46]

The next flight test was Apollo 6, the final uncrewed flight test of the Apollo program, on 4 April. Two minutes into that mission, telemetry received at Bermuda indicated thrust fluctuations of the S-IC first stage engines that caused the entire rocket stack to bounce like a giant pogo stick for approximately 30 seconds. During the "pogo", telemetry also showed low-frequency oscillations reached as high as ±0.6 g inside the CM, exceeding the design criteria of ±0.25 g stipulated for human flight.[47] (This was a flight rule

Protruding from the back of the Apollo Service Module was the spacecraft's autotracking S-band antenna. It was a "quad-feed" system meaning that the system actually consisted of four antennas. Signal strengths of the four were compared so as to allow tracking of the antenna beam to the Earth ground station that the spacecraft was communicating with. In this way, the ground station "drove" the antenna on the spacecraft to keep it always precisely pointed. Shown is the Apollo 16 CSM *Casper* as seen from LM *Orion*. (NASA Image Number AS16-113-18282)

carried over from Project Gemini. This *oscillation* level should not be confused with Apollo launch or reentry loads, which could exceed 8 g and which the spacecraft *was* designed to take.) After the first stage burnt out and was jettisoned, the five Rocketdyne J-2 engines of the S-II second stage came to life.

As acquisition-of-signal occurred over Antigua, telemetry indicated that two of the engines had shut down prematurely. To compensate, the onboard Instrumentation Unit automatically directed the other three engines to fire longer as flight controllers monitored the situation. Even with the extended burn time, the second stage did not reach the desired altitude and velocity before its fuel ran out. Now in order to reach the planned speed, the single S-IVB third stage engine had to burn quite a bit longer than planned. After its shutdown, an orbit determination was made from state vectors received at the Caribbean stations which showed Apollo 6 in a severely lopsided 177 by 367-kilometer (110 by 228-mile) elliptical orbit rather than the desired 257-kilometer (160-mile) circular orbit.[48]

MCC evaluated the situation and decided to continue into the next phase of the flight, a restart of the S-IVB engine to simulate the TLI burn. Command uplinks to the vehicle went unheeded, however. This was verified by telemetry received onboard the *Vanguard* that the simulated TLI burn did not in fact take place. As an alternative, Houston jettisoned the S-IVB and instead commanded the Service Module engine to fire for over seven minutes (which exceeded lunar mission requirements) to simulate the injection burn. The *Vanguard* tracked the CSM out to 22,200 kilometers (13,800 miles) where it was turned around and plunged back into the atmosphere for another reentry test. Because of the extended burn by the Service Propulsion System, Houston expected that the Service Module would not have enough fuel to accelerate the CM to the desired velocity. Network tracking verified this, showing the CM reentering at 35,900 kilometers (22,300 miles) per hour, some 4,500 kilometers per hour (2,800 mph) less than planned.[49]

The period from fall 1968 to the end of 1972 marked the apex of the program, a time in which nine missions were flown to the Moon, landing 12 men on its surface. On 11 October 1968, Apollo 7 was launched with America's first three-person crew: Walter M. Schirra, Jr., Commander; Donn F. Eisele, CSM Pilot; and R. Walter Cunningham, LM Pilot (even though there was no LM). For nearly 11 days, the MSFN tracked the spacecraft as it made 163 orbits around Earth in an engineering flight test to demonstrate the space-worthiness of the new Block II CM, a totally redesigned spacecraft following The Fire. One improvement was a new hatch that could now be opened in just three seconds.

Among the spacecraft's equipment and communication technologies tested was the transmission of live television from the spacecraft, a first for the manned network.[50] The idea of live television had been a topic of debate ever since September 1963, when NASA first directed North American Aviation to install a portable camera in the Block I CM. With weight a constant concern, many engineers viewed the television camera only as a nicety. On occasions when pounds, even ounces, were being shaved from the CM, the camera was usually among the first items to go.

Despite the insistence of most engineers that it was not needed—and the ambivalence of the test-pilot oriented crews—there were those who persistently argued for its inclusion. NASA personnel in Public Affairs, for instance Julian W. Scheer at Headquarters and Paul P. Haney at the MSC, naturally favored the use of television. There were also managers closer to the program who agreed with them. For example, in the spring of 1964, William A. Lee, a MSC engineering manager, wrote to George Low of the Apollo Spacecraft Program Office:

> I take typewriter in hand to plead once more for including in-flight TV. . . . Since [it] has little or no engineering value, the weight penalty must be assessed against a different set of standards. . . . One [objective] of the Apollo Program is to impress the world with our space supremacy. It may be assumed that the first attempt to land on the Moon will have generated a high degree of interest around the world. . . . A large portion of the civilized world will be at their TV sets wondering whether the attempt will succeed or fail. The question before the house is whether the public will receive their report of this climactic moment visually or by voice alone.[51]

Apollo 7 became the world's highest television broadcasting studio in October 1968. The inclusion of television on NASA spaceflights was not reached at in a cavalier way and was due in no small part to the Space Race atmosphere of the Cold War. This picture was from the crew's television transmission on the third day of the mission. On the left is CSM Pilot Donn Eisele; Commander Wally Schirra is on the right. (NASA Image Number S68-50713)

With emphasis on its civilian nature and Kennedy's decision to play out the Moon race on world center-stage, NASA could not avoid the debate. Over the next several years, it continued with persuasive arguments for the case of live television being weighed against technical and operational considerations. Finally, in April of 1968 with the first Block II CSM (CSM-101) ready to be accepted, television became part of Apollo (and, as it turns out, all future NASA human spaceflights) when Samuel Phillips directed George Low to proceed with a camera on Apollo 7.[52] It turned out that television broadcasts on the mission were a huge success, both for NASA public relations and as a technical milestone for the MSFN. The astronauts used television to show (in black and white) views of Earth outside their windows, the uniqueness of working and living in the weightlessness of space, and tours of the new Apollo spacecraft. Lasting seven to eleven minutes each, the broadcasts came to be called "The Wally, Walt and Donn Show," even garnering a special Emmy Award from the Academy of Television Arts and Sciences the following year.[53]

The success of Apollo 7 was followed two months later by what would be the first complete test of the entire Apollo Network. Launched on 21 December 1968, Apollo 8 made the first lunar voyage, carrying astronauts Frank Borman, James A. Lovell, Jr., and William A. Anders to the Moon on a six-day circumlunar flight that culminated with 10 orbits around the Moon. At 10:47 a.m. EST, Capcom Michael Collins relayed through the Hawaii Station, "All right, you are go for TLI," sending men on escape velocity away from Earth for the first time. The mission provided the first true use of the network's large 26-meter (85-feet) USB antennas on an actual human flight; previous activities had involved only system checkouts using Pioneer space probes as TTS.

It also marked a change in the way NASA tracked spacecraft. During Mercury, Gemini and on Apollo 7, communication with the spacecraft was not continuous as the stations could not possibly cover all ground track locations around the globe. However, as Apollo 8 left the confines of Earth towards the Moon, tracking and data acquisition, ironically, became continuous. This somewhat counter-intuitive phenomenon can be explained by simple geometry. As the distance between a spacecraft and Earth increased, the field-of-view required to see it decreased. Also, as a spacecraft sped away from Earth, its motion would appear to an observer on the ground to become more and more stationary. Now instead of the spacecraft racing across the sky in a fast-moving arc, as it would when orbiting Earth, it now traveled on a line (or more precisely, a very shallow arc) slowly away from the observer. As the spacecraft traveled farther and farther away, eventually only a single ground station facing the Moon was needed to communicate with it.

Due to curvature of Earth, the Moon can only be seen comfortably (that is, above the horizon at a fairly high elevation pointing angle) at any one time from locations within a 120° longitude range. Therefore, the three stations 120° apart at Goldstone, Honeysuckle Creek, and Madrid provided continuous coverage to the spacecraft as Earth rotated over a 24-hour period. The only time loss-of-signal occurred on an Apollo mission was when the spacecraft's orbit took it behind the Moon and for those five minutes at the end of the mission during atmospheric reentry when super-heated plasma induced RF transmission black-out.

As successful as live television was on Apollo 7, it paled in comparison to what took place from lunar orbit on Christmas Eve 1968. As a spellbound world glued their eyes to their television sets, the first live images of our planet and lunar landscape as seen by men from the Moon were transmitted from a quarter of a million miles away to the Madrid Station at Fresnedillas, Spain.[54] In a telecast that would forever be etched in the memory of those who were there, black and white images of the Moon and Earth—primitive by today's standards of brilliant high definition television (HDTV)—were shown as each astronaut took turns reading the Creation account from the first 10 verses of the Book of Genesis. As the crew completed their next to last orbit around the Moon, flight controllers—choking back tears by now—looked on as Commander Frank Borman closed the live broadcast with a farewell that reached over a billion people around the world, "We close with good night, good luck, a Merry Christmas and God bless all of you—all of you on the good Earth."[55]

Borman later admitted that he and his crew had not wanted to carry a television camera. Technical reasons aside, they knew that whatever they showed and said from lunar orbit was going to be seen and heard by a whole lot of people. Not a poetic man, Borman, as mission commander, had worried about this the most.

I said 'no' a lot, and the nice thing about it was that NASA gave the commander enough prerogative that they backed him up. I was overruled on one thing and that was because management was a lot smarter than I was. I didn't want to take the damn television camera with me. And they said, 'Let's take it,' and they were right. . . . It turned out to be so important because we could share what we saw with the world. It weighed 12 pounds [5.4 kilograms]. We were cutting out everything, even down to the extra meals, which weighed 16 ounces [0.5 kilograms] or something like that. But I was very short sighted there, and NASA was right.[56]

By including the camera, it made the experience very real to those watching on Earth. "It didn't add a dangerous amount of weight and the camera achieved the purpose for which it was intended: to give all Americans a real feeling for the mission and what it was accomplishing."[57] As it turned out, their broadcast was indeed seen by a worldwide audience, from the Americas to Europe (including East Berlin), parts of Asia and Africa, and even Moscow. Despite some protesting the religious nature of the message, Apollo 8's Christmas Eve broadcast would endure to become one of the most iconic moments in space exploration history.

Having successfully demonstrated the network's 26-meter (85-foot) USB systems, the next mission Apollo 9, went back to again exercise and check out the near-Earth portion of the network. The flight was the first for the LM, the first piloted spacecraft designed exclusively for flying in the airlessness of space. The flight tested, for the first time, MSFN capability to simultaneously track and communicate with both the CSM and the LM. LM USB equipment such as dual-redundant transceivers, the audio center, pulse-code telemetry, central timing, biomedical channels and television were thoroughly tested during this 10-day Earth orbit mission. Communication links between the LM, CSM, and the MSFN ground stations as well as the extra-vehicular mobility unit (the moonwalk spacesuits) were demonstrated.

After 151 revolutions, Gumdrop splashed down on 13 March 1969 near the reentry ship *Huntsville* and was recovered by the carrier USS *Guadalcanal*.[58] Black and white television had worked so well on Apollo 7 through 9 that on the next flight, NASA decided to install a color system in the Apollo 10 CM. Space television had actually come quite far in a short amount of time. During the early Apollo missions, the TV used a slow-scan, black and white camera that was originally intended for development by RCA but, due to procurement delays, was eventually supplied by the MSC as government furnished equipment. That camera yielded a poorly defined, erratically moving image which MSFN stations converted into a standard commercial broadcast format (which after conversion, still exhibited uneven motions). These previous missions had shown to network engineers that there was actually sufficient margin in transmission bandwidth that good quality, *color* television could be attempted in real time.

Weighing "only" 5.4 kilograms (12 pounds), the new Westinghouse color camera could be handheld or bracket-mounted. Its scan rate was at the commercial 30 frames per second, 525 scan lines per frame with a resolution of 200 TV lines at the standard screen aspect ratio of 4:3.[59] What viewers experienced on the ground was a fairly good picture obtained by superimposing the color signals with the imaging (pixel) data. A 7.6-centimeter (3-inch) black-and-white video monitor could even be Velcro-mounted on the camera (or at various locations inside the CM) to aide the crew in focus and exposure adjustment. By Apollo 14, color television capability had been extended from the CM to the LM and onto the lunar surface.

As soon as Apollo 10 splashed down on 26 May bringing to an end the dress rehearsal for the first lunar landing attempt (the G mission), all eyes were on Apollo 11. The historic launch took place before an estimated crowd of one million people on the morning of 16 July 1969. Onboard were Neil A. Armstrong, Commander; Michael Collins, Command Module Pilot; and Edwin E. "Buzz" Aldrin, Lunar Module Pilot. A decade of preparation had been directed toward this mission, and the MSFN now had the responsibility of tracking the three on the greatest voyage ever taken. NASA has flown over 100 more human space missions since Apollo 11 (many much more complex). But historians and grade-schoolers alike still (understandably) look back on this epochal mission as the Agency's high point.

During a visit to the United States in October 1968, John Bolton, Director of Parkes Observatory in western New South Wales, Australia, was approached by Covington's team to consider the possibility of making their 64-meter (210-foot) radio astronomy telescope available to support the historic mission. Although several factors played into this, the driving requirement came down to the fact that Kraft and his team at Houston lacked confidence in the S-band directional antenna of the LM. Specifically, trajectory of the LM on its descent down to the surface was such that after it emerged from behind the Moon, there was a critical but very short period of time to make a "bailout" decision. If the directional antenna was not performing properly, the signal from the lower-gain (much less powerful) VHF omni-directional antenna would be marginal at best using the network's 26-meter (85-foot) antennas.[60]

The way the MSFN stations were spaced also played into this. First, the flight plan had the landing of the Lunar Module *Eagle* taking place towards the end of the viewing window at Goldstone and the beginning of the window at Canberra, Australia. If landing somehow got pushed beyond the Canberra window, however, then Parkes—located some three hours drive west of Canberra—would provide that extra margin to capture the signals. The mission timeline also first drafted by Houston had Armstrong and Aldrin performing the EVA shortly upon landing, with Goldstone being the prime tracking site, and with it, television responsibilities. Honeysuckle Creek, near Canberra, was to track Collins and the Command Module *Columbia* in lunar orbit. In this scenario, the Moon was not due to rise at Parkes until 1:02 pm local Australian time, by which time most, if not all, of the moonwalk would have been completed. Thus, Parkes Observatory was relegated to serve as backup for both the landing and the EVA. To facilitate this setup, the radio telescope would be linked via microwave to Canberra.[61]

This scenario changed about two months before the mission when Flight Operations in Houston decided that, to give the astronauts a better chance to acclimate to the Moon's 1/6th gravity, a sleep period would be allowed before commencing the EVA. Thus, the new plan had the moonwalk starting about 10 hours after landing, which was some 20 minutes after the Moon had set at Goldstone. In the South Pacific, however, the Moon would be high overhead over Parkes. Because of this, the Parkes was redesignated the prime site for receiving the EVA telemetry.[62]

But things changed again. By happenstance, on 17 July—one day after the launch—a fire broke out in the power supply at Tidbinbilla (Canberra) which severely damaged the transmitter on its 26-meter antenna. Despite some quick repair work, GSFC would not take the risk and switched the station's role with Honeysuckle Creek. Thus, the latter would now be the prime station to support lunar EVA, including reception of the crucial bio-medical telemetry from Armstrong's and Aldrin's Portable Life Support System (PLSS) backpacks. This was the top telemetry priority. The 26-meter antenna at nearby Tidbinbilla would be trained on *Columbia* instead.[63]

"Houston, Tranquility Base here. THE *EAGLE* HAS LANDED." The words were said at 4:18 pm EDT on Sunday afternoon 20 July 1969 by Armstrong as Apollo 11 landed on the pristine surface of the Sea of Tranquility. With all LM systems checking out fine and the crew's adrenalin pumping, it would have been incredibly anticlimactic (and probably a little unrealistic) to expect Armstrong and Aldrin to simply just go to sleep for six hours. They had, after all, just landed on the Moon! After discussions with Mission Control, Armstrong exercised his command prerogative and decided to forego the rest period and begin EVA preparations immediately. This began a chain of events from a network perspective that would ultimately decide how telemetry was received and how the world would see humankind's first steps on the Moon.

By skipping the rest period, the EVA would begin five hours before the Moon was to rise at Parkes. However, Goldstone was in a good position. For a while, it seemed as if the Apollo Station in California would have the responsibility of televising the historic first moonwalk as originally planned. But delays kept dragging on as Armstrong and Aldrin prepared for their EVA inside the cramped quarters of the LM. By the time they were ready to egress the ship, moonrise had occurred at both Parkes and Honeysuckle.

While this was going on, a violent wind squall happened to hit the telescope at Parkes while the dish was in its most vulnerable position, pointed at the horizon awaiting moonrise. In this "zero-elevation" position, the face of the dish caught the full force of the two, 112 kilometers-per-hour (70 mph) gusts, subjecting the large antenna to 10 times the force that it was considered safe to withstand. Other structures were also batted around in the swirling winds and the weather remained bad. But in a stroke of good fortune, the winds abated just as the Moon broke horizon at Parkes.[64]

So, because the sleep period was skipped and EVA preparations took longer than expected, no less than three tracking stations—Goldstone, Honeysuckle Creek, Parkes—received telemetry of the incredible first steps on the Moon. Although this was a good thing (plenty of redundancy), it also engendered a dilemma: Which of these TV signals would the world see?

In Australia, signals from both Honeysuckle and Parkes were sent to Sydney by microwave links, where a NASA officer selected between the two to forward on to Houston via the NASCOM. Since moonrise occurred at Parkes just as the EVA was getting underway, the telescope was at a very low elevation angle. As a result, it had to use its less sensitive "off-axis" detector and the received signal strengths were very poor. Antenna elevation angle at Honeysuckle was higher and the resulting signal was better. This meant that its signals were passed on to Mission Control. There, a controller then selected between the Goldstone and the Honeysuckle TV signal. This selected signal (ostensibly the best of the three) was then sent to a media pool television monitor. But this was still not the TV picture that the world saw; there was one more step. The image displayed on this NASA monitor was then filmed lived by a media pool camera for transmission to individual domestic and international TV networks. As a result, what people saw in their homes that evening was of slightly lower quality than what flight controllers and VIPs saw inside Mission Control.[65]

During the first nine minutes of the broadcast, NASA alternated between TV from Goldstone and Honeysuckle, searching for the best one. Neither was very good as they both came from 26-meter antennas (as opposed to the 64-meter dish at Parkes). Because of this, they could only accommodate blurry images using what was called 'slow-scan television'—a picture transmission method used mainly by amateur radio operators to transmit and receive black and white pictures. There was one more thing. Not only was the TV picture grainy and blurry, it was upside-down!

This was because as Armstrong began his 2.4-meter (8-foot) descent down the ladder, he pulled a D-ring which dropped open the Modular Equipment Stowage Assembly (MESA) containing the television camera. Due to the way the camera had to be mounted, however, when the MESA dropped opened, it was upside-down. Avoiding what could have been a major embarrassment forever recorded, technicians at the stations quickly flipped an incon-

Chapter 5 \ The Apollo Years   177

At top is the slow-scan television image from Honeysuckle Creek of Armstrong placing his left foot onto the surface of the Moon. Twenty minutes later when Aldrin came down the ladder, coverage had switched to the 210-foot (64-meter) radio telescope at the Parkes Observatory. The image improved noticeably. The lower picture shows Aldrin checking his jump back up the ladder before stepping onto the surface. Note Armstrong is overexposed in the background from where he stood and took pictures of his crewmate's climb down to the surface. (Scans courtesy of John Saxon. Also available at *http://www.honeysucklecreek.net/msfn_missions/Apollo_11_mission/index.html*)

spicuous toggle switch called the 'Scanner Converter Reversing Switch', just in time to see Armstrong's final descent down the ladder. Although NASA initially began the telecast with Goldstone, by the time Armstrong reached the foot of the ladder, Mission Control had switched to the transmission from Honeysuckle Creek. In this circuitous way, the Australian station was bestowed the privilege of transmitting to the world Armstrong's "one small step."[66]

In a little known vignette of history, the way the camera was mounted in the MESA and the way the compartment dropped opened caused the camera to be slightly tilted with respect to the true horizontal-axis of the LM. What this meant was that an even more harrowing appearance was added to Armstrong's already dramatic climb down the ladder. In reality, although the incline of the ladder was indeed quite precipitous at 65°, it was not as steep as seen on TV, which gave the illusion like it was almost vertical.[67]

Eight minutes and fifty-one seconds into the broadcast, the Moon had risen sufficiently high over Parkes that the telescope could now capture lunar transmissions with its main detector. Normal television scans rates could now be accommodated and the picture quality improved. Houston quickly switched to Parkes. Thus, the world saw Buzz Aldrin's descent down the ladder much clearer than his commander's 20 minutes earlier. NASA stayed with the Parkes television for the remainder of the two and a half hour telecast.[68]

Twelve hours later, the Madrid Station tracked *Eagle* as it lifted off the surface of the Moon to successfully rendezvous and dock with *Columbia*. After rejoining Collins, Apollo 11 made its critical TEI burn for home. On the morning of 24 July 1968, humankind's first journey to the surface of the Moon came to an end as Hawaii and the *Huntsville* tracked *Columbia* to a perfect splashdown less than five kilometers (three miles) from the recovery ship USS *Hornet*.

Ozzie Covington, who had been so instrumental in smoothing the lines of communications between Houston and Goddard, would recall years later the almost surreal feeling after it was all over.

> When we finally landed on the Moon on July 20 1969, I was grateful that our cooperative efforts had paid off. However, during the event, I was in the Mission Control Center in Houston. Some of the data from the lunar excursion module became sporadic and I really became uptight. NASA Administrator Thomas O. Paine happened to stand nearby and noticed my nervousness. He urged me to take it easy. We had come this far and would make it fine, he assured me. Well, we did![69]

This sense of tension followed by great relief was echoed by Bill Wood, who by then was the head of the Manned Network. On Apollo 11, he spent the entire eight days at the GSFC Network Control Center, working, eating, and sleeping there. "When I eventually got home," Wood said, "there was a big sign

Armstrong's photographic counterpart to the television image of Aldrin descending the ladder as seen in the previous figure. (NASA Image Number MSFC-6900937)

'Welcome' greeting me. However, [by then] I was emotionally exhausted and it took me quite a while to really comprehend as to what had happened, even though for years, I had been deeply involved in preparing for this event."[70]

As someone in the "trenches" at the field station, Mike Dinn, who was Deputy Director in charge of Operations at Honeysuckle, framed the accomplishment of the historic mission in a somewhat different perspective. To him, Apollo 11 was a simulation that went well. "The station had reached a point of capability whereby it was comfortable not only with a nominal mission, but comfortable that the station could cope with just about anything nonstandard," said Dinn.

> We had thought through and tried to simulate as many different things as could happen, and so I was comfortable with it. You knew you had the next pass to cope with. Every pass was crucial and critical, even though it might have ended up routine and nothing happened. You had to be, almost, literally on your toes, organized and prepared and staffed to cope with any anomaly. I was also comfortable with the management aspects of it. It was a very good operational philosophy that Chris Kraft had brought whereby everybody in the organization knew the success of the mission depended on them doing their bit properly and correctly, and that the person in the next station was going to do his bit correctly. We were all so busy that it took all your effort and energy to do your part well. And so it was very satisfying and rewarding that we didn't

have Goddard and Houston micromanaging—in great contrast to later years.[71]

As somewhat of a reality check, Dinn told his shift that morning (Australian time) that the most important data coming from the Moon that day was not going to be television but the bio-medical telemetry of Armstrong and Aldrin. Said Dinn:

"If you're there doing a job, you should be concentrating on the job at hand and the data at hand. . . . The luxury of 'whooping it up' doesn't fit in there. That is the least time you'd be whooping it up is when something critical just occurred. After Apollo 11 landed, you heard Gene Kranz say something like 'Right, we've got to stay or no-stay'. There wasn't time there to be whooping it up. I fully recognize this doesn't fit in with what the colloquial media, books, and the like want to say. But I'm afraid that's what it was. Yes, we were pleased and satisfied with what we achieved, but we were only a small cog in the machine. And yes, we'd done our bit well, but we weren't as tested as we were in simulations. I used to say that a nominal Apollo mission used about 5 percent of our capability because we had lots of redundancies. . . . When it came down to it, there was an enormous amount of onboard redundancies. They didn't need the network all the time . . . and to me, that wasn't a negative; that was a positive. It showed a lot of clever, intelligent, management and design of the mission and the hardware. You had so much redundancy and so many backups and so many options. They were all designed into the mission planning. Yes, there was satisfaction. It was the culmination of what we had trained for, and everybody performed. The satisfaction for me was to help bring the station from this state of not being very competent to one of the best in the network, as Bill Wood told me years later."[72]

The greatest challenge for the network during the Apollo years occurred in April 1970 when the flight of Apollo 13 had to be aborted as the spacecraft approached the Moon. Fifty-six hours into the mission with the spacecraft some 322,000 kilometers (200,000 miles) from Earth, damaged wires and insulation inside the Number 2 oxygen tank caused it to explode during a routine tank "stir". The explosion ruptured a line and damaged a valve in the Number 1 oxygen tank, causing it to also lose oxygen. The entire Service Module oxygen supply boiled away in less than three hours, which led to the loss of water, electrical power and use of the Service Propulsion System.

With the lunar landing now scrubbed, the mission turned into a race against time, one of saving the crew before all the life-support consumables expired. Astronauts James Lovell, John L. "Jack" Swigert, Jr., and Fred W. Haise, Jr. quickly powered up the Lunar Module, still attached to the CSM, as a lifeboat. All spacecraft systems except for life support were turned off to save power. Only a low power transmission link tethered the crippled spacecraft to Mission Control. Robert L. Owen, the MSFN Associate Chief for Network Engineering at Goddard during the mission, recalled the network improvising and adapting in real time to the situation. "There was a transponder on board the S-IVB (third-stage of the Saturn launch vehicle) which operated on exactly the same frequency as the transponder on the LM. In our planning, we had never considered powering up the Lunar Module until after the S-IVB had expired. However, when the power failure forced our astronauts to get out of the CM into the LM, we faced the problem of having the S-IVB floating nearby, utilizing the same communication frequencies. This was no good, and we quickly had to work out a scheme which would enable us to capture the signal from the Lunar Module. Eventually, the S-IVB crashed into the Moon, but in the meantime, we had to have reliable communications. We succeeded by working out a configuration we had never anticipated. Apollo 13 presented us with a frightening situation, which luckily, we were able to meet."[73]

To save power, telemetry had to be transmitted back to Earth using low power transmitters on the LM. Here, the 64-meter radio telescope at Parkes Observatory once again entered the picture. Originally, the Moon was too far north to be seen very well from the observatory and the telescope was not scheduled to support Apollo 13. But as soon as the accident occurred, NASA quickly recognized that Parkes could and would in fact be needed to track the failing spacecraft on its altered free-return trajectory.

The Australians, led by observatory director John Bolton, immediately began to prepare the station. While astronomy equipment was carried down a ladder from the antenna pedestal, the NASA antenna feed was taken up in a lift, installed and checked out at the center of the dish. In a job that usually took one week, the facility was reconfigured in 10 hours after receiving the go-ahead from Goddard.[74] Since Parkes was not slated to support this mission, microwave links which had been established for Apollo 11 and 12 were not operational when the emergency occurred. With urgency, a team of engineers from Honeysuckle and Tidbinbilla arrived at Parkes within hours to reestablish the links to Sydney before the next pass of the spacecraft.

Parkes inclusion was critical owing to the interference of the S-IVB as the 26-meter (85-foot) antenna did not have a narrow enough beamwidth to discriminate between the Saturn third stage and the LM *Odyssey*. When Parkes moved out of view, the 64-meter (210-foot) dish at Goldstone was able to do the same, and together, the two were able to track and communicate with Apollo 13, saving the flight from turning into a disaster. From a Mission Control perspective, it was NASA's finest moment.[75]

Eighty-seven hours after the explosion, Apollo 13 splashed down southeast of American Samoa with Lovell, Swigert and Haise safely strapped into their couches to bring to an end the only aborted lunar mission of the entire program. In retrospect, Apollo 13 represented a constructive failure that highlighted not only the coordination and preparation of network engineers at GSFC and stations around the world, but also the teamwork and cooperation between the various NASA centers and, more broadly, with the space agency's international partners.[76]

Lyn Dunseith, Director of the Data Systems and Analysis Directorate at the JSC recalled years later that:

> Throughout the entire program, Goddard provided us with the data we so critically needed. The quality of this support is best evident by the lack of a crisis in a crisis situation, such as the ill-fated Apollo 13 mission. Even during the flight, we had command and voice capability to handle a very serious condition. Our astronauts returned safely thanks in large measure to superb communications and tracking capabilities provided by the Goddard team. Its members are as much a part of manned space flight as anyone in Houston or at the Cape.[77]

Moon landings continued to unfold after Apollo 13, becoming more ambitious and complex with each mission. Scientific exploration of our nearest neighbor began in earnest on Apollo 12 and moved forward until Apollo 17 concluded the program. Compared to the life and death drama of Apollo 13, these missions went relatively smooth, though not totally trouble free. On Apollo 14, for instance, a malfunctioning abort switch gave flight controllers real trouble. The MSFN enabled Houston to send commands to reprogram the computers aboard the LM directing it to ignore that particular signal. Without this capability, the mission would have had to be aborted since the crew would not have been able to separate from the CM and a lunar landing would not have been possible. Former Flight Director Chris Kraft would say that, "On virtually every flight, the network and its people, while in the background, were 'under the gun'. We relied on them in every critical situation."[78]

These landings left Apollo Lunar Surface Experiments Packages (ALSEP) in five geographical locations across the lunar surface. ALSEPs were a combination of experiments which the astronauts deployed at a site sufficiently far from the LM to collect lunar surface experimental data. There was a central processing station to which all of the peripheral experiment and the Radioisotope Thermoelectric Generator (RTG) were attached. The ASLEPs provided power and data with network stations through its own transmitter and antenna. With several packages in place, these ALSEPs, connected as a network, returned more data than any could on its own.

Take the seismometer network emplaced by Apollo 12, 14, 15, and 16. It enabled the location of impacts and moonquakes to be determined very precisely. The network of three Lunar Surface Magnetometers enabled the study of solar wind plasma movement by tracing its magnetic field. Closing out the program, Apollo 17 carried an enhanced package of surface experiments. With nuclear power from the RTGs, ALSEP transmissions were received by

Apollo 15 Lunar Module pilot James B. Irwin loads-up the "rover", Lunar Roving Vehicle, with tools and equipment in preparation for the first lunar extravehicular activity at the Hadley-Apennine landing site on 31 July 1971. A portion of the Lunar Module Falcon is visible on the left. St. George crater is about five kilometers (three miles) in the background. Clearly seen is the one-meter (three-foot) steerable Unified S-Band (USB) antenna of the rover through which Houston could remotely control the vehicle if needed. This photograph was taken by Mission Commander David R. Scott. (NASA Image Number AS15-86-11602)

NASA's Spaceflight Tracking and Data Network for years after the last astronauts had left the Moon.[79]

Having surpassed President Kennedy's goal of landing a man on the Moon and returning him safely to Earth by the end of 1969, the final three Apollo flights that took place between July 1971 and December 1972 were conducted with scientific exploration in mind. The last of the Apollo lunar flights (the so-called "J-missions" with their emphasis on science), featured the Lunar Roving Vehicle (LRV), a 210-kilogram (460-pound) battery-powered car manufactured by Boeing-Delco. It was essentially an all-terrain vehicle designed to operate in the low-gravity, vacuum, dusty environment of the Moon. The Rover could carry 490 kilograms (1,080 pounds)—allowing for 180 kilograms (400 pounds) for each astronaut, his suit and the portable life-support system—a total distance of 92 kilometers (57 miles) to survey and sample considerable stretches of the terrain.[80]

Communicating with the rover posed a number of new challenges to the MSFN. For example, incorporating it into the television transmission scheme created a special set of problems. One issue in particular was how to control the motion of the LRV color television camera. Houston's method of operating the camera was to issue start/stop commands relayed through the network computers at the respective ground station. There was, however, a time lag of 2.5 seconds in the time it took to start and stop the rover camera from the time a command was issued at the MCC. This meant that if the Flight Controller operating the camera wanted to turn it by 5°, the "Stop" command would have to be dispatched before the "Start" command reached the Moon! To compensate, network engineers designed a fix to the ground station computers that staggered start/stop commands thereby allowing the camera to function without having to modify its control format.

Another potential obstacle to successful LRV television transmissions stemmed from voice and telemetry sub-carrier interference into the video portion of the rover's USB signal. Because the telemetry transmission spectrum overlapped the voice and video data frequencies, the interference left annoying herringbone patterns on TV. To solve this problem, engineers from GSFC, Johns Hopkins University's Applied Physics Laboratory, and the Goldstone Communication Complex produced a band-pass filter that removed the interference while preserving the video transmission to produce crystal clear images from the rover camera the quality of which would not be surpassed until HDTV became available 25 years later on the Space Shuttle and International Space Station.[81]

Introduction of the rover also increased the number of transmission sources that the network had to keep track of. The MSFN now had to synchronize all the activities of the Command Module orbiting the Moon, the Lunar Module parked on the surface, the LRV moving around on the surface, and finally, the two astronauts who may each be walking around in different directions. Keeping track of just where the rover was with respect to the LM was obviously important. Needless to say, a more reliable technique was needed than to simply allow the astronauts to visually follow their tracks back to the LM.

The solution—a rather novel one devised by Goddard engineers—was to pinpoint the rover's position with respect to the LM by extracting differential Doppler data from the two separate S-band transmissions coming, respectively, from the LM and the rover. By observing the Doppler shift, the network could precisely track the rover to provide the necessary navigation data. Mission Control then passed the data to the astronauts who then charted their course, enabling them to venture great distances, even after losing sight of the LM.

Proper coordination of lunar surface activities also required communication between the two astronauts on the surface and the CM Pilot in orbit. Support from MSFN stations was needed since direct line-of-site communications between the two lunar parties was limited to a brief overhead pass on each orbit. Since the ground network could see the CM for just about 50 percent of each orbit and because it was in continuous contact with the astronauts on the surface, MSFN stations served as relay points between the two parties. In this way, real-time voice communications between the surface and the orbiting CM were made possible for about half the time that the astronauts spent on the Moon.

When the Apollo 17 CM *America* splashed down on 17 December 1972, it marked the end of the first epic journeys to the Moon, a lasting tribute to the 400,000 men and women whose skill and determination placed 12 Americans on the surface of our nearest celestial neighbor. The tremendous sense of pride and accomplishment that came with Apollo deeply affected those who worked on the program, some, on a very personal level.

Robert Barnes, who first worked with Ozzie Covington at White Sands and later joined him in Greenbelt, saw the potential of the MSFN to accomplish something rather unique, something historical. Reflecting years later, Barnes said:

> My own involvement with this activity lasted 20 years, more or less, and it was not unlike having a front seat on a roller coaster: you wonder why in hell you got on, but somehow, would not have wanted to miss a chance for such a spectacular ride! With NASA, each of us saw a chance to fulfill a dream. However, in retrospect it must be concluded that all dreams were not the same. Certainly the work that led ultimately to the communications support of Apollo satisfied a host of dreams and was the work of a very dedicated group of people. It stands as an accomplishment for which each member can be justly proud.[82]

Lyn Dunseith, whose team was instrumental in integrating the Goddard network with the MCC, reflected:

> It is fortunate that the computer and communications technology kept pace with the needs of the space program. Indeed this program greatly accelerated the state of the art. Surely, without these tools and the men operating them, we would not have been able to get to the Moon. When we finally landed there and returned our astronauts safely to Earth, I could not fully comprehend what actually had occurred. It really seemed incredulous. For months after Apollo 11, I was somewhat in a daze and found it difficult to believe that we had made that lunar landing, even though

Skylab consisted of four major modules: the Orbital Workshop, Airlock Module, Multiple Docking Adaptor and Apollo Telescope Mount (ATM). The Orbital Workshop was a converted S-IVB third stage of a Saturn V. The ATM could not be accessed from the rest of the space station and a spacewalk was required to reach it. Launched in 1973, three crews visited the station between May 1973 and February 1974. Skylab remained in orbit until 1979. (NASA Image Number MSFC-72-SL-7200-110)

I had been personally involved in this dramatic event.... Yes, I have the book with all the equations and procedures, but I still find it difficult to believe, as I now look at the Moon, that men actually walked and worked there! It was incomprehensible.[83]

Even before these dramatic flights took place to the Moon, NASA was already thinking about what would be next. Beginning in 1964, exploratory studies were initiated under various names such as Extended Apollo (Apollo-X) and Apollo Extension System (AES) to investigate options for space projects that would come after the lunar missions. The next year, these initiatives were consolidated under the Apollo Applications Program (AAP), which by 1966, had narrowed the scope of the potential projects down to one of Earth orbit application; namely, a space station.

NASA had originally planned 20 Apollo lunar missions. But on 2 September 1970, Administrator Thomas Paine announced that due to a $42.1 million congressional cut in FY 1971 NASA appropriations, Apollo 15 and 19 were to be canceled; the remaining missions were redesignated Apollo 14 through 17. This disappointing cut left space-qualified hardware, which had already been made, immediately available for an AAP, specifically, an Orbital Workshop for a space station. On 17 February 1970, the NASA Project Designation Committee officially designated the project Skylab.[84]

Network response was required from the start, as Skylab encountered a number of difficulties. On 14 May 1973, the first two stages of a Saturn V launch vehicle placed America's first space station into low-Earth orbit. At over 86 metric tons, Skylab was at the time the most massive object ever successfully delivered into space. But this almost did not happen.

Sixty-three seconds after liftoff while the first stage was still burning, a crucial micrometeoroid shield on the exterior of the Orbital Workshop designed to protect Skylab from harsh solar heating and micro-impacts, was torn away by aerodynamic forces, carrying with it one of the station's two solar panels. Even the second solar array, as it turned out, did not fully open upon reaching orbit. The overheated and underpowered space station seemed doomed as NASA scrambled to decide whether or not to even attempt launching a crew to inhabit Skylab.

Over the next week, engineers at Goddard, Houston, and Marshall poured over telemetry that revealed the health and status and the extent of damage to the station. Houston remotely maneuvered the massive spacecraft via a series of command uplinks into a position which minimized excessive solar heating. Having bought some time, engineers poured over the telemetry data to come up with the appropriate fixes. A solar shield was taken up with the first launch of Skylab astronauts on 25 May. Upon reaching the station, Commander Charles "Pete" Conrad, Jr. and his crew (Paul J. Weitz, Pilot and Joseph P. Kerwin, Science Pilot) found that although metal surfaces were hot to the touch, internal conditions were much better than expected. The team lost no time deploying the parasol heat shield which produced a rapid drop in temperature and a spacewalk was done to fully release the stuck solar panel. By the fourth day, conditions had improved dramatically to the point where the three were able to settle into their flight plan. (Kerwin later served as the NASA Headquarter's OTDA Representative in Australia from 1982 to 1984).[85]

The project pushed network requirements to new heights. Skylab and its numerous scientific activities created a flood of telemetry that threatened to overwhelm the NASCOM circuits connecting the ground stations to Mission Control. It was the familiar problem of the difference in the data-capturing capability of the sites (now able to receive telemetry at a rate of 250,000 bits-per-second) and the NASCOM line transmission rates (still at a much slower 19,200 bits-per-second). Although 19.2 Kpbs reflected improvement over the recently concluded Apollo lunar flights, ways had to be found to accommodate the discrepancy linking the network stations to Houston. To this end, GSFC network engineers designed a data compression software that enabled each station computer to interrogate and filter-out redundancies and static data that, for instance, had not changed from previous downlinks. The station could then pass on only new or changed (dynamic) data. The modification worked well and was efficient in providing the MCC with all its data need without introducing a time lag.

Even with constant improvements like this, the network was not immune to occasional "glitches." The fixes were usually simple though. On Skylab 2, the ship *Vanguard* picked up and transmitted to the crew interference sound bursts coming from cars and fishing boats near the port at Mar del Plata, Argentina. The solution on that particular day was simple: take the *Vanguard* further out to sea.[86]

As a true testament to the value of humans in space, Skylab overcame its somewhat inauspicious start to serve as home for three crews, each on progressively longer durations: 28 days for the first mission, 59 for the second and a then record-breaking 84 days for the third. The last group returned to Earth on 8 February 1974. Even though the last crew left the station in 1974, network activities continued on Skylab until, quite literally, its last day in orbit. For several years after the last crew had left the station, commands were uplinked so as to maintain the spacecraft's orbit in hopes of preserving it long enough so that one of the early Space Shuttle flights could boost it into a higher and more stable orbit. But when the first Shuttle mission was delayed into 1981, it was apparent to NASA that Skylab was not going to survive its slowly decaying orbit. Like it or not, Skylab was coming down.

With its fate sealed, NASA had to make sure that it would reenter the atmosphere without scattering debris in populated areas. Thus prior to reentry, the station's drag characteristics were altered by uplinking commands that changed its attitude in an attempt to place the impact in the south Atlantic or

Indian Oceans. Skylab finally reentered on 12 July 1979, but it ended up scattering debris over Western Australia. A post-mission review of the telemetry showed that incorrect breakup altitude prediction, uncertainties in the ballistic coefficient and atmospheric density caused the impact area to shift downrange to Australia. The reentry demonstrated just how difficult it really is to perform a controlled reentry, even with good telemetry and an active command capability.[87]

Ed Lawless, who was the NASCOM Voice Network Manager, was in the Goddard Control Center when Skylab reentered. In an interview in 1989, he recalled:

> We did a lot of special tracking to make sure we had very good numbers on where it would most likely reenter at the time it was going to happen.... We knew that it had come down in the Australia area, and we had just started taking all the circuits down. I had broken the circuits to NASA Headquarters and all of a sudden the network got a telephone call in from our switching center in Australia. They had a pilot on the line with a very vivid description of the reentry.[88]

Henry Iuliano, who headed Goddard's Network Operations on Skylab, gave a vivid description of the pilot's encounter:

> The pilot was 100 miles [160 kilometers] east of Perth, flying at 28,000 feet [8,500 meters]. He said he saw this aircraft coming at him [and] thought it was a new type of aircraft that looked like blue metallic steel. It was about 5° above the horizon slightly off to his left, and as it approached him, it turned from steel blue to gray. Then the pilot realized it was turning red, that this was the Skylab. It began to break up in large pieces, with a tail at least 100 miles [160 kilometers] long of smaller pieces behind it, and it disappeared behind to his right 7° below the horizon. From the looks of the path, he estimated that it landed about 300 miles [480 kilometers] in back of him somewhere near Alice Springs, and that's exactly where most of the parts were found! Just before we heard the pilot's report, when Skylab went by the Ascension Island tracking station, they were still receiving telemetry data. They gave us a reading and said it was in a stable condition—actually flying! Instead of tumbling like we thought it would, it was actually flying at 66,000 feet [20,000 meters] and still giving good telemetry. Somewhere between Ascension and Carnarvon, Australia, when the pilot saw it, was when it began to break up.[89]

East-West rivalry had led to the United States planting six flags on the Moon, but it also prevented and forestalled any effort for human space cooperation between the U.S. and the Soviet Union. Without being overly dramatic, imagine a spacecraft stranded in orbit unable to return to Earth. Its crew may be injured or in peril as oxygen slowly runs out. What were the chances of another country sending up a rescue team to bring them home? Unlike today, only two countries possessed that capability in the 1970s.

At a meeting with veteran Soviet space scientist Anatoly Blagonravov in New York in April of 1970, NASA administrator Thomas Paine raised the idea of linking the Apollo and Soyuz spacecrafts on a joint mission. The idea interested the Soviets enough that the two countries reached an agreement on 28 October 1970 to conduct a joint study of a US/USSR rendezvous mission. The official intent of such a mission was to create a space rescue capability that would be available to aid astronauts who might become stranded in Earth orbit. These discussions culminated nearly two years later on 24 May 1972 when—with great satisfaction to the international community at large—U.S. President Richard M. Nixon and USSR Prime Minister Alexey N. Kosygin signed a space pact officially endorsing the project. This first-ever international space mission was officially named the Apollo-Soyuz Test Project, or ASTP, on 30 June.[90]

ASTP was based on a, 17-point technical agreement negotiated in Moscow on 4 through 6 April, 1972. This agreement highlighted the level of international cooperation—with clear requirements on network activities—needed to make the project work. Joint requirements included:

1 Control of the flight of the Apollo-type spacecraft will be accomplished by the American Control Center and that of the Soyuz by the Soviet Control Center, with sufficient communication channels between centers for proper coordination.

2 In the course of control, decisions concerning questions affecting joint elements of the flight program, including countdown coordination, will be made after consultation with the control center of the other country.

3 Joint elements of the flight will be conducted according to coordinated and approved mission documentation, including contingency plans.

4 In the conduct of the flight, preplanned exchanges of technical information and status will be performed on a scheduled basis.

5   The host country control center or host country spacecraft commander will have primary responsibility for deciding the appropriate preplanned contingency course of action for a given situation in the host vehicle. Each country will prepare detailed rules for various equipment failures requiring any of the preplanned contingency courses of action.

6   In situations requiring immediate response, or when out of contact with ground personnel, decisions will be taken by the commander of the host ship according to the preplanned, contingency courses of action.

7   Any television downlink will be immediately transmitted to the other country's control center. The capability to listen to the voice communications between the vehicles and the ground will be available to the other country's control center on a preplanned basis, and upon joint consent, as further required or deemed desirable.

The Soviet ground network on the Apollo Soyuz Test Project consisted of seven stations spanning 125° in longitude across Asia and Europe. (Adapted from Map of the Commonwealth of Independent States from the United States Air Force, link *www.af.mil/art/index.asp?galleryID=193* [accessed 9/22/2007])

8   Both sides will continue to consider techniques for providing additional information and background to the other country's control center personnel to assist in mutual understanding (including the placement of representatives in each others control centers).

9   As a minimum, flight crews should be trained in the other country's language well enough to understand it and act in response as appropriate to establish voice communications regarding normal and contingency courses of action.

10 A public information plan will be developed which takes into account the obligation and practices of both sides.[91]

Apollo-Soyuz presented a new challenge to the GSFC tracking and communications team. The challenge was one of providing links between two orbiting spacecraft with two control centers with two entirely different protocols. The mission was unique in that the NASA network had to, for the first time, function in coordination with a Soviet network. Each had its own communications protocol which now had to "talk to each other." Arrangements reached between the two sides stipulated that each control center could receive all voice and television communications transmitted to either spacecraft. Either crew could be contacted by voice from any station, whether American or Soviet.

Some 2,300 men and women at field stations and 500 at Goddard were assigned to the mission (more than that assigned to the later Apollo flights). The NASA stations that supported ASTP were a subset of the 9-meter USB sites that supported the lunar missions, plus a handful of STADAN sites:

Ascension (ACN)
Bermuda (BDA)
Guam (GWM)
Hawaii (HAW)
Madrid (MAD)
Newfoundland (NFL)
Orroral (ORR)
Quito (QUI)
Rosman (ROS)
Santiago (AGO)

Coverage from Orroral, Quito, Rosman and Santiago indicated the considerable progress that was made in the early 1970s in drawing on STADAN stations to assist in human spaceflight operations. In addition to the land stations, the venerable *Vanguard* was stationed off the Argentine coast near Mar del Plata. Three ARIA aircraft also supported launch and reentry operations in the Indian Ocean and South Pacific, taking off from airbases in South Africa and Australia.

The Soviet network consisted of seven stations stretched across the vast expanse of the USSR. In addition, the Soviets deployed two ships, the *Korolev* (ASK), positioned off Canada, and the *Gagarin* (KYG), near Chile. The Soviet stations were:

>Dzhusaly, Kazakhstan (DJS)
>Eupatoria, Ukraine (EUT)
>Kolpashevo, Russia (KLP)
>Petropavlovsk-Kamchatskaya, Russia (PPK)
>Tbilisi, Georgia (TBL)
>Ulan-Ude, Russia (ULD)
>Ussuriysk, Russia (SDK)[92]

It was during ASTP that a new dimension in space tracking and data acquisition was added. In a harbinger of things soon to come, NASA added for the first time, a specific space element to the network. The newly developed Applications Technology Satellite-6 (ATS-6), made by Fairchild, was used to relay communications from the orbiting spacecraft to ground stations. This increased coverage dramatically, from approximately 17 percent to 60 percent (an increase from 15 to 52 minutes) of each 87 minute orbit.

ATS-6 was the second generation of the GSFC Applications Technology Satellite program. Its predecessors, ATS-1 through 5 launched between 1966 and 1969, was the first generation in the series. Originally designated ATS-F, the program had included a second, very similar satellite called ATS-G, but it was canceled for budgetary reasons. Eight of the experiments on ATS-6 were explicitly designed for communications relay studies to prepare for the next generation TDRSS.

But use of ATS-6 on Apollo-Soyuz was not orginally planned. Bill Wood explained. "We at Goddard were very reluctant to commit the use of the ATS except on the basis of a test and not to meet ASTP requirements. This was another example of the camel's nose in the tent. The very nature of the ATS was as a test program. The closer we got to launch, the more important it seemed to get. We wound up putting a lot of effort into putting equipment in Spain to interface with ATS. Thank goodness it worked, but I for one was nervous."[93]

At nearly 1,360 kilograms (3,000 pounds) with a span of over 15 meters (50 feet), ATS-6 was quite the imposing bird. It included a 9-meter (30-foot) diameter parabolic antenna, an Earth-viewing module located at the focus of the parabola and two solar arrays for power. Not only big, it was also quite complicated for its time. All the communication experiment was located in a section of the Earth-viewing module with feeds for the large antenna mounted on top of the module and Earth-pointing ancillary antennas populating the bottom side of the satellite.[94]

Launched out of the Kennedy Space Center atop a Titan III-C on 30 May 1974, GSFC had a list of performance objectives that they wanted to see from the satellite:

> Demonstrate the feasibility of using a nine meter diameter, deployable, steerable, high-gain antenna with good RF performance in the 6.5 GHz range.
>
> Provide spacecraft fine pointing to within ±0.1° accuracy.
>
> Demonstrate precision interferometer attitude measuring technology.
>
> Provide an Earth-facing, stable spacecraft at geosynchronous altitude for experiments to be selected by NASA Headquarters.

Originally placed in geosynchronous orbit at 94°W over the Galapagos Islands, the big satellite was immediately used to test operational compatibility with the network ground stations. In June 1975, Goddard controllers, transmitting through Rosman, commanded the satellite to 35°E over Lake Victoria, Africa, to support the Indian government's Satellite Instructional Television Experiment (SITE). From this vantage point, ATS-6 could also participate in "millimeter-wave" communication experiments with several European ground terminals as well as relay ASTP data to ground receiving stations.

To do this, it pointed its antenna towards the horizon and generated a signal for the Apollo spacecraft to lock onto as it moved into view. Upon establishing contact, Apollo transmitted telemetry, voice and television to the satellite. ATS-6 then relayed the signals to a 30-meter (100-foot) antenna at the Buitrago ground station outside Madrid. Madrid then acted as the ground terminal, relaying the spacecraft's data via commercial Intelsat to the United States.[95] After supporting ASTP and the one-year Indian experiment, it was slowly moved by a series of ground commands to the Western hemisphere where it was stationed at 140°W over the Pacific until it was deactivated in July 1979. During its final trek as it was being repositioned in July 1976, ATS-6 demonstrated the social benefits possible of data relay by providing temporary (and goodwill) communication services while passing over 27 countries on the way to its final destination.

Engineers and scientists at Goddard conducted a series of space communication experiments using ATS-6 in its five year life. One of them, the "ATS-F Tracking and Data Relay Experiment," designed by F. O. "Fritz" von Bun and exercised in conjunction with a Nimbus weather satellite, was designed specifically as proof-of-concept testing for the upcoming TDRSS (see Chapter 7). ATS-6 also relayed television signals to remote areas of Alaska, the Rocky Mountains and the Appalachians. This operation, beginning in August 1974, brought live, public education television programming to those areas of the United States for the first time.[96]

☆ ☆ ☆

Although preparations leading up to ASTP broke new ground in terms of cooperation between the two countries, the Soviets still found it difficult to break with their veil of secrecy. On 2 December 1974, seven months before the scheduled launch of ASTP, Soyuz 16 was launched from the Baikonur Cosmodrome—completely unannounced. NASA had known that a dress rehearsal was coming, but only when the Soyuz spacecraft reached orbit did Moscow bother to inform the Americans that it was in fact already underway! The Agency was able to put the mission to some use through a quickly organized, 15-hour joint tracking exercise at the behest of the Soviet Union. This even included a simulated launch so that the Soyuz crew had something to "aim" at in a mock rendezvous. Data recorded by NASA ground stations were relayed to Goddard and, after the mission, compared to data received by Soviet stations during the same time period. This comparison merely verified what NASA already knew: the network was ready for the mission.[97]

All this took place in the Cold War. The United States had just pulled out of an unpopular war in Southeast Asia, one which pitted the country face-to-face against communism half a world away. It had cost 50,000 American lives. While U.S. preparations for the mission were done in the open, Soviet preparations, although more open, were still for the most part veiled in secrecy (as Soyuz 16 so clearly illustrated). It was not surprising, then, that NASA went about preparations for the mission with a certain sense of trepidation. While international cooperation was what ASTP was all about, NASA kept finding itself in situations asking "How does one cooperate without giving away too much from a technology standpoint."

A case in point was the technology needed to physically dock the Apollo with the Soyuz. The two not only had different docking mechanisms but also different atmospheres inside the spacecraft (Apollo operated at a cabin pressure of about 0.3 atmospheres while Soyuz operated at 1 atmosphere, or standard sea-level). The technology imbalance led to the U.S. developing with help of the Soviet Union, the Docking Module, the central, critical piece of equipment without which the mission could not have succeeded. The Docking Module turned out to be purely U.S. technology in the end. Technology, however, was not the only thing that changed hands during preparations for ASTP. There was also a language barrier. While it is well known that both flight crews had to learn each other's language during training, what is lesser known was that there were actually classes conducted in Russian at the GSFC to train the network engineers who would be communicating with their counterpart in the Soviet Union. The direction from NASA management was, "If we're going to deal, we have to learn to speak the language."[98]

At 1220 hour GMT, 15 July 1975, Commander Aleksei A. Leonov and Flight Engineer Valeri N. Kubasov blasted off aboard Soyuz 19. It was the first time that a Soviet space launch was seen live on television by its own people and others around the globe. The communication link traveled in a circuitous route: Moscow to Helsinki, Stockholm, Copenhagen, Prague, Hamburg, Frankfurt, and then to a Comsat ground station at Raisting, near Munich, West Germany. From there it was sent via the Comsat satellite to the United States. The routing was requested by the Soviets since an AT&T ground station planned for this flight was not finished in time. According to Charles J. Goodman, Goddard's technical manager for television on ASTP, the routing involved some seven relay points on both the East and the West. It also required conversion of signals from the Russian color system protocol (SECAM III) to the European PAL color system and finally to American standards National Television System Committee (NTSC).[99]

Despite the complexity, communications never showed any noticeable degradation. "Just about everybody broke his back to help make it happen," Goodman remembered. "We had some 50 hours of virtually flawless television transmission. Our arrangements began on November 25, 1974 and everything was in place for the launch some eight months later."[100] Goodman specifically pointed out a first-rate relationship with the European Broadcast Union headquarters in Brussels, Belgium, and its technical personnel. Apollo-Soyuz was seen by more people in more countries than even Apollo 11.

Seven and a half hours later, Commander Thomas P. Stafford, CM Pilot Vance D. Brand and Docking Module Pilot Donald K. "Deke" Slayton were launched atop a Saturn IB rocket from pad 39B at the Kennedy Space Center. After a series of orbital maneuvers—the most complex of its kind during the Apollo era in which the American CSM chased the Soyuz—the two spacecraft began station keeping at 1551 hour GMT on 17 July. They docked 24 minutes later. After Slayton and Stafford equalized the atmosphere inside the Docking Module with that of the Soyuz, the hatches were opened and the now celebrated "space handshake" between the two mission Commanders was televised live to the world.

Over the next two days, the crews exchanged mementos and conducted (token) zero-g science experiments; a second docking was also performed. They also exchanged cuisines, with the Americans offered a choice of hot soups from the different peoples of the USSR—Ukrainian beetroot and cabbage soup, a piquant Georgian mutton broth and Russian sorrel and spinach soup. In return, their western colleagues offered up such delicacies as applesauce, spaghetti, apricot pudding, and bacon squares.[101] After two days, the two vehicles undocked for the final time. After a fly around photography experiment in which the Soyuz was used to block out the Sun simulating an artificial solar eclipse, the two spacecraft went their separate way. Two days later, Leonov and Kubasov de-orbited their spacecraft, bringing it back to Kazakhstan on 21 July. As with the launch six days earlier, their landing and recovery was seen by a live television audience for the first time. Apollo stayed in orbit for another three and a half days, splashing down four and a half miles from the recovery ship *New Orleans* near Hawaii on the afternoon of 24 July to bring to an end the first international space venture and the final Apollo splashdown.[102]

Chris Kraft would reflect years later on the uniqueness of the ASTP experience and what each side was able to learn from the other:

> Getting to know the Russian management approaches, their thoughts and objectives, both in a national and personal sense, was an extremely interesting experience. The Russians are very different and their motivation is certainly not the same as ours. Their pride is very important and their engineering skills are very good. They are just as smart as we are. They did a superb job of building parts of the machinery and in the planning. They needed a great deal of help from us particularly in getting the job done within the management confines that existed in the Soviet Union. Here they needed help, and they told us so. Certainly the Russians do not do things in a manner even closely resembling our approaches. They are more secretive and I am not sure that we really learned how they do things internally. For instance, I do not remember ever having seen an organizational diagram. It was a long and protracted process. In the beginning, I thought a joint project might just not be possible. There was a great lack of credibility and trust between us and our ability to communicate. But slowly, primarily due to the tremendous efforts Glynn S. Lunney, the American Technical Director for the mission and his Russian counterpart, Professor Konstantin Davydovich Bushuyev, their associates and the respective space crews, we found a way to get things on the right track. They deserve all the credit for this. It was a fantastic achievement for both sides when we finally flew this mission.[103]

The Apollo-Soyuz Test Project stood as a symbol of the Nixon-Ford era of détente. This atmosphere of cooperation was short lived however. Soviet-American relations soon deteriorated, reaching a new low in 1979 after the invasion of Afghanistan. Cooperation in space exploration turned tepid and would stay dormant for the next two decades. For the balance of the 1970s, human spaceflight practically disappeared from the American public's eyes. While atmospheric Approach and Landing Tests of the developing Space Shuttle were conducted as the new decade approached, NASA would not return a person into space until 1981. In a way, Apollo Soyuz marked the swansong for the first era of human space presence, one driven by the intense rivalry between the two Cold War superpowers. How fitting then that this era, which began in 1961 with Alan Shepard's 15 minute response to Gagarin's flight, concluded with a handshake in space between those same superpowers.

NASA's Spaceflight Tracking and Data Network stood out during this time to make possible this success story and America's victory in the space race. The role it played led to a much deserved recognition by Congress when in September 1974, the House declared that

> After completing an investigation which took nearly a year, it [has] concluded that the Tracking and Data Acquisition Program is being managed and operated in an effective and efficient manner. The people working in the program—both government and contractor, both U.S. and foreign—are doing an excellent job, and are to be commended for their contributions to the success of the U.S. space program.

As one committee member put it, "They are the unsung heroes of the space program."[104]

Noel W. Hinners, who retired as Director of the Goddard Space Flight Center in 1989, echoed this sentiment when he recalled the uniqueness of the time and place that was the Space Race, and how NASA's tracking and communication networks met the challenge. "There was a unique contest: the dream of man's quest to explore space and the harsh technical realities which had to be faced if these dreams were to come true. The area of space tracking, communications and data acquisition from orbiting satellites and eventually from the Moon and beyond, was an important part of this odyssey. A dedicated team of men and women, both in and out of government, helped to make these dreams a reality. They were the first generation of 'space trackers' whose electronic links tethered the spacecraft to its controllers and scientists. The Goddard Space Flight Center, as a member of the NASA-industry team, [was] proud to have contributed to these expeditions in space."[105]

CHAPTER 6

ERA OF CHANGE

NASA's annual budget was $330 million in 1959. Just six years later, it had ballooned to $5.25 *billion*.¹ Over the next seven years, however, even as the space agency was putting 12 men on the Moon and busy pushing the envelope in launching a plethora of new generation of space probes, science satellites and application spacecraft, it saw its funding gradually being cut. By 1974, it had bottomed out at $3 billion.² In the FY 1973 NASA budget hearings, Gerald M. Truszynski, Associate Administrator for the OTDA, announced plans by the space agency to merge the STADAN and the MSFN into a single, more streamlined network.

Networks were developed under a certain sense of urgency in the early years of the space program. The need to respond to the Soviet Union and to put the American space effort on the fast track sometimes took priority over such matters as coordination of effort and minimizing of cost. It was, after all, a time of pioneering work with many unknowns. By the early 1970s, while the major emphasis on meeting program requirements had not diminished, coordination of these requirements, economic efficiency, and tighter management controls were being given a much higher priority. This fundamental shift to the pragmatic was felt—and felt hard—by those running its spaceflight tracking networks. As one NASA manager recalled, "There wasn't as much money there, wasn't as much activity, [and we began] closing Apollo tracking stations, cleaning up after Apollo."[3]

It was in this atmosphere of renewed fiscal awareness that NASA merged the two networks to form what would be called the Spaceflight Tracking and Data Network, or STDN. There were also other reasons besides budget for a network consolidation. With the decline in scheduled human space activities after Apollo, the argument for a separate, manned-flight network became less compelling. For the engineers, technicians, managers and even the astronauts—the very men and women who had just put Americans on the Moon—there was a definite sense of let-down when it dawned on them that what seemed like an adventure which had just begun was now suddenly over. Bill Watson, the Program Executive at Headquarters who today oversees NASA's Ground Network, was fresh out of school and just starting his career at the time. Reflecting back, Watson said:

> There was a sense of what's next, what we should do next after Skylab and Apollo-Soyuz. It was hard for guys to get excited about scientific, robotic satellites to the extent that they were excited about the manned flights. There was a large hiatus there until the Shuttle program came along, and . . . a lot of folks left the program during that gap.[4]

The numbers reflect this. In 1970, the Agency's fulltime, civilian workforce stood at 31,223. By 1979, this had dropped to 22,633, a reduction of almost 30 percent. NASA cut its workforce by 7 percent in 1972 alone.[5]

Besides fiscal constraint and the rescoping of the Agency's mission, there were also good technical reasons for merging the networks. Both the STADAN and the MSFN were growing increasingly sophisticated. The clear separation of crewed versus uncrewed requirements that had so differentiated the two were becoming more nebulous due to the increasing number of high eccentricity (highly elliptical), high apogee observation satellites being launched. This new class of satellites had much in common from a tracking standpoint with an Apollo spacecraft traveling to and from the Moon. Meanwhile, network managers at Goddard thought that implementing the USB concept throughout the STADAN could serve as the common bond needed for a single, overarching, near-Earth network. All these factors served to provide Truszynski and his office with good reasons to merge the capabilities of the two networks. NASA's thinking was that, with a leaner network, fewer stations could actually provide a more flexible capability to support its upcoming workload for all near-Earth missions, both robotic and piloted.[6]

Network engineers understood that the existing geographical distribution of the stations could effectively be modified into a configuration that would be able to handle the total mix of missions which NASA at the time foresaw for the latter half of the 1970s. Before this transition, there were 25 stations (19 STADAN and 6 NASA-owned, primary MSFN sites) spread over five continents (see maps in Appendix 1). The continual operation and maintenance of so many stations were, not surprisingly, expensive and required a great deal of manpower.

As of the mid-1960s, satellites were carrying much more powerful beacons so that telemetry—and not tracking—was now the pacing item. Technology was also advancing such that having fewer but better instrumented stations was now possible. Bill Wood, at the time Chief of the Manned Flight Operations Division and later Associate Director for Network Operations at Goddard, said of the change:

> As the Apollo program began to wind-down, we realized that both manned and unmanned tracking functions had to be consolidated. It simply became too impractical and too costly to maintain separate networks. . . . This was the time to change from the old to prepare for the future—Skylab, Apollo-Soyuz, Space Shuttle and the many unmanned missions also being planned.[7]

At the Directorate level of Goddard, the organization was reworked starting in January of 1971. Ozzie Covington now consolidated all network activities under him including the field stations, the Network Operations Control Center (NOCC), and communications. Under Jack Mengel were all the Project Operation Control Centers, data processing, and the large computers at the Center.

To implement the change, Goddard made sure that several requirements, both new and old, were going to be met. First, the high data rate, real-time TT&C capability of the manned network were retained since they matched well with the increasingly more complex satellite requirements that were then coming online. Many of the satellites were, in fact, approaching the complexity of and taking on the characteristics of human missions in terms of requirements for command and control, downlink data rates, and the higher operating frequencies at the S-band. Foremost among this new generation were "mega" satellites such as the Earth Resource Technology remote sensing satellite (ERTS), the High Energy Astronomy Observers (HEAO), and the International Ultraviolet Explorer (IUE). On the IUE, for example, the onboard telescope had to be moved at regular intervals by means of ground commands emanating from the GSFC. In general, telemetry rates were pushing state-of-the-art capabilities at 150,000 bits per second.[8]

When ERTS-A was launched on 23 July 1972, it was actually supported by the MSFN. Thus, there was an increased need in the unmanned spacecraft community for the type of technical capabilities which already existed in the MSFN. By 1974, work was well underway to modify the telem-

Technicians check out Earth Resource Technology Satellite ERTS-A at the General Electric Company Astro-Space Division in Princeton, New Jersey in 1972. Launched atop a Delta 900 launch vehicle from Vandenberg Air Force Base in July of that year, ERTS-A was the first in the series of Landsat remote sensing satellites, one of the most successful Earth resource application programs ever. Downlinking its data to the Goddard Space Flight Center at a rate of 15 megabits-per-second, ERTS-A was designed to last one year but was not deactivated until 2 January 1978. (Photo courtesy of the United States Geological Survey)

etry and command processing systems at the existing MSFN sites for compatibility to support science spacecraft. At the same time, however, NASA still had many of the less complex spacecraft such as the old Explorer series, which was still returning a healthy amount of data. These were generally the smaller, spin-stabilized satellites which could not accommodate the newer and larger, high-gain, directional antennas, and therefore, still had to operate at the lower VHF frequencies.

With this wide spectrum of needs, NASA required the full range of capabilities offered by both networks. Apollo just came to an end; the time was right for such a merger. By the end of 1974, the number of ground stations (STADAN and MSFN) had dropped to 17. Two years later, it went to 15. Of the eight ARIA, only four were now available for NASA support, the others having reverted back to the Air Force full time. Four of the five AIS were retired in 1969, leaving the *Vanguard* as the only network vessel to remain in service (it too retired in 1978).[9] Indeed, once the transition to the newly organized STDN was complete in 1976, network operations quickly became more standardized. NASA began to see greater returns from the slimmer network, all the while reducing the manpower needed for operations, logistics, and most importantly, cost.

While NASA did not present the consolidation of the STADAN and the MSFN to Congress until 1973, phase-down activities had already been taking place for some time. The first round of phase outs involved the STADAN stations at Blossom Point, Maryland; East Grand Forks, Minnesota; and Woomera, Australia in 1966. This was followed by shutdown of the temporary sites at Darwin and Cooby Creek in Australia; Lima, Peru (transferred to that country's university); and at Mojave, California in 1969 (the remnant of the old San Diego Minitrack station which had moved to Goldstone). A year later, St. John's, Newfoundland, on the eastern-most point of Canada, was shut down for good as was Fort Myers, Florida in 1972. By the time the STDN consolidation occurred, STADAN had, in fact, already streamlined down to nine stations (plus the NTTF in Greenbelt).

On the MSFN side, downsizing began soon after Apollo 11 when requirements for Apollo were carefully reevaluated by Headquarter's Office of Manned Space Flight. With little fanfare, NASA soon began reducing the number of MSFN stations as well, beginning with the shutdown of Antigua in the South Caribbean Sea on 15 August 1970. The Agency had determined that limitations on launch azimuth angles for flights following Apollo 13 would not require data from Antigua and that no increase in risk to mission success would be incurred as a result of the shutdown. In the words of a NASA spokesman, Antigua was simply the victim of "reduced requirements for NASA's worldwide tracking system."[10]

The station had a 9-meter (30-foot) USB system as its centerpiece. After Apollo 11, it was almost immediately relegated to a caretaker status with the 17 Bendix employees and 11 Antiguans put on standby status.[11] Most of the equipment was transferred to other facilities in the MSFN. Although human spaceflight requirements for Antigua were soon deleted all together, the requirement to support other NASA launches out of the KSC still remained and the station stayed open at a reduced level. But the writing was on the wall. Soon thereafter, NASA pulled out of Antigua. The Air Force Eastern Test Range station on Antigua agreed to provide services to NASA as needed—on a cost reimbursable basis.[12]

The review board also showed that either the Corpus Christi Station in Texas or the Guaymas Station in Mexico could be closed. Had all factors basically been equal (including politics), the decision would have come down to fiscal considerations; that is, which one would yield the most cost savings. But Texas had one thing going for it that Guaymas did not. Due to its desirable location to support Earth resource satellites, it was the logical choice to remain open. By utilizing USB equipment from Guaymas, the station would be able to support both crewed and uncrewed programs. Because of this, the decision was made to keep Texas operational and close Guaymas. A meeting was held in Mexico City on 16 June 1970 with Mexican space officials, the U.S. Ambassador, and Gerald Truszynski discussing plans on how best to phase out the station. This was followed by a second meeting two months later in which it was agreed upon that NASA would remove two of the three major station systems for relocation to other parts of the network. The third system would be left in Guaymas to support Mexican space activities and programs of mutual interest to the Mexican science community and the United States.[13]

This was a good way to close a station. In addition to promoting goodwill between the two neighboring governments, the Mexican National Commission for Outer Space (CNEE) and NASA were, at the time, cooperating on two scientific projects. One was to develop a system using weather data acquired from U.S. satellites by using automatic picture transmission equipment. The other was to develop capabilities and applications for Earth observations using advanced airborne remote sensing instruments. The two countries were also completing plans for a cooperative project involving meteorological sounding rockets. After details of the agreement were ironed out, joint press announcements officially closing Guaymas were released by both governments on 12 November 1970.

By the following February, NASA's withdrawal from the station was complete. This brought to an end a decade of association during which America blazed a pioneering trail into space. From John Glenn's first flight into orbit to Apollo 11, Guaymas was there. Commenting on the legacy of the station, Dr. George M. Low, then Acting NASA Administrator, noted most fondly that the "cooperative establishment and operation of the station over the 10 most exciting years in space exploration stood as a tribute to the friendship and understanding between the two countries."[14] In particular, he singled out members of the Mexico-U.S. Commission for Space Observations who first laid the groundwork in 1959 to make Guaymas possible, specifically recognizing: Hugh Dryden, Chris Kraft (Director of Flight Operations at the Manned Spacecraft Center), Ralph Cushman (Special Assistant, Office of the Administrator), and Dr. Eugenio Mendez Docurro (Secretary of Communications and Transport, Av. Universidad Xola). This was quite the fitting tribute to a decade which saw the sleepy little railroad town of Empalme, Sonora (12 miles outside the actual city of Guaymas) thrusted into the international space forefront to become, even today over 30 years later, a source of pride for the Mexican people.

As the transition took place, plans regarding which stations to keep and which to close could change quickly, and often did. Take Canary Island, for instance. In the summer of 1973, NASA Headquarters proposed a five-year extension to the Spanish government that the station be kept open until 1978, when NASA's TDRSS was then scheduled to become operational.[15] The station seemed safe for another five years. Several requirements still needed support including telemetry reception from the Apollo Lunar Surface Experiment Packages that had been left on the Moon by the astronauts and the upcoming Apollo-Soyuz Test Project that would take place in 1975. Ironically though, it was this same requirement to support ASTP that ended up providing the impetus needed to shut down Canary Island.

This twist of fate came about due to the requirement for live television, a critical requirement on the highly publicized ASTP. It had been anticipated (correctly) by NASA that this particular mission, as the first international human spaceflight between the two Cold War rivals, would draw worldwide interest not seen since that of Apollo 11 five years earlier. As early planning requirements for extensive real-time coverage were being developed (jointly by the ASTP Program Office in Washington DC and Moscow), it became apparent to both that this requirement was not going to be met effectively using existing MSFN capabilities. Something better was needed. The Agency would use the ATS-6 to directly receive television signals from the Apollo spacecraft and then retransmit them to a ground station in Spain, rather than depending on the ground stations alone.

Fallout from this decision on Canary Island came quickly. On 22 January 1975, Truszynski sent a letter to the Director General of the Madrid Station (of which the Canary Island station was a part of) that NASA has "regretfully come to the conclusion that both near and long term data acquisition requirements do not support the continuation of the Canary Island station and would desire to close the station as soon as possible."[16] Canary Island's fate was officially sealed two days later by a notification from Truszynski to NASA's Assistant Administrator for International Affairs that services on Canary Island were no longer needed and that the State Department was requested to take appropriate actions as soon as possible to shut down the station. Thus, Canary Island went from being a crucial land station in the eastern Atlantic to "not necessary" in the span of not even a single mission. In a way, it was a harbinger of things to come as ATS-6 tested out the new concept of space communications, one that would rely almost exclusively on space-based satellites to do the job that ground stations once did.

* * *

Soon after Apollo 11, the Guam and Hawaii stations took center stage in a budget fight between the Bureau of the Budget and NASA. During the FY 1971 budget process, the Budget Bureau notified NASA that Guam and Hawaii were going to be phased out and their operations transferred to the DOD satellite control station on their respective island. Each year, with a few exceptions, every department and agency of the federal government has to negotiate the "necessary evil" of the budget process; NASA was no exception. While budget negotiations were an annual ritual, what the Bureau was telling NASA in this instance was considered by the space agency as being somewhat "out-of-line." The Budget Bureau's position was that NASA should shut down these stations, but that in order to "alleviate to the extent possible impact on mission support," the DOD would "give the NASA manned missions highest priority in workload allocation."[17]

In November 1969, Administrator Thomas Paine rejected this proposal outright, making it clear that this was indeed an assumption of fait accompli, one not based on any DOD-NASA discussion after it was proposed at the start of the budget process. A paper was drafted explaining why NASA believed that any such consolidation would be neither operationally feasible nor cost effective. NASA's viewpoint was based in part on a preliminary joint NASA-DOD sponsored study to evaluate the merit of consolidating the NASA and DOD network facilities on Guam and Hawaii. No long-term operational costs were identified which would have offset the substantial immediate cost of modifying and relocating the equipment and expansion of facilities required to handle the high-priority functions of both agencies. Before sending this paper to the Budget Bureau, Paine confirmed that "responsible officials in the Air Force agree with us that the conclusions of this study are still valid."[18]

This did not end the matter however. Three months later, the Bureau of the Budget once again informed NASA that the Guam and Hawaii stations were to be phased out. This time, in a strongly worded letter to Robert P. Mayo, Director of the Bureau of the Budget, Paine voiced the Agency's concern that they now appeared to be under direction, without prior consultation, to take an action which was operationally and economically unsound in the view of both NASA and the U.S. Air Force. Since he was at the time accompanying the Apollo 11 crew in the "Giant Leap" victory tour in the Far

Gerald M. Truszynski (far left) rose through the ranks to become NASA's Associate Administrator for Tracking and Data Acquisition from 1968–1978. This picture shows Truszynski when he was Head of the Instrumentation Division participating in the 27 January 1953 ground breaking ceremony of the NACA High-Speed Flight Research Station (which became the Dryden Flight Research Center) on the northwest edge of Rogers Dry Lake in the Mojave Desert. Pictured with Truszynski were Joseph Vensel, Head of the Operations Branch; Walt Williams, Head of the Station, scooping the first shovelful of dirt; Marion Kent, Head of Personnel; and California state official Arthur Samet. (NASA Image Number E-980)

East, Paine volunteered to change his travel plans so that he could personally look into the situation at the NASA and Air Force stations in Hawaii on his return. In the meantime, he directed Truszynski and his office to review again the requirements on both islands with DOD officials. Drawing a line in the sand, Paine concluded his letter to Mayo in no uncertain terms, saying, "Unless new information is developed in my visit or in the review, I will then formally reopen this matter with you, and if necessary, the President."[19]

Guam and Hawaii went on to survive that year's budget process. In fact, both stations went on to become among the longest-serving STDN sites, remaining operational for another 19 years, finally closing in 1989.

As NASA stations began to close around the globe, none was more of a political target than the Johannesburg Station in the Republic of South Africa. This was one of the few communication complexes where the DSN and the STADAN shared a location. Roots of the DSN go back to the late 1950s. As the United States moved from the realm of Earth-orbiting satellites to begin sending probes to the Moon and beyond, a "World Net" was established by the DOD's Advanced Research Projects Office. This World Net formed the nucleus of what would go on to become the DSN. In order to maintain continuous coverage of space probes departing the planet as Earth rotates, three sites are needed, each situated about 120° apart. The DOD—and later NASA—had placed the first two sites at Goldstone, California and Woomera, Australia. Completing the World Net was the construction of a third station in the country of South Africa, where a government-owned, 4,000 acre grassland valley near the Hartebeestpoort Dam 65 kilometers (40 miles) north of Johannesburg was provided.[20] The station became operational in June of 1961.

To meet tracking requirements in the Southern Hemisphere, a $5 million expansion at the Johannesburg complex was done three years later that brought the number of stations to three. One was run by the U.S. Air Force to control its satellites. Due to its obvious military nature, the station was staffed entirely with Americans. The other was a NASA satellite tracking station. The remaining site, for all intents and purposes, was part of the NASA station but was operated for the Smithsonian Institute, its roots dating back to the IGY and Minitrack.[21] Unlike the Air Force, NASA staffed these two stations with South African workers and normally only had a U.S. liaison officer present onsite. Under a 1960 agreement with the space agency, the National Institute for Telecommunications Research, a part of South Africa's Council for Scientific and Industrial Research (CSIR), had full responsibility for management of the station which they operated so as to meet NASA's technical requirements. The station was fairly extensive. At its peak, the NASA side of Johannesburg employed some 280 South Africans of whom about one in five were black.[22]

Even as NASA began working with the South African government to establish stations there, the potential fallout from that country's racial segregationist policies was not lost on many in the United States. NASA was fully aware that an agreement with a government espousing such policies could become a political flashpoint. But at the same time, it could not just discard the technical merit of such a location. Here's why: for optimal coverage of interplanetary probes launched on trajectories from Florida, an antenna was best placed as far south as possible, preferably deep in the Southern Hemisphere. The Republic of South Africa, being on the very southern tip of the continent, was ideal. As unfavorable as the South African political climate was, it was actually the most democratic and most stable government accessible to the United States on the continent at the time. To keep its options open, even as negotiations were being held with South Africa, NASA still looked at other locations, particularly those in southern Europe. These, in order of preference, were Sicily, Sardinia, south Spain, and south Portugal. Headquarters also looked into a possible cooperative arrangement with France, which at the time was considering the purchase of a 26-meter (85-foot) antenna from the Collins Radio Company to build a ground station of its own on the Normandy peninsula.[23]

These were more than just cursory looks. Site survey teams consisting of members from Headquarters and the Jet Propulsion Laboratory were sent to all these locations as NASA wrestled with whether or not to proceed with South Africa. In the end, it was decided that the geographical location, along with the country's already robust scientific community and the expressed enthusiasm of the South Africans, best advocated putting stations there.

It did not take long for tensions to arise. Accusations centering on the station started to surface even back in 1962, that South Africa might be putting pressure on the U.S. government for NASA to adopt a segregationist policy there. This was a serious concern, so much so that Associate Administrator Edmond Buckley wanted an early evaluation of the matter by asking the State Department to look into the situation.[24] Time did not assuage the tension between the two governments, though. In fact, things only got worse. The situation came to an early head when in May 1965, the United States asked South Africa for permission to have a squadron of advance-planes from the aircraft carrier USS *Independence* land at airports when the ship was scheduled to dock at Capetown. The government granted the Americans permission, provided the planes' crews were white. Up until then, American planes had often landed at South Africa airports and on occasion, there had been mixed-race crews including blacks. However, never before had the South Africans explicitly asked for all-white crews.[25] This caught the State Department totally off guard. In an attempt to clarify the meaning of the South African response, the United States asked if this was a condition or a suggestion. If it was a condition, South Africa was told it would not be accepted. If it was a suggestion, no guarantees could be given. With no clear response from the South African government and not wanting to escalate the already well publicized series of events, the USS *Independence*, in the end, skirted the issue by bypassing Capetown altogether.

American resolve was further tested just a month later when, for the first time, pressure to actually shut down the station officially came from the South African government. This time, Premier Hendrik F. Verwoerd announced in a press release that he had told the United States it cannot employ "negro scientists in the South African stations," and that his government "would not admit American negroes if they were assigned to work at the tracking stations."[26] Verwoerd's comments on the tracking station staff seemed to most observers at the time to have been a condition, deliberately made, so as to provoke an American response. An opinion editorial came out that same week in the *South African Sunday Times* declaring that the United States would have to decide whether it can "afford morally" to overlook Dr. Verwoerd's remarks. The irony was that just three years earlier when South Africa's role in NASA's tracking network was being heralded, the same newspaper headlined "South Africa has Important Part in U.S. Moon-Shot."[27]

On the other side of the Atlantic, the station became a major target of blacks and liberal politicians who protested that the United States should not be putting money into a country with whose racial policies we do not agree. Into the 1970s, numerous congressional inquiries and hearings before the House Subcommittee on Aeronautics and Space Technology were conducted. Led mainly by prominent liberal members of the Democratic Party, these hearings aimed to determine just what exactly NASA was doing in South Africa. To that end, they looked at what the United States was doing to improve the working and living conditions and opportunities for black South Africans employed at Johannesburg. NASA administrators from Headquarters also answered questions before the House Committee on Foreign Affairs regarding the specific racial breakdown of employees, salary breakdown by race, wage practices, and NASA's hiring practices of Black Africans. (The irony was that NASA did not do any hiring in South Africa. CSIR hired African employees from an agricultural group resident in the area of the station while whites were hired through normal CSIR employment channels for technical assignment.)[28]

As hearings progressed through the mid-1960s to the mid-1970s, the issue intensified to the point where heightened scrutiny was placed on even the smallest of details, such as educational assistance, Christmas bonuses paid to whites versus negros, eating facilities and provisions for medical services. On one side of the aisle, members of the House Congressional Black Caucus, led by Representative Charles B. Rangel of New York, viewed the station as an egregious symbol of American acquiescence to apartheid. Others in Congress, led by Representative Olin E. Teague of Texas, Chairman of the House Space Committee, argued that the station was really South African, not American, since NASA did not employ any Americans in South Africa. Information gathered by NASA at the behest of Representative Charles C. Diggs of Michigan showed that, whereas blacks held about 25 percent of the jobs at the station, they received only about 5 percent of the wages paid by NASA through CSIR. In 1972, after returning from a visit to the site, Diggs reported that black employees were barred not only from the station cafeteria but from most of the technical and all of the supervisory jobs and from the technical training programs. Representative Rangel charged that gross disparities existed between fringe benefits given to white and black employees, benefits such as sick leave, vacation time, and medical benefits. To support his case, Rangel presented numbers showing that the highest paid black employee—a "skilled laboratory assistant"—earned $2,005 per year, just barely more than the lowest paid white employee—a "raw trainee"—who earned $1,930 a year.[29]

Even when there was good news for NASA regarding South Africa, it was tainted by what could only be called handwriting on the wall. In May of 1973, a House bill that would have cut $3 million of NASA funding for stations in South Africa was defeated. However, in defeat, more votes than ever before (104 to 294) were rallied. That same month, Massachusetts Senator Edward M. Kennedy introduced an amendment to cut off funds for the station, an amendment he later withdrew but only after the Senate Space Committee's new chairman, Utah's Frank E. Moss, promised to look fully into the matter in the fall session. According to Moss, the unconditional shutdown of the station would have meant that another station would have to immediately be established elsewhere. If this had to be done, the replacement cost would have amounted to around $35 million, something that would have been difficult to justify on the bill that late in the budget process.[30]

Throughout this debate, NASA consistently countered that local improvement programs which accompanied the stations were in fact making a difference. For example, the United States was, at the time, providing approximately $109,000 a year (1973 dollars) on improvement programs for the black station community. Among them was the building of houses for the African staff, at the rate of one completed every two months, and the construction of an elementary school. By 1974, 18 new houses had been constructed plus the school. Under the agreement between NASA and CSIR, the South Africans provided the initial construction funds which were then reimbursed by the United States upon completion. NASA also operated a small medical facility onsite, the services of which were made available to the Black African staff and their families. Although it was only staffed part time—a nurse was on duty three days a week and a doctor visited once a week—it was, nevertheless, one of the very few modern medical facilities in the Hartebeestpoort area that provided services to the black community, and as such, was well used. However, station critics in Congress regarded these improvements as merely cosmetic, noting that South Africa seemed not to think the station important enough to its own interests to justify making exceptions to the rules of apartheid. "The system is so unyielding," said an aide to Charles Rangel, "that if the U.S. had forced the point, South Africa would have just kicked the station out."[31]

As things turned out, Senator Kennedy did not have to wait until the fall session. After more than a decade of defending the station, on 10 July 1973, Administrator James C. Fletcher announced that it would begin pulling out of South Africa the following summer and would withdraw U.S. support entirely by late 1975. The phase-out would be done in two stages, starting first with the

DSN side in June of 1974 followed by the STADAN side after completion of the near-Earth phase of the Viking Mars missions.[32] The decision to phase out Johannesburg did not, however, signal the immediate cessation of all NASA activities in South Africa, just its tracking stations. Meteorological data collection as part of NASA's worldwide program to conduct high altitude air sampling in all hemispheres continued. Data analysis for the LANDSAT-2 satellite (in which the U.S. was one of roughly 50 countries involved) and lunar sample analysis continued for years thereafter, some even to this day.

As NASA pulled out of Johannesburg—and other stations for that matter—what to do with the equipment and hardware usually came down to two options: 1); Remove all or part of it at the Agency's own expense, with the implied, parallel responsibility to restore the site to its original condition; or 2); Dispose of the property, all or part of it, within the host country in accordance with arrangements agreed to beforehand by the two governments. In South Africa, the cost to dismantle the Deep Space portion (DSS-51) would have amounted to $643,500 with an additional storage cost of $11,060 (1975 dollars).[33] Based on this estimate, NASA determined that its real property interests at the tracking station constituted foreign excess property which had essentially no commercial value. Eventually, it was concluded to be in the best interest of the U.S. government to either donate or abandon in place the property to the South Africans. In doing so, it was mutually understood that the assets would be relinquished with the provision that no further U.S. obligation or liability remained. NASA, in essence, washed its hands of South Africa.[34]

Nevertheless, finger-pointing continued. Noting that the Agency had previously closed down two similar tracking stations in just the past year—Fort Myers, Florida, and Woomera, Australia—the Agency said that the South Africa decision was based entirely on technical requirements and was in no way a response to political pressure. Critics in Congress disagreed. "Frankly," said a spokesman for Senator Moss, "I think they just saw the handwriting on the wall, the message being that the station was becoming an embarrassment." Moss himself later released a statement praising NASA for its decision to pull out, adding "Apartheid has always been repugnant to me."[35]

In reality, NASA began planning phase out activities for the station as early as 1971. Its official position was that there would be an absence of requirements for long-period, near-Earth, Southern Hemisphere coverage after Viking left for Mars in 1975. Following that, deep space requirements could be handled by the DSN stations at Canberra, Goldstone, and Madrid. With this plan in mind, discussions were held with CSIR in August of that year to give them as much time as possible to work out staffing plans. A concern at the time for both countries' space programs was to not just abandon the station but rather, retain enough competent staff through the transition period as it moved from being a jointly sponsored site to one that was fully South African.

In the phase out discussions with CSIR, the fate and future of the Black African staff were, in fact, discussed at length, down to the number of Black Africans which might remain employed after NASA relinquished funding. There was particular concern on NASA's part that Black African staff would be declared "redundant" and whether they would be treated equitably relative to the white staff. At Headquarters, Gerald Truszynski, in his discussions with Dr. Frank Hewitt of CSIR, felt that the South Africans appreciated the American position, with Hewitt saying he "reflected a genuine concern for the future of this group."[36]

A legacy of these discussions was that it led the South Africans to implement several policy changes with regards to Black station staff members. One had to do with the pension they were receiving. At the time, the African staff members were covered by a different benefits plan that was generally (and obviously) inferior to that of the white staff. This "Provident Fund Plan" was soon changed so that the same formula was used in calculating the pension for all staff members. In addition, after these changes were made, CSIR allowed the Black African staff who were declared "redundant" to, where appropriate, continue occupying their houses, thereby enabling them to look for other employment before moving their families off the station site. Arrangements were also made for CSIR to provide a vigorous outplacement service and reemployment counseling. On the other hand, the one service which the South Africans did not continue after NASA ceased its funding was the secondary school bursary program which the Agency had started. CSIR deemed this to be outside of their normal responsibility and charter as they had many Blacks employed in their agency's other activities who were not receiving any educational assistance.

In the end, two-thirds of the Black staff (39 out of 59) were released after NASA pulled out of Johannesburg.[37]

※　※　※

As a principal site in the Southern Hemisphere, Tananarive (TAN) had been busy, supporting a host of science satellites as well as all the Gemini, Apollo and Skylab missions. The routine began to change in 1972 when the Malagasy government underwent a series of political upheavals. In May of that year, the president of the ruling Social Democratic Party, which had been in power since Madagascar first gained independence from France in 1960, resigned under political pressure. The unrest continued over the next three years, culminating with the brutal assassination of the military dictator which put the country under martial law in February of 1975. Before long, a new Marxist regime was formed under the leadership of a 38-year-old revolutionary named Didier Ratsiraka. Under President Ratsiraka, known in the region as the "Red Admiral," the government became highly centralized and committed to revolutionary socialistic ideals. (These policies did not change until the 1990s only after the formation of new political parties.)

One of the first foreign policy changes that Ratsiraka made was to impose a rent on the United States for operating a NASA ground station in his country. In the original memorandum 20 years earlier which both countries signed establishing a site near Majunga, it was agreed upon that there would be no exchange of funds and no rent exacted for use of land. But now, Ratsiraka was demanding $1 million per year, *retroactive to 1963* on back taxes. This was a demand that the United States obviously could not agree to.

Negotiations were conducted but to no avail. A few weeks later, the Supreme Council of Revolution of the Malagasy Republic forced the station closed. This action came upon NASA unexpectedly. During that time, GSFC was still improving on the station and in the process of adding a Unified S-band antenna. Under the guise of avoiding "possible maneuvers of sabotage," President Ratsiraka immediately placed it under military control. The Station Director and Bendix workers with their families were allowed to evacuate, but all equipment had to be left behind. At the time of closure, there were the two Goddard appointed NASA employees and 50 Bendix workers, along with their dependents—148 rather apprehensive Americans total—at the station.[38]

With the abrupt shut down, Goddard had to make some quick changes in order that support for Apollo-Soyuz, which was to launch the very next day, would not be disrupted. They improvised by tasking the geosynchronous ATS-6 to serve as a data link. Workload from other satellites was shifted to other stations. ARIA instrumentation aircraft and the *Vanguard* were repositioned to help support other launch activities out of the Eastern and Western Test Ranges. These changes resulted in some temporary scheduling problems but otherwise proved adequate and Apollo-Soyuz went on to be an unparalleled success.

Over the course of the next five years, the Malagasy government periodically allowed NASA back into the country to remove equipment. On 3 April 1980, the last of the remaining hardware that NASA still wanted was removed from Tananarive. By diplomatic note, the remaining U.S. property was turned over to Madagascar the next day. This note, which was actually received by the U.S. Embassy the previous October, expressed essentially an agreement on the list of equipment NASA would remove and the monetary settlement. The removal process, in effect, was the final act that brought to an end five years of negotiations by the State Department to repatriate NASA equipment following the forced closure of the station.[39] Besides the stress and disruption experienced by the staff and their families, the closure also had an effect on NASA in terms of operating cost. After Tananarive was shut down, the *Vanguard* was called on to fulfill some of its requirements. In the mid-1970s, the annual cost to operate a tracking ship was quite high, about $6 million per year. By contrast, a land station like Tananarive cost around $2.8 million, or less than half that of a ship.[40]

* * *

One of the first actions in the reorganization for STDN took place at the Goddard Space Flight Center where management of both networks was consolidated as early as May of 1971. Two new directorates, the Mission and Data Operations (M&DO) Directorate and the Networks Directorate (ND), replaced the Tracking and Data Systems (T&DS) and Manned Flight Support (MFS) Directorates. In this new arrangement, divisions within the M&DO managed the data processing activities and the computing requirements of the network. The ND became responsible for operation of all the STDN elements, from NASCOM to the ground stations and the satellites. In a harbinger of things to come in the 1990s, it was at this time that the Networks Directorate formed the Network Office for International Operations, which allowed GSFC to start handling some of the foreign policy work that up until then had rested exclusively in the domain of NASA Headquarters in Washington, DC. This move seemingly made sense at the time, coinciding with preparations for Apollo-Soyuz which demanded a lot of technical interaction with Soviet Union network engineers at the working level.

Although STDN was considered a new network (or at least a greatly retooled old network), much of the way in which it was run continued as before, including usage of acronyms—a well known hallmark of NASA. The MCC in Houston continued to serve as the focal point during human space missions with the responsibility of directing all ground stations when a flight was in progress. Meanwhile, the NOCC (Network Operations Control Center) in Greenbelt continued in its role of controlling the network including overseeing all network preparations leading up to the launch of a human space mission.

In this capacity, NOCC engineers monitored console displays and established direct voice links amongst all the mission elements such as the launch site at the Cape, the network ground stations and the appropriate Project Operations Control Center (POCC). These POCCs that began emerging during the previous decade were essentially individual operation control centers at the GSFC that were built to specifically control certain types of satellites such as the Applications Technology Satellite (ATSOCC) or the Orbital Astronomical Observatory (OAOOCC). Not to be left out were the multimission "umbrella" centers such as the Multisatellite Operations Control Center (MSOCC) and the Mission Operations Control Center (MISSOC) that scheduled network support for all classes of satellites and assigned each station a weekly list of satellites that were to be monitored.

To accommodate the data downlink from the new generation of satellites that were now nothing short of orbiting laboratories, Goddard enhanced the STDN with increasing centralized capabilities. A typical change was the greater reliance on electronic data transfer methods in the late 1970s with the implementation of systems such as the Telemetry Online Processing System (TELOPS), that eliminated the need to ship magnetic tapes from the field stations. Raw data was, instead, transmitted over communication lines to a dedicated storage system at GSFC.

The POCC themselves also continued to acquire improved technology, permitting scientists stationed at Goddard to manipulate the orbits and attitudes of satellites with greater ease. Take for example, as a progenitor of touch-screen technology, a scientist seated at the IUE Operations Center. This person could, by simply pointing a light-pen at a specified portion of a video display, swing the IUE telescope around to look at another part of the

The Goddard Space Flight Center has been home to many Project Operations Control Centers over the years. Shown here is the Space Telescope Operations Center where commands to the Hubble Space Telescope (HST) originate and where its systems are monitored. The picture was taken in December 1999 as engineers monitored activities during the telescope's third repair mission. Today, command and control of the HST is done mostly from the Space Telescope Science Institute in nearby Baltimore.(Image courtesy of NASA, available at *http://hubblesite.org/gallery/spacecraft/01/*)

sky. In a move exemplifying NASA's continual effort to obtain ever better quality videos, scientists supporting Earth resource and remote sensing satellites also received a special Image Processing Facility (IPF) at Goddard that provided video and pictures that were continuously corrected for distortions introduced by spacecraft equipment or during transmission—a progenitor to today's high definition television transmissions.[41]

Stations reported back each week to the MISSOC on the performance of their satellite coverage. To measure performance, matrices were set up and grades given on how well stations were doing their jobs, whether excellent, good or poor. The success rate for each station was measured by how many passes were supported and how much low bit-error data was captured. This was then compared to how much *could have been* or *should have been* captured. The focus was primarily on the amount of data captured on satellite passes and not on cost of operations. This may not have been the best way to gauge how well a station did. Former Quito Station Director Charles Force said, "I thought at the time that was a mistake. You need to have some kind of a balance between how much you are spending and how well you are doing the job. But it was totally focused on how well the job was being done."[42]

The system was also not without its flaws. Force remembered an example that always puzzled him:

> It was my first exposure to performance evaluation using a matrix and what I learned from that personally, was that you have to be very certain the matrix measures what you want. One example I remember. If a satellite came over the horizon and the station was tracking it and the receiver failed, and they start losing data, they would then be scored down so [that] if they got the receiver up before the end of the pass and covered the balance of the pass, they got a lower score than if they didn't get the receiver back up.

In other words, the station would actually get penalized if it successfully recovered from the receiver failure than if it hadn't. "That was the system," said Force. "That was idiotic why it was set up that way. I have no idea because I was not in on the early days of the STADAN."[43]

While the original STADAN side of the house relied on these metrics to grade the performance of its stations, no such matrix was used on the MSFN side of the house. This apparent dichotomy in the way the two networks operated prior to their merger can, in large part, be traced back to the way the two networks came about—and the competition that followed. For years, while both the STADAN and the MSFN were run by Goddard, they were separate, up through the directorate level. Specifically, "Code 800" ran the MSFN while "Code 500" ran the STADAN.

From the onset, there were cultural differences, and with it, friction. Some would go so far as to call it jealousy. While Code 500 dated back to Jack Mengel and the team that created Minitrack, Code 800's heritage was basically an offshoot of Langley—a lot of the people actually came over to Goddard from Langley. Because of this, the two networks had different heritages and different cultures in everything from the way they operated to how people were used in the field. STADAN stations, for example, were generally more "remote" in the way they operated in the sense that data was gathered and sent back to Goddard where it was then assimilated and processed. Thus, there was a lot of effort by technicians and engineers physically at GSFC running the computers. Unlike today where desktop and notebook computers can be found in every office, this way of centralization made sense at a time when mainframe computers were required to do the massive calculations. These mainframes were expensive, to put it mildly, and required a fair amount of maintenance. Therefore, the STADAN had a lot higher percentage of its technical expertise stationed at Goddard in proportion to the field. This was exactly the opposite of the MSFN, that had more computational capability in the field and, therefore, had more of its share of expertise assigned to the field.

There has been the conception throughout the years that by nature of its mission, the MSFN was somehow more glamorous and had a higher profile (public exposure) than the STADAN, and therefore, got more attention and resources. This was, in all likelihood, exactly what happened. It was an undeniable fact that the MSFN received more attention than the STADAN in the one area where it most mattered: funding. In 1968, for example, two-thirds of the budget for Goddard's tracking operations went to the MFSN whereas one-third went to the STADAN.[44]

It's been said that where the money lies, so lies the priority. This apparent inequity was well recognized and unfortunately, resented within the STADAN system. An "us-and-them" attitude developed in many circles. As Force put it, "They didn't talk to each other that much."[45] The presence of this "sibling rivalry" is probably not too surprising considering the diversity of the people and their talent that was (and is) the Goddard family.

The years have shown that while such differences and strong feelings existed, they were worked out and the STDN moved on. The network that came out of it was a far better and more efficient network than before. Force would later say compellingly of the big picture, "The people did work together [and] the job was done successfully. There were an awful lot of good people that did work together and there was an awful lot done right and successful."[46] Indeed, the ensuing three decades have proven that.

By 1975, the merging of STADAN and MSFN was complete. Table 6-1 is a glance of the reorganized network in the mid-1970s.

Over time, these sites adjusted to the changing demands of the integrated network to support tracking of both human spaceflight missions

Table 6-1: Ground Stations of the Spaceflight Tracking and Data Network in 1975[47]

Station (Location)	Call Sign	Latitude Longitude	Year Established	Year Phased Out	Original Network	Primary Capabilities
North America						
Alaska (near Fairbanks)	ALASKA	64°59'N 147°31'W	1962	1984 (Transferred to NOAA)	STADAN	GRARR, MOTS*, SATAN 40, 45, and 85-ft dish antennas
Goldstone (Mojave Desert, California)	GDS	35°20'N 116°54'W	1967	1985 (Turned over to DSN and re-designated as DSS 16 & 17)	MSFN and STADAN	30 and 85-ft USB
Merritt Island (Kennedy Space Center, Florida)	MILA	28°25'N 80°40'W	1966	Still operating	MSFN	30-ft USB C-band radar
Network Test and Training Facility (Goddard Space Flight Center, Maryland)	NTTF	38°59'N 76°51'W	1966	1986 (Transferred to Wallops)	MSFN and STADAN	30 and 59-ft antennas
Rosman (North Carolina)	ROSMAN	35°12'N 82°52'W	1963	1981 (Transferred to DOD)	STADAN	Two 85-ft antennas GRARR; three SATANs, MOTS ATS telemetry and command
White Sands (New Mexico)	WHS	32°21'N 106°22'W	1961	Reestablished as nearby TDRSS ground terminals	MSFN	C-band radar VHF voice
Pacific						
Guam	GWM	13°18'N 144°44'E	1966	1989	MSFN	30-ft USB
Hawaii (Kokee Park, Kauai)	HAW	22°07'N 157°40'W	1961	1989	MSFN	Two yagi command 14-ft antennas C-band radar 30-ft USB

*While the MOTS cameras remained at the stations, they were no longer required for calibration after phaseout of Minitrack. They continued to be tracking devices for geodetic research, photographing the Pageos spacecraft in the mid-1960's. By the 1970's, MOTS was no longer an operational network.

continued on the next page

Station (Location)	Call Sign	Latitude Longitude	Year Established	Year Phased Out	Original Network	Primary Capabilities
South America						
Quito (Ecuador)	QUITO	00°37'S 78°35'W	1957	1982	MINITRACK	40-ft antenna SATAN, three Yagi command, MOTS
Santiago (Chile)	AGO	33°09'S 70°40'W	1957	1989	MINITRACK	40-ft antenna GRARR SATANs (2 receive, 1 command) Yagi command, MOTS
Atlantic						
Ascension Island	ACN	7°57'S 14°35'W	1967	1989	MSFN	30-ft USB (Also had a 30-ft DSN antenna which was phased out in 1969)
Bermuda	BDA	32°15'N 64°50'W	1961	1998	MSFN	C-band radar 30-ft USB
Europe						
Madrid (Spain)	MAD (RID after 1984)	40°27'N 4°10'W	1967	1985 (Transferred to DSN and re-designated as DSS 66)	MSFN	85-ft USB
Winkfield (England)	WNKFLD	51°27'N 00°42'W	1961	1981 (Turned over to the British)	STADAN	14-ft antenna SATAN, MOTS Yagi command
Australia						
Canberra (Honeysuckle Creek)	CAN	35°24'S 148°59'E	1966	1984 (Moved to Tidbinbilla, transferred to DSN and re-designated as DSS 46)	MSFN	85-ft USB
Orroral Valley	ORR	35°38'S 148°57'E	1965	1984 (Turned over to the University of Tasmania)	STADAN	85-ft antenna Two SATANs Yagi command MOTS

A "Spanish Watchman" on the hills overlooking NASA's Madrid Spaceflight Tracking and Data Network Station and its prominent 26 meter (85 foot) Unified S-band antenna. The station at Fresnedillas, some 50 kilometers(30 miles) west of the city of Madrid, was the NASA ground station that tracked *Eagle* to the lunar surface on the historic flight of Apollo 11. The station was phased out and transferred to the nearby Deep Space Network site at Robledo in 1985. (Photo courtesy of Larry Haug and Colin Mackellar.)

and applications satellites. From a purely technical standpoint, the augmentation of former STADAN stations with Unified S-band hardware was the biggest single improvement that NASA took to provide a common capability across the STDN stations. By the mid-1970s, this had been done at the Goldstone Apollo site (1972), Fairbanks (1974), Orroral Valley (1974), and Santiago (1974).

* * *

How busy a station was depended on how many spacecraft it was assigned to track. It was not necessarily true that the largest and best-equipped stations were the busiest. It all depended on where a particular ground station was located. For example, it may not be surprising that Fairbanks, Rosman,

Canberra, and Goldstone—all home to one or more 26-meter (85-foot) antenna—were normally on four-shift, 24/7 operations. Contrast that with the Madrid Station at Fresnedillas. It was also among the best-equipped stations in the network boasting its own 26-meter system, but it only operated on a two-shift basis. Moreover, older stations like Quito, Santiago, and Winkfield—which could all trace their roots back to the old Minitrack days—still operated on four shifts, even well into the 1970s.

This variation in scheduling of workload had everything to do with the numbers and types of Earth science satellites that a station was called on to monitor. When one talked about the largest number of different satellites that a given station supported, Johannesburg and Orroral immediately came to mind. Each monitored 30 or more satellites in the mid-1970s. Fairbanks and Winkfield were not far behind at 24 and 22, respectively. Overall, NASA's STDN provided coverage to some 50 different satellites in the 1970s.[48] In December 1975, the GSFC made a familiar move by awarding Bendix a two year, $104 million contract—with provisions for three additional one-year extensions—to continue its role as the prime operator of the network into the decade of the 80s and the age of the Space Shuttle.[49]

Even though American human space presence clearly saw a period of quiescence in the mid to late 1970s, science and application satellite activities continued to flourish. One area of research in which Goddard satellites and the STDN played a leading role was in tectonics—the study of the structural deformation of Earth's crust. On 4 May 1976, LAGEOS 1—the LAser GEOdynamics Satellite—was launched on a 50-year, high inclination Earth-orbit mission to study the geophysical behavior of our planet. The idea behind the mission was that long term data received from the satellite could be used to monitor the motion of Earth's tectonic plates, for example, and to measure the gravitational field and nutation (wobble) in the axis of rotation.[50]

In this activity, the exact position of the STDN on the surface of Earth was itself a piece of scientific data. Most tracking is done under the presumption that the location of the ground station is known and that tracking determines where the satellite is with respect to the ground station. In laser tectonics, the logic is reversed: the location of the satellite is known. What is desired is the exact location of the ground station. The LAGEOS satellites, covered with tiny "retroflectors," reflect laser beams transmitted from various ground stations. Covering the two-foot spherical satellite were 426 cube-corner reflectors made of fused silica glass and the heat tolerant element of germanium. By measuring the time between transmission of the beam and reception of the reflected signal from the satellite, stations on Earth can thus precisely measure the distance between themselves and the satellite. The accuracy obtained is extremely high, with distance measurements correct to within one to three centimeters, or about an inch.[51]

To return the laser data to Earth, Goddard Space Flight Center came up with the Goddard Laser Tracking Network, or GLTN, in 1975. The GLTN functioned as somewhat of a "mini network," same as the STDN but using laser instead of radio frequency signals—essentially an optical system. A laser at a GLTN station would emit a beam to the satellite which would then be reflected and returned to the station. The interval between the start of the transmission and the receipt of the return signal was recorded at the GTLN station and multiplied by the speed of light to obtain the precise distance between the station and the satellite. In this way, ever subtle changes over time in the satellite-to-station distance painted a picture of motion in Earth's tectonic plates. As a result, significant information concerning fault line movements and the dynamics of earthquakes, for example, could be deduced.

Because of tectonic science requirements, the design of GLTN stations also had to emphasize mobility. During the 1970s, most of these sites were configured into Mobile Laser Ranging Systems (MOBLAS), with each MOBLAS housing the required hardware in three instrumentation vans. By the end of the decade, the pace of laser ranging activities had picked up to where MOBLAS units had been deployed to diverse locations such as Bear Lake, Utah; Quincy and San Diego, California; and places in Australia and around the Pacific and Indian Oceans. A fixed site, meanwhile, operated on the grounds of Goddard in Greenbelt, Maryland. Like its STDN counterpart the Network Training and Test Facility, this fixed station essentially acted as a test site for development of additional MOBLAS units. To achieve even more mobility, Goddard engineers soon developed a second generation system called the Transportable Laser Ranging System, or TLRS. Instead of large instrumentation vans, the TLRS consisted of relatively small, box-like transportable units that were readily borne by trucks and aircraft. Their performance was even better than that of the MOBLAS. By 1990, these laser tracking systems had achieved astonishing ranging accuracies, down to the sub-centimeter level, a must for measuring the slow and virtually indiscernible movement of Earth's crust over time. The work continues today with the MOBLAS and the TLRS terminals still operating at Goddard.[52]

* * *

In 1977, the NASA began a series of low altitude, atmospheric glide tests of the Space Shuttle. These Approach and Landing Tests (ALT) took place at the Dryden Flight Research Center near Edwards Air Force Base, California from February through November. The ALT was the first step in the flight qualification of the Shuttle Orbiter, verifying its flight worthiness as it glided to a landing after returning from orbit. Testing began with three ground taxi runs of the Shuttle *Enterprise* mounted atop a Boeing 747 Shuttle Carrier Aircraft—a highly modified Boeing 747-100—to determine loads, control characteristics, steering and braking of the mated vehicles. This was followed by eight so-called "captive-flights" of the *Enterprise* (five uncrewed, the last three crewed) attached to the 747 to evaluate the structural integrity and aerodynamic performance of the mated pair in the air. Five free-flights of the Orbiter concluded the test program.

To support the ALT, Goddard set up a special mobile telemetry station in 1975 at Buckhorn Lake (a dry lake) on a hill overlooking the landing strip. Buckhorn (BUC) was a fairly simple ground station, with transportable equipment consisting of two 4.3-meter (14-foot) antennas and C-band radar, equipment in part used previously at Grand Bahama during Apollo. Trailers housed UHF air-to-ground voice and S-band telemetry equipment. After supporting the ALT and seven Shuttle orbital missions, Buckhorn was closed-out in 1983 following the STS-8 night landing. One of its 4.3-meter antennas was permanently transferred to the nearby Dryden Flight Research Facility while the other hardware was put back into the STDN equipment pool.

Three and a half years after ALT ended, the STDN tracked the first Space Shuttle into orbit. After a six year hiatus, America finally returned to space, this time ushering in a new era in space transportation with the launch of the Shuttle *Columbia* on STS-1 the morning of 12 April 1981. As Commander John W. Young and Pilot Robert L. Crippen lifted off from pad 39A at the KSC, long-time Flight Director Christopher Kraft called it the most tense moment in all his years at Mission Control. Never before had NASA flown a crew on the very first launch attempt of a new rocket. Boosted by the largest solid rocket motors ever made, the entire Shuttle stack cleared the launch tower within seconds, a surprise to those at Mission Control who remembered the painstakingly slow liftoff of the mammoth Saturn V just a decade earlier.

The 2,500 men and women of Goddard's STDN had prepared six years for the launch.[53] Even though in the eyes of the public, little activity had come from the space agency since the mid-1970s, it was quite a different story behind the scenes. Much had improved. Station equipment had been upgraded to accommodate the new multi-channel S- and Ku-band communication system of the Shuttle. Telemetry rates from tracking stations had increased to 128,000 bits per second (128 kbps) in real time versus the 14 to 21 kbps of the 1960s. Telemetry streams were transmitted to the JSC in real time on three 56 kbps circuits.[54] A key communication change was the implementation of S-band air-to-ground voice circuits in addition to UHF radio capability. The Shuttle continued to use the UHF air-to-ground voice system but the USB system of tracking, telemetry, and command developed for Apollo was now expanded to include two-way voice. This required the development and installation of equipment to digitize and multiplex voice on the command channel and de-multiplex voice from the telemetry channel. Voice was now multiplexed with commands for uplink and downlinked with telemetry. The links were then de-multiplexed, converted to analog voice and redirected to the MCC in Houston. The more than two million circuit miles of NASCOM lines continued to be upgraded, relying on domestic and international land lines, submarine cables, commercial satellites, and microwave radio systems (not unlike relaying of cell phone signals today) to interconnect the overseas stations with the launch facility at Kennedy and the MCC in Houston.[55]

In contrast to earlier human space missions when the need for voice contacts steadily declined, the Space Shuttle program imposed a goal of 30 percent voice contact on each orbit to monitor things like critical ascent and reentry events, payload delivery tasks and on-orbit crew science activities. Studies done in the late 1970s, however, showed the STDN could provide voice support for only about 23 percent of the time. Gaps in voice coverage had to be addressed. As with earlier renditions of the network, international cooperation once again held the key.

The Dakar Station in Senegal on the western most point of the African continent looking towards the Atlantic was an ideal location to track the Shuttle's ascent into orbit on eastern launch azimuths. The arid setting was typical of west Africa as were the facilities. Clearly visible are the 4.2-meter (14-foot) USB antenna on the right and the quad-helix command antenna on the left. During Shuttle missions, DKR was staffed by about a dozen NASA contractors and Senegal workers. (Photo courtesy of Gary Schulz)

Discussions and diplomatic negotiations resulted in new stations at several important locations, each equipped with UHF air-to-ground voice systems to fill coverage gaps. As B. Harry McKeehan, former Chief of International Operations at Goddard who spearheaded many of these talks put it, "These countries gave us their full support," even in places such as Pakistan, where NASA operated a Landsat station near the industry center of Rawalpindi.[56] In June 1982, a station was added in Dakar (DKR), Senegal—the western most point of Africa—to support the Shuttle's Orbital Maneuvering System first burn (OMS-1), a critical event in the ascent to orbit timeline where a decision as whether or not to continue onto orbit or to initiate the Abort-Once-Around (AOA) sequence had to be made. DKR also provided an additional contact point for each orbit once the Shuttle was safely in orbit. In addition, Dakar served as the early Transatlantic Abort (TAL) landing site.

Also established in Africa was the Botswana Station (BOT) at Gaborone. Also called Kgale, it was added in 1981 primarily to cover the OMS-2 circularization burn. Since the Johannesburg Station was closed-out in 1975, Botswana assumed many of the functions formerly handled by the South Africa station. In the archipelagos of Seychelles, NASA called on the Air Force, tasking their 18-meter (60-foot) antenna to serve as the Indian Ocean Station (IOS) some 1,100 kilometers (700 miles) northeast of the Madagascar coastline.[57] Yarragadee (YAR) in Western Australia, was added in 1980—just prior to the launch of STS-1—to provide coverage for the Shuttle's deorbit burn and reentry.[58]

Rounding out the changes to the STDN required for Shuttle support was a 4.3-meter (14-foot) antenna atop the 1,980-meter (6,500-foot) Tula Peak (TULA) in 1979. Situated on the grounds of Holloman Air Force Base just outside the gates of White Sands Missile Range, New Mexico, TULA alleviated the approximate, 10-minute communications gap during each Shuttle orbit over the southern United States.[59]

By July of 1982, the STDN reached its zenith in terms of the number of ground stations (20 stations). For Shuttle support, it had acquired new outposts in the United States, Africa, the Indian Ocean and Australia. Network engineers had greatly enhanced its capabilities by incorporating the latest data processing and transmission innovations of the 1970s. The result was an unprecedented network with 10-fold increase in telemetry and data handling capacity over that of Apollo just a decade ago.

Despite the tide of innovation and streamlining that went into the leaner and more efficient STDN of the 1980s, NASA could not totally offset the rising cost of station operations. The network remained manpower intensive. This made operations highly susceptible to inflation, not only in the U.S., but even more so overseas, where the impact of wage escalation was even more stifling. Remote, often isolated locations were especially burdensome on cost, staffing and maintenance. Spiraling cost at certain locations—particularly Spain, Australia, Goldstone and Alaska—could no longer be ignored. The prime culprit—wage increases—had led to a 15 to 20 percent jump in operating cost at Madrid and the Australian stations, and increases of 13 and 25 percent, respectively, at Goldstone and Fairbanks. This equated to a 6 percent across the board cost increase over the entire network. While this was not huge, in the cost-conscious days following Apollo, and with double digit inflation of the late 1970s, it was enough for the space agency to begin closing down more ground stations.[60]

One of them was Rosman in North Carolina. Among the best-equipped of the original STADAN sites, Rosman had been supporting ATS-6 which was no longer operating by December of 1979, and the OAO which had completed its mission in November 1980. Although NASA pulled out of Rosman in January 1981, five years earlier it had been the target of a well-publicized (at least among the locals working there), rumored-closing. The 1976 events did not stem from technical reasons though. That year, a labor dispute arose between employees of Bendix and their company with respect to a collective bargaining agreement.

What happened was this: the station employees, who were represented by the International Brotherhood of Electrical Workers (IBEW), on 25 February of that year commenced negotiations with Bendix on their labor agreement which was soon to expire. But despite numerous meetings, an agreement could not be reached and so the union went on strike. While labor disputes were not uncommon, this strike caught the attention of the local North Carolina residents who began to feel that NASA may be thinking of closing down "their" station because of the dispute. Although the Agency really had no such intentions and (by law) had to leave IBEW and Bendix to work out their differences, state and U.S. representatives from North Carolina soon, perhaps in somewhat of a panic, got into the fray and began questioning NASA on its "true intentions."

Only after an official letter from NASA Headquarters was sent to Congressman Roy A. Taylor clarifying the Agency's position that the strike would not affect the status of the station did the rumors begin to fade. In the end, the dispute was resolved when IBEW accepted a new labor offer from BFEC which included a 15 percent wage increase—not bad considering that the average wage settlement for all major collective bargaining agreements negotiated in the U.S. during the first quarter of 1976 was just 8.8 percent.[61]

Six years later, though, dwindling pass requirements did cause NASA to really leave Rosman. But instead of just shutting it down, the DOD received authorization from Congress to assume operations of the tracking facility. As for the workers, all 119 Bendix employees assigned to the station were offered jobs elsewhere within the company. Some remained, others did not. Of the 119, 30 transferred to other Bendix locations while 34 were retained by the DOD at Rosman. Fifty-five others declined to accept employment elsewhere and were terminated.[62]

Like Rosman, other stations were also reassigned, either within NASA or to another agency. In 1974, the NTTF became part of the operational network—Greenbelt or BLT—and expanded to take on responsibilities for NASA's IUE. This ended in 1986 when the Center decided, after the deactivation of IUE, to align all support activities at Wallops Island off the coast of Virginia. With the decision made, the 12-meter (40-foot) antenna used on IUE was given to the nearby United States Naval Academy and the 9-meter (30-foot) USB system moved to Wallops. Following this decision, this rather unique facility reverted back to its original role of serving the network as a test bed and training center.[63]

Another case in point was in Alaska after LANDSAT-3 went inoperative in 1983. On 30 September 1984, NASA operations at Fairbanks, Alaska ceased when it granted the National Oceanic and Atmospheric Administration (NOAA) a temporary-use permit to operate the station to track weather satellites. The polar orbiting Landsat was by now using TDRS-1 for support and Goddard no longer had any pressing requirements for Alaska. Being in Alaska, the station was one of the most expensive stations to operate. NASA nevertheless continued to provide operations and maintenance support to NOAA for the next four years (at a cost of $1,920,000) until a permanent transfer was finally granted by the Bureau of Land Management.[64]

In 1985, Alaska's remaining 26-meter (85-foot) antenna was transferred to the Jet Propulsion Laboratory for continued support on Nimbus and the Dynamic Explorer satellites. At the time of the station's closing, NASA had invested $12 million of capital equipment in Alaska which, after transferring to NOAA, brought to an end 26 years of NASA operations there.[65] The Agency's absence from the state, however, would be rather short-lived as it would soon return to the area, this time to conduct scientific research, activities which continue to this day.

Other sites were closed out in a more permanent way. In November 1981, one of the most venerable stations in the network came to an end. During STS-2, the second flight of the Shuttle *Columbia*, the Quito Station in Ecuador was shut down as planned. Fiscal belt-tightening and steep foreign inflation rates often overpowered the international cooperation value of keeping an overseas station open. (Another consideration was the balance between international cooperation and the desire for more "U.S. territory-based solutions.") One of the original Minitrack sites, the station was located near Mount Cotopaxi, the highest active volcano in the world, 56 kilometers (35 miles) south of the Ecuadorian capital and had served faithfully as a key Southern Hemisphere station dating all the way back to 1957 and Sputnik. Bendix had operated Quito, along with the Ecuadorian Services Company

(ESCO) who provided subcontractor services, since 1961. Closing it saved NASA an estimated $4 million annually.[66]

Quito exemplified the "international value" of a NASA overseas station. It showed how the seed of a NASA station in a foreign country germinated to eventually became a technological national resource for that country, a resource that endures to this day. Ecuador's main product is agriculture. In the late 1950s, there was an economic need for companies willing to enter into other fields such as technology and oil. In 1960, a group of Ecuadorian executives—visionaries in hindsight—led by Carlos H. "Polo" Cadena founded ESCO. BFEC soon awarded ESCO a subcontract to help operate the Quito STADAN station.

The station grew as NASA grew. It upgraded from Minitrack to a three-link station in the mid-1960s. The compliment of Ecuadorian nationals bloomed from 50 to 220. With the consolidation of STDN, it expanded from supporting only application satellites to human spaceflight support on ASTP continuing on to the Space Shuttle. This was a giant step forward for an overseas station, one that required a dynamic and joint managerial effort by Bendix and ESCO. A transition from American station staff to Ecuadorians took place and Goddard implemented its training and certification program with outstanding results. Cadena himself was a strong proponent of "station nationalization" and firmly believed that it was in the best interest of his employees.[67]

The Ecuadorian government had designated the Esceula Politecnica Nacional as NASA's cooperating agency, responsible for facilitating and monitoring the Agency's activities in that country. In the early 1970s, its Director, José Rubén Orellana, expressed dissatisfaction with NASA's integration of Ecuadorian nationals into the station staff, as provided for in the international agreement. In response, NASA brought in new station management: Charles Force was transferred from Guam while Bendix named Cliff Benson as their new Senior Manager. They quickly determined that Orellana's charges were valid and moved aggressively to remedy the situation. Over the next two years, over 50 Bendix personnel were replaced with Ecuadorian nationals. The willingness with which people like BFEC Logistics Supervisor Harry Bailey trained Fabian Mosquera as his replacement, for example—not knowing where he himself would go next—was impressive! A year or so later even Benson himself was replaced with a national: Julio Torres. Open animosity, while it did exist, was infrequent as Bendix management understood their role and made every effort to place their workers in other positions with other parts of the company. While operational performance of the station had previously been quite good, it improved even more under the new personnel, and operations costs were simultaneously reduced.[68] In 22 years, Quito provided over half a million hours (578,160 hours to be exact) of direct mission support, one of the highest in the STDN. Numerous performance awards were bestowed by both NASA and Bendix.[69]

A panoramic view of the Quito Station in 1973. The station was located at 3,650 meters (12,000 feet) elevation 69 kilometers (43 miles) south of the Equator, at the base of Mount Cotopaxi. A herd of llamas that frequented the station is slightly visible grazing just left of the 12-meter (40-foot) USB antenna. The deactivated Minitrack antenna is visible in the background between the two larger buildings. The photograph is unusual because Cotopaxi is cloud-free, and because of the rare vantage point—from the top of a communications tower along the nearby Pan American Highway that was accessible only by climbing 30 meters (100 feet) up an open ladder. (Photograph by Charles Force)

At 7:04 a.m. local time on 14 November 1981, as astronauts Joe H. Engle and Richard H. Truly passed over Mount Cotopaxi for the final time, they expressed their appreciation to the 75 station employees. Words of bittersweet thankfulness also went to the Quito crew from a host of Agency officials, including: Robert E. "Ed" Smylie, NASA's Associate Administrator for Tracking and Data Systems; John H. McElroy, Deputy Director of the Goddard Center; Richard S. Sade, NASA's Director of Networks; Mike Stevens, the Shuttle Network Manager; Walt LaFleur, Deputy Director of Networks; and Daniel A. Spintman, Chief of the Goddard Network Operations Division.[70]

The last formal agreement with Ecuador came on 4 December. On that day, the State Department authorized the U.S. Embassy to exchange

As large as the 26-meter (85-foot) antennas of the STDN were, they were dwarfed by the 70-meter (230-foot) dishes that the DSN uses to communicate with spacecraft at the outer reaches of the Solar System. This photographic rendering drives home the size of these dishes at Canberra, Goldstone and Madrid. (Photograph courtesy of NASA)

notes with Quito extending the agreement for another six month to allow NASA to perform "cleanup work" completely closing-out the station. Station equipment was transferred to Dakar, Senegal, which at the time was just being established as the Transatlantic Abort emergency landing site for the Space Shuttle. On 1 July 1982, the facility was transferred to the government of Ecuador who, in turn, assigned the CLIRSEN agency the responsibility for its operations. It has been used since to support Earth science data acquisition and regional land management and development. A number of nationals who started at the Quito Station have gone on to play an important role in the industrial development of Ecuador.[71]

Also closed during this time, with no fanfare, was the small station atop Tula Peak near Alamogordo, New Mexico. Its relatively light workload allowed it to be phased out and its responsibilities reassigned to other STDN sites. From a scheduling perspective, its impact was small, less than 5 percent coverage for most scientific application satellites. The loss to Space Shuttle support was even less at 3 percent and none involved mandatory or mission critical events. TULA had only been operational for less than three years but closing it would save the Agency half a million dollars a year. Just four and a half months after its closing, however, Tula Peak had to be quickly reactivated—literally overnight—to support a contingency landing of the Shuttle *Columbia* at White Sands. Due to wet ground conditions at Edwards in California, STS-3 was diverted to New Mexico (the KSC was still unavailable for Shuttle landings in 1982). Getting TULA up and running in just over 24 hours was a rather impressive feat of logistics and field engineering, a feat that once again demonstrated the "badgeless" teamwork of those who made the STDN possible.[72]

This steady phase out of the ground network continued through the 1980s. In 1981, NASA transferred ownership of its only station in England—the Winkfield Station at Berkshire—to the British, who having operated it since its establishment in 1961, continued to use it for radio research. Also realized in the big picture was the long-planned consolidation in 1985 of STDN stations at California, Australia, and Spain with their DSN counterparts. Under the reorganization, STDN capabilities were retained but now as part of the DSN. They would still be used to support the Agency's near-Earth and highly elliptical orbiting spacecraft but would be run out of the Jet Propulsion Laboratory. The thinking was that by combining the capabilities in each geographical location, more efficient use of the facilities could be realized.

First to be realigned was Goldstone. Of all the ground stations at Goldstone, only one, the Apollo Station (GDS) built in 1967, was originally part of the STDN; all others were original DSN equipment. To meet tracking requirements on the Apollo program, DSN assets were used as a wing-station, modified for USB operations and tasked to support the primary Apollo antenna. At Goldstone, the wing-station was the Pioneer Station (DSS-11), the first of the DSN sites constructed back in 1958 (It is now a National Historical Landmark). In general, a wing-station was not equally equipped as its STDN counterpart, but it provided the redundant systems (transmitters and receivers) that were needed under Apollo mission rules. This was a technically sound requirement. At lunar distances, the very narrow beamwidth of the 26-meter (85-foot) antennas (0.43°) meant that one was needed to track the Command/Service Module circling the Moon while the other was needed to focus on the Lunar Module as it made its way down and back up from the lunar surface. Under the Goldstone consolidation, Apollo GDS was reassigned to the JPL and redesignated DSS-16. A smaller 9-meter (30-foot) USB antenna was also transferred, redesignated DSS-17.[73]

An essentially parallel move was made in Spain at the Madrid Station (originally abbreviation MAD, which was changed to RID in 1984) built in

1965 at Fresnedillas 50 kilometers (30 miles) west of the capital city. The station had operated under bilateral agreements signed by the U.S. and Spain on 29 January 1964 and 11 October 1965 to establish mutual cooperation in the scientific investigation of outer space. There, the 26-meter (85-foot) STDN antenna used on Apollo was moved to Robledo by GSFC workers, placed under the auspices of JPL and redesignated DSS-66 as part of the Madrid Deep Space Communication Complex (MDSCC). Like Goldstone, MAD also had a wing-station assigned to it during Apollo. DSS-61 was just eight kilometers (five miles) away at Robledo de Chavela. It was modified for USB operations. In 1971, MDSCC became one of the first NASA tracking facilities to be turned over completely to a foreign government. Under the agreement, INTA, the Spanish National Institute of Aerospace Technology, today operates Madrid on behalf of NASA.[74]

Finally, the STDN stations half a world away in Australia were phased out. Today, mobs of wild kangaroos freely roam the abandoned grounds of Orroral Valley where one of the busiest stations once stood. ORR as it was

Abandoned in 1985, site of the Orroral Valley Station is today home to hundreds of kangaroos and their joeys in the serene valley. Shown here are remnants of where the main Operations Building used to be. (Photograph by the author)

In this picture from early April 1970, former Honeysuckle Operations Coordinator John Saxon (left) and Deputy Station Director Mike Dinn man the Ops console during pre-mission simulations for Apollo 13. Saxon holds the distinction of being the only person to have talked with an astronaut on the Moon from the Southern Hemisphere. During the Apollo 16 EVA, an earthquake knocked out the Los Angeles NASCOM node which caused Mission Control in Houston to temporarily go off the air. Since HSK was in communication with the crew at the time, Saxon chatted with Mission Commander John Young as Houston slowly got back on the air. The two agreed to share a toast if they should ever meet. They finally did—22 years later when the former mission commander visited Australia in 1994 to commemorate the 25th anniversary of the first Moon landing. ("Long Time Between Drinks," *Canberra Chronicle*, 16 July 1994. Photograph courtesy of Colin Mackellar, *www.honeysucklecreek.net/people/at_work.html*)

known, located 58 kilometers (36 miles) southwest of Canberra, was established as a STADAN facility in 1965. It was used mainly to support science and application satellites until its closure in 1985. In addition to its 26-meter (85-foot) antenna, ORR also had the Minitrack and the old Smithsonian Baker-Nunn optical cameras transferred from Island Lagoon, Woomera. In 1984, NASA shutdown the station and the USB antenna was donated to the University of Tasmania. The next year, it was moved to Mount Pleasant, east of Hobart, Tasmania, where it stands today. The Baker-Nunn camera was also donated, but to the University of New South Wales. The remaining Minitrack control equipment was handed over to the Commonwealth Department of Territories and was used for a while by the Australian Department of Transport and Communications for monitoring small satellites.[75]

Goddard had also put in a laser ranging facility at the station 1972 for geodetic research which was operated on behalf of NASA by the Australian Land Information Survey Group (now called Geoscience Australia) from 1975. This laser tracking facility operated at Orroral until 1998 when it was shut down and the equipment moved back to the United States. Continuing the work started at Orroral, a new laser ranging facility was established at Mount Stromlo, near the Australian Capital in 1998, wholly operated by Australia. But five years later, a devastating wildfire erupted in the hills surrounding Canberra which reached the outskirts of the city and destroyed the facility (as well as over 500 homes). A replacement facility was built in mid-2004 which continues to operate today.[76]

Just a few miles north of Orroral on Apollo Road was Honeysuckle Creek (HSK), perhaps the most historical of all the Australian sites because of its unique role on Apollo 11. In November 1981 after the second flight of the Space Shuttle, HSK closed its doors and simply faded away. Hamish Lindsay, who worked the consoles at the station, said in his book that "There were no farewells, no speeches, no parties, no wakes. All the equipment was removed, we pulled the last of the cables out, and walked out the door. During its short but glorious life, Honeysuckle Creek distinguished itself as a top station around the world in two completely different spheres as a Manned Space Flight Station and then as a Deep Space Station DSS."[77]

NASA transferred the HSK antenna to the Canberra DSN station at nearby Tidbinbilla where it has served as DSS-46 since 1983. Planned for phase-out in the coming years, the fate of the "old Honeysuckle antenna" as it is affectionately called, is nebulous. Those who worked at Honeysuckle would hate to see this piece of history simply scrapped. To this end, space enthusiasts, former station workers, and local residents in the area have banded together to form an ad hoc, private, "Save the Antenna" campaign. Their hope is that perhaps one day the historic antenna which received telemetry and video of mankind's first steps on the Moon will be restored, maybe even to stand once again at its original location in the hills of Namadgi National Park. Whether or not there will be a concerted effort by NASA or the Australian space agency CSIRO to preserve the legendary antenna in someway remains to be seen.

From a goodwill perspective, the closing of Guam was perhaps the most difficult. If ever there was a station outside of the 50 united states that could be called family, it was Guam. From the time of its ground breaking in 1966 to the later operation of the TDRSS ground node on the island, the Guamanians consistently strived for that close association with NASA, and vice versa. An important objective in originally establishing a station on Guam was for the United States to contribute to the economic growth of the island and to serve as an educational catalyst on the territory.

In late 1988 when it became apparent that the station was going to be closed, the Guam government pleaded with NASA to keep it open. At the time, it employed 91 people, of whom almost two-thirds were hired in Guam, at an annual payroll of $3 million. In an effort to save the station, Guam Governor Joseph F. Ada formally requested that NASA Administrator James Fletcher reconsider the decision, saying "the station has lent luster to the territory of Guam and has been a great source of pride for our people."[78]

The station at Dandan, establishment of which had been such the personal campaign of Governor Guerrero, was put into caretaker status in 1989 and closed out the following year. Some equipment was left in place at the request of the State Department, who was interested in using the facilities, while the remaining equipment was transferred to the government of Guam. While many stations may have simply ceased operating without any fanfare when they were shut down, this was definitely not the case at Dandan. At the conclusion of the final pass of the Solar Maximum Mission (Solar Max) at 10:30 a.m. on 30 June 1989 Guam time, simultaneous farewells took place on the island and on the other side of the globe at NASA Headquarters. Present at the ceremony were one time Guam Station Director Charles Force; Robert Spearing, former Director of Goddard's Mission Operations and Data Systems Directorate; and a host of other NASA and contractor employees who had worked the station over the years.[79]

Going back to even before the establishment of NASA and the Minitrack days when the network was set up by the Army Corps of Engineers, the United States always tried to bring local people into the operation. The station at Santiago, Chile, was an example of where this policy worked to near perfection, even if it were at times the target of anti-American political demonstrations. The station was a remarkable example of the long-standing goodwill engendered by the networks' activities. It was eventually operated entirely by Chileans. (Even the Station Director was Chilean, working for the University of Chile under NASA contract.) Wes Bodin, the former Associate Chief for Ground Network at Goddard, explained. "This policy created a cooperative spirit with the countries NASA dealt with, created a mutual relationship. And as we phased out a station, we transferred the equipment in total to the local government. At Santiago, we transferred the entire operating entity over to the University of Chile. The University kept the Station Director and part of his crew to operate as a space tracking station."[80]

After NASA left the station, it still bought services from Chile. In the late 1980s, the university reconfigured the station to support the COSPAS-SARSAT project, a multilateral, cooperative project sponsored primarily by the United States, Canada, France, and the former Soviet Union. (COSPAS was an acronym for the Russian phrase "Cosmicheskaya Sistyema Poiska Avariynich Sudov" meaning "Space System for the Search of Vessels in Distress" while SARSAT stood for "Search And Rescue Satellite-Aided Tracking.") The program used satellites to help search and rescue efforts by detecting signals emitted by airplanes, boats, and others in distress. Even today, the European Space Agency (ESA) and the National Space Development Agency of Japan (NASDA) continue to use such services provided by the Santiago Station.[81]

The year 1989 also saw the end of NASA operations on Ascension Island. The most isolated location in the network, Ascension ended up as one of the longest serving stations, operating without interruption for close to 25 years.[82] This streak was nearly broken, however, in 1982. From March to June of that year, the United Kingdom and Argentina engaged in a military conflict over the Falkland Islands to the south. During this brief but intense conflict, Ascension Island was used by the British for logistical support and as a result, commercial communications on and off the island were heavily disrupted. Fortunately for NASA, technical support was able to continue for the most part as the Agency maintained its own communication lines on and off the island for direct mission support. But the situation was not without its share of tense moments. Even though military action took place almost 10,000 kilometers (6,200 miles) to the south and Bendix workers on Ascension were at no time in any real danger, concerned family members back in the U.S. nevertheless had plenty of difficulty placing commercial telephone calls to their loved ones. Much of the problem was resolved when the company made available special circuits, routing telephone traffic through its headquarters in Columbia, Maryland to reach their families.[83]

Seven years later, operations at Ascension would be interrupted, this time for good. While the technical reasons to shut down Ascension Island were clear, how to close the site and what to do with it afterwards were not as obvious. Here, international cooperation with the international space community once again came to the forefront. What happened was that a series of events occurred as NASA was deciding to phase out the station, events that ended up involving three parties: the island government on Ascension, NASA, and the Europeans. Before the *Challenger* Space Shuttle accident in 1986 broke NASA's stride in constructing its Tracking and Data Relay Satellite System, or TDRSS, the space agency had planned to transfer ACN to the Air Force's Eastern Space and Missile Center (ESMC) when NASA operations ceased there in October of 1985.

To this effect, in a memorandum of understanding between GSFC Director Noel Hinners and the ESMC commander, authority and terms of the transfer were laid out in which Goddard had the responsibility to provide logistical support to ESMC for supplies and materials. Conversely, ESMC was to reimburse Goddard for contractor support provided to them during this transition period. But by 1986 when it was evident that TDRSS was going to be delayed, NASA quickly extended its agreement with British Cable and Wireless, who provided the only way to transmit wideband data off the island. (The Air Force station, ASC, did not rely on cables but rather used high frequency radio transmissions to the Eastern Test Range as their primary communications link.)

Although the commercialization of space would not reach a full swing until the following decade, even in the 1980s it was not difficult to see that a fundamental shift in the space landscape was already taking place. This change was the movement of space from the realm of government sponsorship to commercial commodity. With the Reagan administration being a strong proponent of privatization, as the nation's space agency, this paradigm shift was not lost on NASA. In fact, the Agency had already been operating from Ascension for a few years under an agreement with the Europeans.

During this time, U.S. dealings on the island, in the words of French program officials, "has been excellent."[84] For it to work, international cooperation had to have flexibility, and on occasion, some good fortune. A case in point was the handling of coverage for Ariane's 9 November 1985 launch, which happened to coincide with Shuttle mission STS-51A. The ESA Ariane carried the GTE Spacenet 2 and the European Marecs B2 satellites while the Shuttle mission included deploying two satellites and the recovery of Westar 6 and Palapa-B2. The Shuttle launch was originally planned for 7 November and would have required Ascension Island tracking support two days later for satellite retrieval operations. It ended up, however, being delayed 24 hours causing Ascension support for the Westar and Palapa recovery to now occur on the 10th. So, because of the Shuttle delay—not an unusual occurrence—the station was now free to cover Ariane without conflict.

But according to NASA mission rules, Ascension Island was not available for Ariane support for a 48-hour period before and after a Shuttle launch, and for a similar period before and after a scheduled landing. In practice, though, flexibility in scheduling was not uncommon so as to accommodate international partners' needs. Commenting on the series of events during STS-51A in 1984, Clet Yven, a Station Chief for the French Centre National d'Etudes Spatiales CNES said:

> What we have seen in practice is that Ascension Island availability is handled on a case-by-case basis, and the periods blocked against our use depend upon the mission. They [NASA] have demonstrated excellent flexibility and have said they could free their facilities for short periods in certain cases, even when there may be general scheduling conflicts with Shuttle missions.[85]

Hence, private commercial space launch industry officials were at the time especially concerned that the station closings (not just Ascension but the others as well) would leave them without the ability to receive data from their boosters. It was logical for the United States to consider commercializing some sites. In discussing the fate of the Ascension Station with NASA Headquarters, it was the perspective of the Office of General Counsel that "commercialization is feasible and would be consistent with other efforts to commercialize the space industry."[86] Such an arrangement could generally benefit everyone involved. In addition to the potential financial gain to the commercial operator, such an arrangement would also address these concerns, keeping the station operational while providing a continued source of revenue to benefit the local economy.

Recognizing the European's need for downrange launch support out of Kourou, French Guiana, officials from NASA and CNES met at the KSC in January of 1986. On the day of the *Challenger* launch, David W. Harris, at the time Manager of Space Network Operations at GSFC, happened to be leading a contingent from Goddard to discuss with CNES these issues. Breaking the meeting to view the launch—in one of those moments indelibly etched in one's memory—the team immediately recognized the horror of the situation. Still numb from what they just witnessed, the group disbanded that morning and agreed to reconvene at a later date.[87]

Three long years would pass before the group met again on 12 January 1989 to finish their talks. NASA already had plans to transfer the facility to the Air Force ESMC at the end of the fiscal year under an existing memorandum of understanding with the DOD. Under the proposed agreement with ESA, the Europeans would in turn install their own equipment on Ascension by March of 1990, to be operated by British Cable and Wireless personnel. The station was perfect for ESA since from its spaceport at Kourou, equatorial launches of the Ariane rocket flew almost directly over the island. Therefore, ESA requested that NASA continue operating the Ascension Station just a little bit longer, on a monthly reimbursable basis, until April 1990, when their equipment would be installed and become operational.[88] Thereafter, ESA would assume full operations on its own to provide tracking services to its international customers.

An agreement was thus signed on 21 February 1989, extending NASA operations on the island on a cost reimbursable basis, one that would have ESA pay NASA $283,000 per month to keep the station open.[89] As for the facilities that ESA did not want to use, NASA was requested by the Island Administrator to restore the site to its preexisting condition. In this cleanup, the Operations Building was transferred to the Ascension government for use by the local community. All other buildings were demolished and the rubble hauled off. A significant restoration effort was the cleaning out of Devil's Ashpit which had been used as a trash pit for a quarter-century. As one can imagine, this was no easy task since The Ashpit was quite large—30 meters deep by 40 meters wide by 90 meters long (100 by 125 by 300 feet)—with sheer, fragile walls. The cleanup took a year.[90]

In this rapid succession of station closings, perhaps no other group of people was more affected on a day-to-day basis than the contractors and their families. In 1989, BFEC for example, employed over 300 people at Ascension, Dakar, Guam, Hawaii, Santiago and Yarragardee, a good portion of whom had established families at these remote outposts (except for Ascension, which was "singles only").[91] This meant that by shutting down these stations, a few thousand people were going to be uprooted, some from the only homes they knew. Of course, there were some places where it was easier to leave than others.

Take Ascension versus Hawaii, for instance. While it may not have been all that difficult for folks to walk away from a place called Devil's Ashpit—recall that "If you can't go to the Moon, the next best place is Ascension Island!"—it was quite a different story for those who were stationed in the tropical settings of Hawaii. Located on a 25-acre site at Kokee State Park, the Hawaii Station was near Waimea Canyon on the west side of Kauai—one of the most scenic sights in the world. Since its establishment in 1961, the station had supported every U.S. human spaceflight with the exception of the first two sub-orbital Mercury missions. With its lush, green settings and surrounding hillside, the area is often used for motion picture and television location shots. Thus, it was not surprising that once assigned to Hawaii, one usually stayed in Hawaii.

Many of the employees at Kokee had been there for 20 years or more and had established roots there. When NASA announced that the site would be shut down at the end of the fiscal year on 30 September 1989, most of those at the station were offered positions elsewhere by Bendix. Few wanted to leave though. In the end however, with limited job opportunities on the island for skilled technicians, most took the offers and reluctantly left the island. Those who did not left the company. After 29 years of service which saw the station track John Glenn around Earth and bring back 27 astronauts from the Moon, much of the land was returned to the state of Hawaii. The station was turned over to the U.S. Navy's Pacific Missile Range. The 9-meter (30-foot) USB antenna system continued to be used for years on the Goddard Crustal Dynamics Project and is still being operated for science—tracking radio stars and studying plate tectonics by the University of Hawaii.

Finally, Botswana, Dakar and Yarragardee—the early UHF air-to-ground Shuttle voice stations—were closed in 1986, 1995, and 1991, respectively. Hardware from these stations were transferred to other STDN sites or mostly just donated to the host country. In the case of Botswana, the legacy of having hosted a "space station" in their country was preserved as NASA donated the surplus equipment to the Botswana National Museum, who made an exhibit commemorating their involvement and contribution to the success of the early Shuttle flights.[92]

★ ★ ★

Chapter 6 \ Era of Change

It was clear by the end of the 1980s that the era of NASA's worldwide, ground-based network had come to an end. Goddard's once sprawling STDN had been reduced by over 75 percent. Deemphasis had come a long way in just a few years. If one were to ask what the largest structures ever assembled on the face of Earth is, answers might range from the Great Wall of China to the Great Pyramids of Giza. From an infrastructure point of view, NASA's family of tracking networks—NASCOM, Minitrack, STADAN, MSFN, STDN, as well as the DSN—put together comprised one of the most wide-reaching infrastructures of the twentieth century, a true testament to the men and women who engineered it, built it, and made it work. Eventually though, technology and better access into space would supersede the need for such an extensive ground network. Instead of being tied to the surface of Earth, this new kind of network would now literally be based in space. It would change the STDN from a network using many ground stations into one using only a handful of satellites called the Tracking and Data Relay Satellite System, or TDRSS. TDRSS would enable Earth orbiting spacecraft such as the Hubble Space Telescope, the Space Shuttle and the International Space Station, to continuously communicate with control centers on the ground without an elaborate and expensive network of stations.

This fundamental change in spaceflight communications from primarily a ground-based network to a space-based network was something that NASA had in fact been working on since the early 1970s.

In other words, the revolutionary change to this new kind of network did not take place overnight.

CHAPTER 7

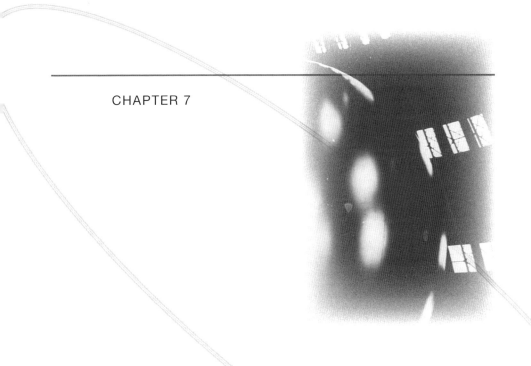

A NETWORK IN SPACE

The roots of a satellite-based communications network can be traced to 1945, when a Royal Air Force radar specialist and member of the British Interplanetary Society, Arthur C. Clarke expounded on his concept of what is known today as the "geosynchronous satellite." The reason geosynchronous communication satellites are needed is really very simple: The curvature of our Earth limits how far we can see. Consequently, a network of tracking stations, even when spread around the world, can only see and communicate with an orbiting satellite about 15 percent of the time, only when it passed within the station's field-of-view. In his article, Clarke accurately hypothesized that a satellite placed into orbit at an altitude of 35,900 kilometers (22,300 miles) over the Equator would circle Earth at the same angular rate that Earth rotated. In such an orbit, it would appear to an observer on the ground to be hanging motionless over the Equator. Thus, he concluded that a stationary satellite at geosynchronous altitude would be in an excellent position to relay communications around the globe. To this end, he suggested that use of three *manned* satellites in orbit could be used to relay programs for the newly invented medium of television.[1]

Clarke's article apparently had little lasting effect, however, in spite of the story being repeated in the 1951–1952 publication *The Exploration of*

Space. Lying dormant for several years, it was not until 1954 when a scientist named John R. Pierce at AT&T's Bell Telephone Laboratories carefully reevaluated the various technical merits (and the potential commercial windfall) of Clarke's proposal. Since the terms geosynchronous and geostationary had not yet been invented, Pierce, in a 1954 speech and 1955 paper, elaborated on the utility of a communications "mirror" in space. Along these lines, he added the concept of a medium-orbit "repeater" and a 24-hour orbit "repeater." In comparing the communications capability of a satellite, which he roughly put at 1,000 simultaneous telephone calls, with the capacity of the first trans-Atlantic telephone cable (TAT-1), which could then carry only 36 simultaneous telephone calls at a cost of $40 million dollars—Pierce wondered if such a "repeater satellite" could be worth over a billion dollars to his company![2]

Within 10 years, Clarke's and Pierce's concept would be translated into reality as communication satellites enabled viewers from around the world to enjoy the 1964 Tokyo Olympics live on television.

Spurred on by the 1957 launch of Sputnik 1 and later the Explorer satellites, the use of artificial satellites for communications quickly became a high-interest item in academia, the fledgling space industry and in the government. Many in the military saw its obvious strategic potential. NASA too, understood its incredible potential towards global communications. However, due to Congressional fears of "duplication" and in keeping with NASA's civilian charter, the Agency pretty much confined itself to experiments with passive, reflective, mirror-like satellites such as Echo 1 and 2. These were essentially nothing more than gigantic, shiny, Mylar balloons that bounced radio signals from one point on Earth to another. Meanwhile, the DOD dabbled in the more "active" satellites which actually amplified the signals received, providing much higher quality and stronger returns.

Government agencies, however, were not the only ones involved. In 1960, AT&T filed with the FCC for permission to launch an experimental communications satellite with the full intention of following it up with an operational system. The U.S. government was caught somewhat off guard since there was really no policy in place to regulate the decisions needed to implement the AT&T proposal. (This is somewhat akin to the situation that the FAA found itself in during the 1990s with respect to the commercial space launch market. Many laws and policies were in effect, but the FAA found itself having to quickly adapt them into guidelines for a cottage industry interested in this new commercial arena.)

The pace quickened. By the middle of 1961, NASA had awarded a competitive contract to RCA—who won the contract over AT&T and Hughes Aircraft—to build the medium-orbit (6,500 kilometer or 4,000 miles high), Relay communication satellite. Undeterred, AT&T would soon build its own satellite, the Telstar, which NASA launched for them on a cost-reimbursable basis in July 1962.

Chapter 7 \ A Network in Space   245

Echo, America's first communication satellite, was a passive spacecraft based on a balloon design created by engineers at NASA's Langley Research Center. Made of highly reflective Mylar, the satellite measured 30.5 meters (100 feet) in diameter. Once in orbit, residual air inside the balloon expanded, and it would begin its task of reflecting radio transmissions from one ground station to another. Satellites like Echo 1 shown here during an inflation test generated a lot of interest because they could be seen with the naked eye from the ground as they passed overhead. (NASA Image Number GPN-2002-000122)

On 25 May 1961, President John F. Kennedy spoke to the nation, committing to an American Moon landing by the end of the decade. But in another, long forgotten, portion of that speech, the President also committed the country to build a global satellite communications network. To this end, NASA and the Hughes Aircraft Company began developing a small, experimental, geostationary satellite called Syncom. Its first launch in January

Antennas for communicating with satellites have come a long way. "The Horn" antenna at Bell Telephone Laboratories in Holmdel, New Jersey was built in 1959 for pioneering work in communicating with the NASA Echo satellites. Made of aluminum with a steel base, it was 15 meters (50 feet) in length and weighed in at 18 metric tons (40,000 pounds). Used to detect radio waves that bounced off Echo, this primitive antenna was later modified to work with the Telstar Communication Satellite. In 1990, The Horn was dedicated to the National Park Service as a National Historic Landmark. (NASA Image Number GPN-2003-00013)

1963 went successfully, but unfortunately, the satellite failed to operate after injection into geostationary orbit. The second attempt in July 1963, though, was a complete success. These pioneering experiments soon paved the way for the semi-private, U.S. government subsidized Communications Satellite Corporation, COMSAT, that was formed as a result of the Communications Satellite Act of 1962 (a fallout from Kennedy's commitment), to pave the way for the world's first commercial communications satellite.[3]

Not surprisingly, the United States was not the only country in the West interested in this new realm. Understanding full well the global nature of the endeavor, NASA began negotiations with the Europeans to build ground stations on their soil (negotiations which AT&T had begun two years earlier in preparation for its Telstar experiment). Soon, Earth stations existed in Great Britain, France, Germany, Italy, Brazil, and Japan. Further negotiations over the next two years eventually led to a new international organization, one which would ultimately assume ownership of the satellites and responsibility for management of the new commercial space communications network.

On 20 August 1964, INTELSAT (the International Telecommunications Satellite Consortium) was officially formed with America's COMSAT as a majority owner. INTELSAT would eventually come to have more than 140 member nations, becoming the world's largest commercial satellite communications service provider. In this cooperative, owners contribute in proportion to usage of satellite services and receive a return on their investment. On 6 April 1965, the consortium launched the Early Bird from Cape Canaveral, and the age of international satellite communications was born. Today, INTELSAT operates a fleet of more than 20 geostationary satellites, providing television, telephone, and data services to literally billions of people worldwide. To manage the system, the consortium establishes technical and operating standards for ground stations which all users must comply with. Using antennas as small as 1.5 feet in diameter, users such as television and telephone companies, along with data service providers around the world, can access the system on a 24/7 basis to support their customers.[4]

But back in the 1960s, much of the early use of the COMSAT/INTELSAT system was to provide circuits for NASA's communications network NASCOM, relaying data back and forth between ground stations and their respective control centers. By the end of the decade, fortuitous timing led to the INTELSAT-3 series completing the global network just days before a billion people watched on live television mankind's first steps on the Moon on 20 July 1969.

During this time, communication satellites were fairly simple and not very big. Like Syncom, they were all spin-stabilized. In order to keep proper orientation in the weightlessness of space, an object (any object) has to be stabilized, either actively with an attitude control system consisting of small thrusters, or passively by spinning so as to conserve angular momentum (like how a bicycle wheel or a top stays upright when spun). By the 1970s, three-axis stabilization using gyroscopes had matured to the point where they could be used to reliably maintain the orientation of a satellite in orbit.[5] This made a huge difference. Since a satellite would no longer have to be spinning, it could now accommodate large directional antennas to support high data rates and deploy very large solar arrays for power. With more power came more equipment, sophistication, and more capabilities.

Technology steadily improved through the 1960s and 1970s. Perhaps an even more important improvement than new stabilization techniques was the increase in the amount of power that RF signals can be transmitted at and the utilization of higher frequencies in the RF spectrum. At the heart of signal amplification is a device called the Traveling Wave Tube, or TWT. Invented by Austrian born physicist Rudolf Kompfner and his colleagues at

Started in 1964, the International Telecommunications Satellite or INTELSAT ushered in the era of communication satellites for everyday use. Today, the consortium consists of over 140 member nations. This photograph shows Intelsat IV in an anechoic (sound-absorbing) test chamber in 1972. Built by the Hughes Aircraft Company, NASA placed it in geosynchronous orbit over the Atlantic with a then state-of-the-art capacity of 6,000 voice circuits or 13 television channels. (NASA Image Number 72-H-872)

Bell Laboratories, the TWT amplifiers date back to the beginning of the space communications era. Early tubes had power output only in the one-watt range (less than a common household nightlight). By the early 1970s, though, TWTs with a couple hundred watt capabilities were becoming available. What this meant was that ground stations no longer needed large dish reflectors costing millions of dollars to build. Antennas for satellite services quickly and dramatically shrank to the point where a 3-meter (10-foot) dish costing around $30,000 could now do the job that once required a 26-meter (85-foot) dish.[6] Advancements have continued in this field to where today, direct-broadcast application satellites have TWTs in the 300 watt range, requiring receive antennas that are only a foot or two (0.3 to 0.6 meters) in diameter and which cost less than a hundred dollars a piece. This has resulted in a huge leap in the amount and types of services available to everyday users literally anywhere in the world—as evidenced by the boom in the number of satellite television subscribers in recent years.

These sweeping strides in communications satellite technology provided NASA with the technology it needed to turn the TDRSS (Tracking and Data Relay Satellite System) from a concept on the drawing board into reality.[7] In fact, it would not be an overstatement to call TDRSS a national resource, one that has totally transformed the way space communications are done. In its planning and conceptual stage for about 10 years, implementation of the TDRSS in the 1970s and 1980s was, without a doubt, the biggest evolution in NASA tracking and communications during that time. So different was TDRSS that, to put it simply, it made the sprawling network of global ground stations a thing of the past.

Of the many communication satellites launched prior to the Shuttle era, only one—ATS-6 on Apollo-Soyuz—played a key role for tracking and data acquisition on a human space mission. Its success in 1975 took place at an important juncture. By this time, the Agency had completed Apollo and had already conducted several years of feasibility studies on a space-based communications network. ATS-6 underscored the unique ability of a communication satellite to serve as an orbiting platform, greatly enlarging the field-of-view capable from a single location.

Thus, the timing seemed right to establish a completely new kind of network, one based in space. Cost-benefit analysis done by GSFC drove the point home. By the 1970s, the sheer number of American spacecraft requiring network support had exceeded 50 and the cost of running ground stations was rising. Moreover, the STDN, as a ground-based system, had inherent weaknesses. Each station, for example, could monitor only two spacecraft at the same time and all stations working together could only hold a spacecraft in view for a small percentage of each orbit. TDRSS changed all that. "The network will take on a whole different complexion, becoming primarily a satellite-to-satellite network. But the big advantage that we'll get from that is the amount of coverage we'll get. That's the big benefit of TDRSS," said Henry Iuliano in comparing the expected performance of the TDRSS to the STDN.

> We have a very, very reliable network out there right now but it has wide gaps of information, compared to what we will get from TDRSS. There's no comparison. During the aborted Apollo 13 lunar mission, voice contact was very important because you would have to wait sometimes 20 to 25 minutes between contacts. We had to fit as much communication as possible into that short span, whereas now, once the TDRSS system is fully deployed, we'll have absolute coverage for a Shuttle mission. You can call and talk just about any time you want to.[8]

Its implementation greatly slashed the number of ground stations, saving NASA an estimated $500 million dollars in network operating expenses alone while providing this almost seamless communication capability.[9]

The original plan envisioned three satellites, each placed in geosynchronous orbit: one over the Eastern Hemisphere, one over the Western Hemisphere, and a spare positioned between the two. They would be connected to the ground at a single ground terminal. In this way, TDRSS could provide 100 percent viewing of spacecraft orbiting between 1,200 and 5,000 kilometers (745 and 3,100 miles) altitude. Craft orbiting above this altitude would be assigned to the DSN while for those orbiting below 1,200 kilometers, TDRSS could provide 85 percent coverage for—not perfect but still a far cry better than that offered by traditional ground stations.[10]

These hard facts were compelling and NASA's commitment to TDRSS was firm by the mid-1970s. Originally intended for inauguration with the Space Shuttle in the 1979 to 1980 time frame, implementation of the new system experienced many frustrations, and unfortunately, a tragic setback as well. This series of events was to prevent the TDRSS from meeting its full potential for nearly the entire decade of the 80s. Even though by the late 1970s, when the Agency knew that the new Space Network (SN) was not going to happen for a few more years, there was nevertheless optimism on the part of planners that there would not be too much of a delay between the inception of Shuttle flights and when NASA would have an operational SN in place. Even as late as December 1979, Goddard was counting on TDRSS taking over all tracking and data support of near Earth-orbiting spacecraft by 1982.[11]

Though it took the better part of the decade to complete, by 1989 NASA finally had what it had been waiting for. With TDRSS now available, the size of the ground network indeed shrank dramatically while communications coverage grew, from some 15 percent to over 85 percent, a six-fold increase. On top of that, network complexity was greatly reduced. TDRSS does not perform processing of user traffic but rather, operates simply as a

Chapter 7 \ A Network in Space   251

A drawing of the TDRS-3 spacecraft. The Tracking and Data Relay Satellite System (TDRSS) provides nearly uninterrupted communications with the International Space Station (ISS), the Space Shuttle and Earth orbiting satellites, replacing the intermittent coverage provided by the Spaceflight Tracking and Data Network (STDN) ground stations. (NASA Image Number MSFC-8893551)

"bent-pipe" repeater, one in which signals and data between spacecraft and ground terminals are relayed but not processed in real time. One Goddard network manager prognosticated (correctly, as it turns out) in 1989 on the future of space communications on the eve of a fully operational TDRSS network.

> We're certainly not going to go out of business. We'll start exchanging data through international programs, and there'll be increased contact with the universities and foreign space programs. We'll still maintain a NASCOM presence in Europe and Australia through the DSN and the domestic network is going to continue to grow in communications capabilities through the universities and scientific project control centers. As the tracking stations go away and the Shuttle flies on a regular basis, we'll have more and more satellites and more and more scientific data to exchange, so we'll be changing the network. Instead of linking up to tracking stations around the world, we're linking up users to the data we're getting back from the spacecraft, and that's going to continue to grow.[12]

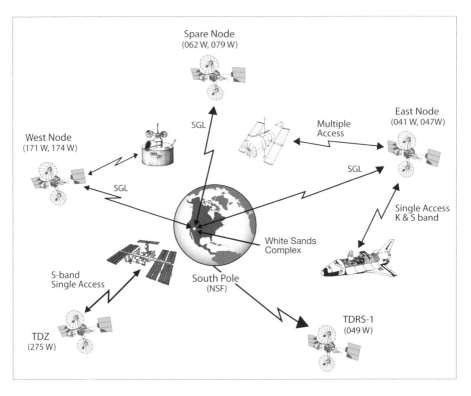

The Tracking and Data Relay Satellite System, TDRSS. The first generation Space Network used S and Ku-band to relay communications from up to 20 satellites at the same time. Ka-band capability was added beginning with three second generation satellites in the early 2000s. (Adapted from Roger Flaherty, Satellite Communications, Goddard Space Flight Center, May 2002)

This is how the system works. Data, voice, and video acquired by the constellation of satellites are relayed to a centrally-located terminal on the grounds of NASA's White Sands Test Facility in southern New Mexico— the White Sands Ground Terminal—or on Guam. From there, the raw data is sent directly by domestic communications satellites to control centers at the JSC, the GSFC or wherever it may be needed by independent users. In this way, nearly continuous communications with the ISS, for example, is allowed. This permits far greater flexibility in mission operations than had been previously achievable with a network of stations on the ground. To carry out the commercial side of the program, TDRSS also serves the space and science community at large by providing near-continuous coverage for over two dozen low-Earth orbiting spacecraft all at the same time. As one former manager put it, this new kind of space-centric network "focuses on the total data system, from instrument to scientist."[13]

All this ties back to just how the TDRSS came about. Studies for a tracking and data system that would rely on satellites rather than on a network of ground stations date back to the early 1960s. It was the DOD, not NASA, who first planted the seed. In the interest of controlling the "high ground," the United States Air Force knew that a space-based communications network could be the key. To this end, they held discussions with the Lockheed Missiles and Space Company and General Electric to investigate the feasibility of putting into space a so-called network of "Instrumentation Satellites."

NASA, however, was not far behind. In 1964, tracking personnel at the GSFC requested that Headquarter's OTDA consider funding an "orbiting tracking and data station" as a research and development project. OTDA managers in Washington were intrigued with the idea and put it on the agenda for Future Advanced Studies. Two years later in April 1966, the RCA Astro-Electronics Division and Lockheed were both awarded six-month contracts to define the characteristics of what was by then called an "Orbiting Data Relay Network."

By fall of the following year, OTDA was convinced that the space-based concept had a future. Goddard was thus tasked to establish a Data Relay Satellite System (DRSS) Requirements and Interface Panel, which included specialists from human spaceflight and science applications offices from around NASA. This panel's assignment was to oversee the definition and startup of such a system.[14]

The DRSS focused on a basic plan that called for a two satellite network in geosynchronous orbit over the Equator. In this configuration, an "East" satellite would be placed off the northeast coast of Brazil and a "West" satellite placed southwest of the Hawaiian islands. The goal was to have a system that could "be developed to augment and, to the extent practical, replace certain of the facilities that [comprised] NASA's tracking and data acquisition network."[15] The Agency was hoping for an operational network in orbit in the 1974 to 1975 time frame. To do this, Goddard had to expend considerable effort designing a system that would meet user needs at a time when most of the users were not even around yet. In other words, how did NASA know that this system it was designing would meet the needs of a future user community for the next 15 years?

To answer these questions, network planners developed what was called "loading analysis" computer programs. These programs evaluated whether the designs would satisfy user demands and determined how changes to staffing and closure of ground stations would affect the existing users. Meetings were held to identify the needs, understand onboard recording capabilities, data dump requirements, antenna design, and orbit planning.

For instance, it was through such analysis that Goddard came to understand that two so-called Single Access (SA) antennas and an array of 30 Multiple Access (MA) antennas could be used to satisfy those needs. (The number of spacecraft that can be supported by the MA system is determined by the phasing equipment on the ground, not by the number of antennas on the spacecraft.) From a station closure standpoint, loading analysis was used to help phase-down ground station shifts and closures in anticipation of each successful TDRS launch.

By May of 1971, Goddard was ready to issue Requests for Proposals to the industry for design of what was now officially called the Tracking and Data Relay Satellite System—TDRSS. An open competition led to Hughes Aircraft and North American Rockwell both being awarded two-year design contracts. However, before the contractors could finish their studies, NASA management realized that a budget conscious Congress would likely not fully fund development from the ground up of an effort that was still at a minimum four or five years down the road.

NASA had to think about new ways to procure TDRSS.

In what could only be termed a radical departure from the way it had operated up until then—and in an effort to get the project started without committing the Agency to a future purchase of a suite of satellites—OTDA decided to *lease rather than buy* a satellite system. In other words, rather than proceeding with a government-owned and operated system, NASA would, in essence, negotiate with private industry for a long term contract, one that would have the latter sell communication services back to the government. Since TDRSS was categorized as a support program rather than an agency research and development program, NASA considered leasing to be a viable option. Besides, all the technology required to implement the system was labeled as either off-the-shelf or in a high enough technology readiness level that leasing was considered no riskier than buying.[16]

In a flip-flop of the traditional customer-client relationship, the space agency was now a customer of private industry. Again, the impetus for this fundamental departure in the way NASA did business was rather simple and as usual, came down to economics. By obtaining this capability from industry on a long term, fixed price service basis, the Agency hoped to save money, and at the same time, spur on the commercial space sector.

In September 1973, Administrator James C. Fletcher wrote to individual members of Congress advising them of the Agency's budget needs for FY 1975. Among the new programs listed was TDRSS. Regarding the crucial role that the new system will play in Space Transportation System (STS) (Shuttle) operations, he wrote:

> Our studies have shown that the only way to meet our future tracking and data acquisition needs with reasonable expenditure of

funds will be through a ...TDRSS. Such a system will improve our Earth orbital tracking and data acquisition capabilities and meet the high data rates anticipated when the Space Shuttle is in operation, while at the same time, permitting the elimination of most of the ground stations in the present.[17]

Fletcher's statement to Congress captured *the main reason* for TDRSS: it was cheaper than augmenting the ground stations to meet Shuttle requirements.

The Agency had already identified six companies that were interested in the project, but, in this case, needed the assistance of Congress to develop the necessary legislation to authorize NASA to enter into such a contractual arrangement, since something like this had never been done before. Congress debated the wisdom of such a relationship through the spring of 1974, but finally authorized the go-ahead in May.[18]

Looking back over the last 50 years, the transfusion of technology from the government-borne space program to the private sector has occurred in many areas. Nowhere has this been more visible than in the realm of communications. Even in 1977, Gerald Truszynski summed it up rather succinctly when he testified before Congress, saying "The TDRSS contract, we think, is a good example of government developments moving into commercial applications."[19]

NASA now had the authority it needed to proceed with this leasing venture. It was at this time that the Agency's Headquarters made the (fatalistic in hindsight) decision that NASA had no basis to preclude telecommunications companies from bidding. The fallout of this decision was that by October 1974, no less than 27 companies or teams of companies had indicated their interest in bidding for the design, fabrication and operation of TDRSS. On 7 February 1975, Goddard issued a Request for Proposals for Phase I studies which would detail the system design and cost. In June, awards went to two contractor teams: RCA Global Communications, Inc., and Western Union Space Communications, Inc. A separate contract was awarded to Hughes Aircraft to define the user antennas systems that would be required by customer satellites.[20]

By 15 January 1976, Western Union and RCA had completed their six-month Phase I studies. Both were now intensely competing for the Phase II production contract, the winner of which would actually build and operate the system. These two were not the only ones busy. Throughout the year, announcements came of awards for several smaller, support contracts. One was given to Hughes, as expected, for the company to continue on the user (customer) antenna system. Others were awarded for building various support hardware. The big announcement for the TDRSS prime contract itself did not come, however, until the end of the year.

On 12 December 1976, in what could only be called a shock to the aerospace industry, NASA awarded the lucrative, 10 year, $800 million prime contract to Western Union Space Communications, Inc., otherwise known as Spacecom—a wholly owned subsidiary of the Western Union Corporation headquartered in Upper Saddle River, New Jersey. It ranked among the largest contracts ever awarded by the Agency, even dating back to the big procurement days of Apollo. Western Union, while a leading communication services provider (it continues to be one of the largest wire service companies in the world), had virtually no experience in the aerospace world.

Under the Western Union team, TRW's Defense and Space Systems Group in Redondo Beach, California would build the satellites and provide the computers and software for the ground terminal at White Sands. Unlike its prime, TRW *was* a leading satellite manufacturer for the DOD and NASA, and thus provided the valuable experience of working on large aerospace projects that Western Union so sorely lacked. In addition to TRW, the Harris Corporation's Government Communications Systems Division in Melbourne, Florida, was on the team. Harris, a leader in communications and information technology, was responsible for $60 million of the contract to build and integrate the system's antennas at the White Sands terminal.[21]

After the network was up and running, terms of the contract called for 10 years of services to be provided by Western Union Spacecom to NASA in both the space and ground segments. This included six spacecraft with components for a seventh. But here is where the contract was different. Unlike traditional procurements where the government provided funding from the onset, no money would be forthcoming to Western Union until the system was operational. Since no funds would be forthcoming from NASA until TDRSS became operational, the development of the project was financed with loans provided to Western Union by the Federal Financing Bank, an arm of the U.S. Treasury. To make this work, Congress had to actually pass a law, which they did on 30 July 1977. Under the terms of Public Law 95-76, NASA would make loan repayments to the bank once services began.[22]

Unfortunately, and almost from the beginning, the contract with Western Union ran into problems. While large, government procurements on this scale are already difficult enough to handle, the TDRSS procurement had an added level of complexity. More specifically, the space agency was trying (for the first time) to build what was known as a "shared system." What this meant was that TDRSS would actually serve two purposes: It would be designed and built to provide NASA with a new communications network, but it would also be designed and built to provide commercial communications. Part of this venture called for one satellite to be dedicated exclusively for use by Western Union to provide domestic communication services once the constellation was complete.

This sharing of the system introduced some technical complications into the system. But that was not the main problem. What really became an issue to NASA was that Western Union was unable to market the commercial part of it. Bob Spearing, NASA's Director of Space Communications who was at Goddard during this time, explained what happened:

> In a sense, they [Western Union] were ahead of their time. They were designing a commercial satellite package that worked at Ku-band. Ku-band was not a household word at the time. It was a new emerging capability and they just weren't able to get traction. So that created some difficulties in terms of how they were going to proceed with NASA. The idea of the shared system was that it costs less because the satellite would serve two purposes. When that started to go down the drain, there were a lot of contractual issues that transpired with NASA to try and resolve that problem leading eventually to NASA actually buying out the commercial side of the system.

By buying out the system, the Agency in essence changed TDRSS from a shared system back to one that was basically dedicated for NASA use. However, the commercial capability remained on the satellites. As Spearing said, "The design was far enough along at that point that it would have been much more costly to scrap the design and start over, so we actually built the satellites with the commercial capability."[23]

In 1980, in the first of a succession of moves, the TDRSS operations contract was transferred to a partnership of Western Union, Fairchild and Continental Telephone. Then three years later, in July 1983, Western Union got out of the contract all together by selling its 50 percent of the business to the other two partners. The buyout continued. In 1985, Fairchild, sold its share, leaving Continental Telecom—better known as Contel—as the sole owner of Spacecom and the TDRSS contract. This continued until 1990 when a new contract was negotiated which finally transferred ownership of the system back to NASA. Contel remained onboard but was now the space agency's contractor that operated the system for NASA.[24]

The failure of Western Union in their role as the TDRSS prime contractor can be traced in large part to the nature of the company itself. Unlike its subcontractors, TRW and Harris, Western Union was not a major player in the aerospace industry. As such, it operated in the highly regulated environment of the telecommunications industry where it was not unusual to find four lawyers and managers for every engineer. As a communication services provider, it knew how to get the most out of a network. However, it lacked the experience to actually build one. From a technical standpoint, the concurrent development of the TDRSS with the Space Shuttle in the late

1970s also meant that Western Union had to work closely with Rockwell (the Orbiter prime contractor) as well as the JSC. This was again something that the company did not do successfully. Western Union's function on the contract thus became more and more administrative than technical, even to the point where TRW ended up assuming most of the systems engineering and integration role.

Former Associate Administrator for the Office of Space Operations Robert O. Aller presented to senior Agency management and Congress in 1989 a "lessons learned" workshop from the TDRSS procurement process. Aller gathered 30 NASA and industry people who were closely involved in the process to review its successes and its problems. The eight lessons learned concisely addressed the heart of the matter:

1 *Shared Service Concept.* The concept of combining a commercial need with an established NASA need is valid, and may offer significant savings to the government through shared costs; however, the rights and operational utilization needs, availability, and privileges of each party must be clearly established in advance.

2 *Leased-Service Concept.* A leased-service concept should be based on the use of available commercial services or existing system technology if service is mission-critical.

3 *Interdependency with Government-Provided Services.* The interdependency of government-provided services to the establishment of a shared-lease service should be avoided or minimized to avoid government impact to the enabling of the leased services.

4 *Fixed-Price Contract for Developmental Work.* A fixed-price contract is not appropriate for development of a mission-critical support system where significant technology development may be required or where substantial changes to requirements may occur.

5 *Government Control Under Leased Service.* Under a leased-service arrangement, NASA must accept some loss of control over physical assets and accept risks of system outages or failures.

6 *Operational Interface.* In a fixed-price environment, establish the government/contractor operational interface at a point where changes in requirements affect only the government side, so far as possible.

7   *End-to-End Engineering and Operations Analysis.* In a leased-service approach to obtaining a mission support capability, it is just as essential initially to establish a comprehensive end-to-end systems engineering analysis and an operations and testing plan as would be done in a conventional NASA space system development program.

8   *Considerations for Prime Contractor.* The prime contractor must be one who has an extensive background in the business at hand.[25]

Spearing elaborated on these lessons and what happened:

In a sense, it was like NASA does today. In other words, if NASA lets a contract today, we would be in that oversight and management role and we would have a group of contractors handling the various elements, usually one lead contractor with some subcontractors associated with it. So we had this extra layer in there, if you will, with Western Union, driven principally by this shared system concept.[26]

Despite these challenges, work on TDRSS pressed on. Entering its final year of development in 1979, hardware fabrication continued in both the space and ground segments.[27] In the space segment, manufacturing of the high precision spacecraft antennas was the main item. Other activities included finishing up work on the propulsion system, specifically, qualification testing of the propellant tanks and acceptance testing of the Reaction Control System (RCS) that will be used to maneuver the satellites. In the ground segment, the Operations Building and ground antenna installation at the White Sands Ground Terminal (WSGT) was completed while hardware checkout and software development continued.[28]

Since nothing like TDRSS had been built before, technical challenges were expected. They were essentially the kind of things expected with building a brand new system, both in the design, and in particular, with the software. One way to describe the nature of a networked system involving many components such as the TDRSS is that it is "tightly coupled." This means that the software is such that if there is an anomaly in one part, it is going to affect a lot of other parts of the system. Along these lines, TDRSS was not only a tightly coupled system but an integrated system as well, with many subsystems that all had to work together. As a forerunner to today's so-called lights-out operation, TDRSS was envisioned by its designers to be capable of around-the-clock, unattended automatic operations.

For example, an operating schedule could be uploaded to a TDRS spacecraft. From there, it was up to the software to control the system, both on the ground and aboard the spacecraft—to configure links and acquire a given user satellite at the appropriate times. This was not at all trivial considering that each TDRS might be accessing 20 satellites at the same time, each in their own orbit while entering and exiting the spacecraft's field-of-view. With scheduling now automated, the number of ground controllers and the operational cost could be greatly reduced, to the point where personnel were needed only to monitor the system and implement changes. This move to systems automation was a major intent of the TDRSS. Thus for TDRSS to work, the software simply had to work.

Eventually, TRW engineers, working with Goddard, ironed out the problems. To demonstrate its capability, Spearing recalled that

> One day, just to show off a little bit, when we got it working, we had the operations team actually get up from their consoles and walk out of the room. We actually watched from a monitor to see how the system did. It went right through the whole process, acquired the spacecraft, got the signal and the data flowed out the back door. We wouldn't do that normally—just sort of a little showoff thing that we did for the local folks. We were not tracking the satellite operationally, just using it as a target of opportunity.[29]

But it proved the point: TDRSS was ready.

Another key hurdle that had to be cleared by the Agency as the initial operating capability of TDRSS approached was to make sure that there was not going to be radio frequency interference with other transmissions. A FCC Electromagnetic Compatibility Analysis was done in Annapolis to make sure that NASA's new system would "operate on a not-to-interfere basis with other services" operating in the 13.25 to 15.35-GHz regime.[30] At the heart of this analysis were classified DOD assets that operated in the same frequency range.

With the FCC analysis showing no serious radio frequency conflicts—and with the planned initial operating capability of the Space Shuttle quickly approaching—GSFC, that same year, made some rather significant decisions. The most important of these had to do with how the new satellites were going to be launched. Instead of sticking with the original decision to use a combination of expendable launch vehicles like the Atlas/Centaur and the reusable Shuttle, it tied TDRS launches exclusively to the latter. To do this, the spacecraft would be mounted atop an Inertial Upper Stage (IUS) rocket and the whole stack loaded horizontally inside the Shuttle payload bay. Once in Earth orbit, the TDRS/IUS stack would be raised up and gently deployed (literally pushed away) from the Orbiter. After it had moved a safe distance, the IUS would be ignited placing the TDRS on a course to geosynchronous orbit. The launch mode was officially tied to cost but ostensibly made the Shuttle that much more indispensable as it would now be the TDRS's only ticket into space. NASA, in essence, became a key supplier to its own satellite contractor. It was a watershed decision, one that would end up directly affecting the fate of TDRSS for years to come.

Other modifications had more to do with the capabilities of the satellite itself. Provisions for increasing spacecraft weight, reliability and station-keeping fuel reserves were added. Its tolerance in high radio frequency interference environments up in geosynchronous orbit was improved. (A Spacecom analysis done the year before had indicated that pulsed interference signals emanating from ground radar systems could create substantial TDRSS system upsets.) Overall, the value of these modifications added about $80 million to the project, which brought the total value to $866 million, plus award fees. When it became apparent that the Shuttle was not going to fly until after 1980, NASA slipped the schedule and delayed the launch of TDRS-1 until December 1980. As it turned out, it would not fly until 1983.[31]

From 1983 to 1995, NASA launched seven (TDRS-1 through 7) first generation TDRSS satellites. At the time, they were the largest and most advanced communication satellites ever made, weighing 2,270 kilograms (5,000 pounds) each and measuring 17.4 meters (57 feet) from one end of the solar panels to the other (equivalent to the height of a five-story building). In fact, the spacecraft was so large it would collapse under its own weight and could only be opened in the weightlessness of space.[32] Physical attributes aside, the heart of the spacecraft is its data handling capability. Operating in the S- and Ku-band, each satellite's electronic relay system could handle up to 300 million bits of information per second (300 Mbps), unheard of at the time considering 150 Mbps was considered high-rate service. Since eight bits of data make one digital word, this capability was somewhat akin to processing three and a half, 20-volume sets of encyclopedias every second.[33]

Looking somewhat like a giant, robotic bird out of a science fiction novel, the TDRSS spacecraft had several distinguishing, easily recognizable features. Foremost among them were the two huge, wing-like solar arrays which provided the satellite with over 1,800-watts of electrical power. The total array consisted of six (three on each side) 3.8 by 1.3-meter (12.6 by 4.2-foot) panels weighing approximately 130 kilograms (288 pounds) with a total photo-cell area of 30 square meters (317 square feet). These wings were movable so they could be kept pointed to the Sun. To do this, the arrays rotated about a common axis by two identical electro-mechanical drive assemblies which were individually controlled (Sun oriented) by the onboard Attitude Control System (ACS).[34]

Solar energy converted by the photo-voltaic cells was then used to charge the onboard nickel-cadmium (NiCd) batteries. These were capable of producing a power output of 1,440 watts and were housed in the hexagonal

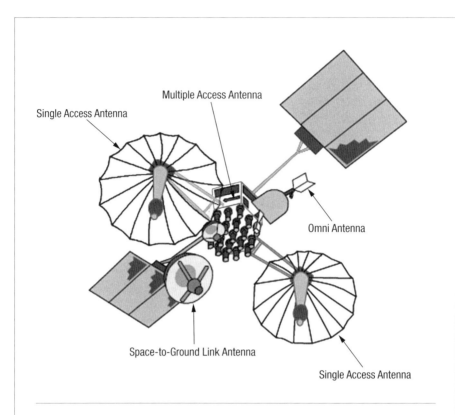

The first generation Tracking and Data Relay Satellite (TDRS-1 through 7). (Adapted from Space Network User's Guide, SNUG-450, Revision 8, NASA Goddard Space Flight Center)

equipment module of the main body of the spacecraft. Since electricity was a precious commodity (true of any spacecraft), TDRS battery usage was carefully monitored and controlled via the ground at White Sands. To maintain spacecraft weight symmetry, these batteries were configured in two assemblies, each comprised of 36 sealed NiCd cells. With each assembly weighing 66 kilograms (145 pounds), they were quite heavy, but had good electrical capacity at 40 amp-hours each, about that of an automobile battery.[35]

To show just how far technology had come over the years, the 10 milliwatt mercury battery that powered the transmitter on the old Vanguard satellite was designed to last 10 to 14 days. Since the TDRS batteries were not self-contained but rechargeable via solar power, their design life was 10 years minimum. Since the spacecraft had four major power busses, electricity was routed from the solar arrays and batteries to the spacecraft systems using an onboard Power Control Unit (PCU). As its name implied, the PCU controlled the charge and discharge rates of the batteries.

All active space vehicles require some type of ACS, or Attitude Control System—unless the spacecraft is purely passive like the Echo. On the TDRS, the onboard ACS contained all the equipment necessary to control its orientation and stabilization. In addition, it served to point the antennas, drove the solar arrays and controlled thruster firings for precise, three-axis station keeping. Like most modern control systems, the ACS used a combination of miniature momentum wheels, gyroscopes, and accelerometers to precisely measure its inertial attitude and position in space (exactly how it was oriented with respect to the stars and the horizon). An important capability that the ACS provided was to recover the satellite should there be a loss in attitude control—for example, a spin from a highly unlikely, nondestructive impact with space debris.[36]

Since the spacecraft was designed to stay in orbit for at least 10 years, it carried its own fuel to provided impulses for maneuvering and precision station keeping. Onboard were 680 kilograms (1,500 pounds) of hydrazine propellant, enough to operate the spacecraft for 10 years.[37] Like electrical power, the propellant budget was also carefully monitored on the ground. Rounding out the ACS was a solar sail which compensated for the effects of solar wind against the asymmetrical body shape of the satellite.

In addition to power and propulsion, a critical requirement for any spacecraft is the thermal protection needed for it to survive the extreme temperatures of space. On the surface of Earth, we are protected by the atmosphere so that temperature changes are relatively gradual. But outside the atmosphere, temperatures can swing by more than 280°C (500°F) during each orbit. When TDRS was in daylight, the temperature could reach 117°C (243°F); when it was on the night side, the temperature dropped to -173°C (-279°F). This is why spacecraft are often seen wrapped in gold thermal protection blankets. TDRS's Thermal Control System (TCS) maintained its temperature within acceptable limits during all prelaunch, launch, orbit insertion, and on-orbit activities for the duration of its mission. To control the temperature, the TCS used a combination of insulation blankets, radiator panels, thermostatically controlled heaters, and special reflective surface coatings. For example, radiators were located on the upper and lower faces of the equipment compartment to help reduce solar heating effects. Components with nonradiating external surfaces were covered by aluminized Mylar or Kapton insulation blankets which were electrically grounded together to the main spacecraft structure so as to prevent any on-orbit static charge build up.[38]

Along with the solar arrays, the antennas of the spacecraft were undoubtedly its most prominent features. In fact, TDRS carried five antennas. Particularly noticeable were the two 4.9-meter (16-foot) diameter, high-gain parabolic antennas which resemble giant parasols after unfurling. These were the so-called Single Access (SA) antennas, providing dual frequency communications at both the S-band (2.025 to 2.300 GHz) and Ku-band (13.775 to 15.0034 GHz). They were called SA because they tracked and relayed communications only with a single user spacecraft at any one time, in response to ground commands. The two SA antennas were steerable in two-axes and could be slewed for this purpose, following an object as it moved below, crossing TDRS's field-of-view.[39]

The high-rate service provided by these antennas was available to different satellite users who wanted to use the TDRSS on a time-shared basis. While the antenna may only be pointed at a single position, it was capable of supporting two users if they were operating at the different S- and Ku-Band frequencies. In other words, with the SA antennas capable of handling dual frequencies, each could actually be used to support two user satellites at the same time—one on S-band and one on Ku band—if both were within the antenna's field-of-view. To keep design complexity at a minimum and to reduce circuit cable loss (that is, loss of radio signal strength as it travels through a finite length of wiring), the SA receivers and transmitters were actually mounted on the back of these large antennas.

Since every pound that is launched into space drives up the cost, materials are usually selected with as high a strength-to-weight ratio as possible, and as durable as possible; exotic manufacturing techniques are thus not uncommon. This was particularly true with something as big as the SA antennas. In this case, the primary reflector surface was made of a molybdenum wire mesh, woven like cloth, on the same type of machine used to make material for women's hosiery. For RF reflectivity and thermal tolerance, it was clad in 14-carat gold. When unfurled, its 18.9 square meters (203 square feet) of mesh was stretched tightly on 16 high-strength tubular ribs by fine, thread-like quartz cords. In this way, the antenna looked somewhat like a large, glittering, metallic spiderweb. Despite the size, the entire antenna structure weighed only about 23 kilograms (50 pounds) on Earth. To help explain their lightweight sophistication, NASA liked to publicize the following fact: Because of the support and structure that would be needed to counterbalance the effect of gravity, an antenna of similar capability and size based on Earth would need to weigh about 2,270 kilograms (5,000 pounds).[40]

Mounted on the lower side of the spacecraft's main body was the MA antenna. It was an electronically steerable, 30-element, phased-array antenna used to relay communications for multiple customer satellites simultaneously. To relay signals, 12 of the elements—called helices—were diplexed (split) for transmit and receive while all 20 were used as receive elements. Signals from each helix antenna were received at the same frequency, multiplexed or combined into a single composite signal and transmitted to the ground. In the ground equipment, the combined signal was demultiplexed and distributed to 20 sets of beam-forming equipment that discriminate among the 30 signals to extract signals of individual users. So a TDRS functioned somewhat like a celestial switchboard, receiving data from up to 20 different satellites while transmitting to 12, all at the same time. (The 12 that it was transmitting to could be other satellites or be the same ones from which it was receiving data.)[41]

From its vantage point at geosynchronous altitude, the 13° field-of-view of the MA meant it could see all spacecraft in orbits of 1000 km (620 miles) or below—the majority of low-Earth orbit spacecraft. Not only could it track all spacecraft below this altitude, it could also track many aircraft simultaneously. The MA service was attractive because it was very reliable, and for TT&C and low science data rate functions, it could provide user support everywhere and at any time. By contrast, the SA service was attractive because it could handle high data rates (300 kbps for S-band or 25 Mbps for Ku-band SA forward service versus only 10 kbps for MA service). Another difference was that the MA antenna operated only in the S-band. More specifically, it forwarded signals at 2106.4 MHz and received return signals at 2287.5 MHz.[42] When the system was being designed in the 1970s, this S-band only capability was deemed sufficient by most communications experts for handling the commercial satellite traffic then envisioned for the coming decade. This is only partially true now 35 years later.

While the Single and Multiple Access antennas were fine for communicating, tracking, and relaying data between the TDRS and other satellites, they could not be used to actually link the spacecraft with the ground. This was done with a separate Space-to-Ground Link antenna, or the SGL. It was a pointable, 2-meter (6.6-foot) diameter dish whose only purpose is to provide the uplink and downlink between the TDRS and the ground terminals at Whites Sands and Guam. Signals were relayed with the SGL using the more bandwidth efficient Ku-band (13.4 to 15.25 GHz). The SGL antenna, unassuming in appearance compared to the pair of SA antennas, handled all

The Tracking and Data Relay Satellite (TRDS) stowed in the Shuttle payload bay is raised to a vertical attitude in preparation for deployment from low-Earth orbit. Shown here is TDRS-6 being deployed from the Shuttle *Endeavour* on STS-54 on 13 January 1993. Clearly visible is one of the Single Access (SA) parasol antennas seen folded at the top. The solar arrays are also in the stowed position. The Inertial Upper Stage(IUS) is visible below the satellite. (NASA Image Number STS054-71-025)

customer scheduling and service requests as well as NASA's own TDRSS command and telemetry. It was, in essence, the customers' only electronic link back to Earth.

Finally, there was the Omni Antenna which supported the spacecraft's TT&C system. The TT&C collected data from the various onboard subsystems and transmitted the telemetry down to White Sands so that the spacecraft's health and status could be ascertained (for example, how fast were the batteries discharging). Conversely, it processed and implemented commands uplinked from the ground (for example, initiate thruster firings to rotate the craft). The TT&C system provided range and range rate information by computing precise turnaround-and-retransmission delays in signals to-and-from the ground.

Once the TDRS was operational in orbit, TT&C was normally done at Ku-band through the SGL antenna. However, there were exceptions and that was where the Omni came in. Looking rather inconspicuous—an oddly-shaped polygon—this omni-directional antenna mounted on the side of the main structural body operated in the S-band and was used strictly by NASA for command and control. Specifically, it was used during deployment from the Shuttle and, if necessary, during system recovery in the event of an emergency. It supported no customer services. With the Omni, TDRSS control on the ground could switch satellite operations to failsafe mode at any time for a variety of reasons: prevention of command lockout caused by failure of the primary SGL equipment, anomalous spacecraft attitude or pointing errors, and something that NASA hoped never happens—remote (hostile) takeover of the spacecraft. To put it simply, if one thinks of the SGL as the spacecraft's normal link back to Earth, then the Omni was, for all intents and purposes, the spacecraft's last-chance lifeline.[43]

* * *

When STS-6 left Pad 39A at the KSC on the afternoon of 4 April 1983, it had a few firsts. It was the first flight of the new Shuttle *Challenger*. It was the first use of the improved, lightweight External Tank and the lightweight SRB casings. The mission had the first spacewalk (EVA) of the Shuttle program, one that lasted 4 hours and 17 minutes to check out the new generation of spacesuits that will be used by Shuttle astronauts. And finally, it launched the first Tracking and Data Relay Satellite, TDRS-1.

The launch, originally slated for 20 January 1983, was delayed several times due to leaks discovered in *Challenger*'s main engine fuel lines while it was on the pad. But in an unfortunate turn of events, as engine repairs were being made, a severe rain storm swept through the Cape that caused TDRS-1 to be contaminated while it was still in the Payload Changeout Room (PCR) at the pad. As a result, workers had to take it back to its checkout facility, have it cleaned, rechecked and remounted into the Shuttle payload bay. (The PCR and the payload bay first had to be cleaned out also.) With this temporary roadblock cleared, STS-6, commanded by Skylab veteran Paul Weitz, lifted off without further delay at 1:30 pm. EST on 4 April, sending the crew of four on their five-day mission to deploy the first TDRS.[44]

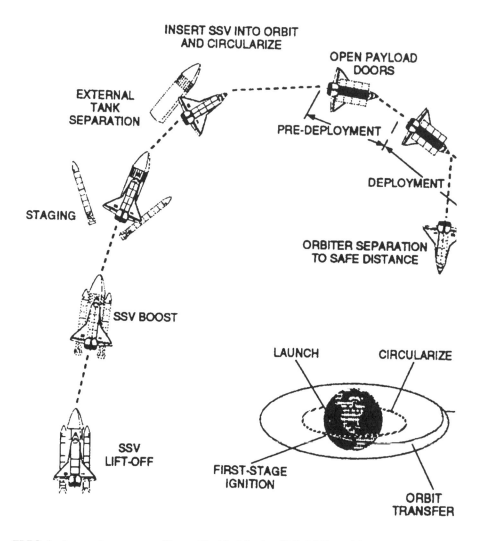

TDRS deployment sequence. (Space Shuttle Mission STS-54 Press Kit, January 1993, NASA Headquarters)

Deployment of a TDRS from the Space Shuttle is a well orchestrated series of events. After reaching orbit, the Shuttle's payload bay doors are opened and its Ku-band antenna deployed. This antenna—stowed on the right, forward side of the payload bay—was crucial for checking out and communicating with the new satellite. As efficient a bandwidth as Ku-band is, one drawback of having to operate in this high frequency is the inherently narrow

Chapter 7 \ A Network in Space

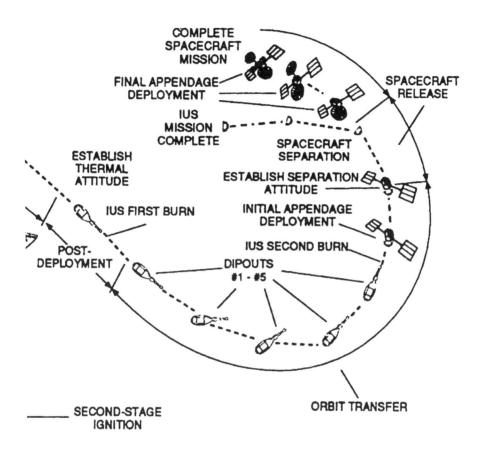

pencil-like beam needed to focus the signals.[45] This makes it somewhat difficult for the SGL antenna to lock onto the signal in order to communicate with the Shuttle. However, since an S-band system can get by with an inherently larger beam, the Omni antenna is first used to lock the Ku-band antenna into position after the satellite is deployed from the Shuttle—a process known as acquisition. For anyone who has ever looked for an object in the night sky using a telescope, this is not unlike having to first use a finder-scope to point the main telescope in the vicinity of the star.

Once the Omni has locked on, the Ku-band system is turned on. To perform the acquisition, the Shuttle's Ku-band antenna is gimbaled so it can acquire the TDRS by executing a preprogrammed search. In this search, if the satellite's SGL signal is not detected within the first 8° of a scan, the search automatically expands to 20° and is repeated. The entire search typically takes only about three minutes. The scanning stops once the acquired signal strength meets a given threshold. At that point, the Ku-band system becomes operational.[46]

About an hour after release, having moved sufficiently far from the Shuttle, the IUS first-stage rocket motor is ignited. Built by Boeing Aerospace for the U.S. Air Force, the two-stage IUS solid-rocket boosts the TDRS into its 35,900 kilometer (22,300 mile) geosynchronous orbit since the Shuttle itself cannot go that high. This is then followed by a second-stage motor burn. Once this burn is successfully completed, the TDRS—still attached to the IUS—is well on its way to geosynchronous orbit and the Shuttle and her crew have essentially done their job.

There is, however, still more to do, this time by the ground. First, there is the geosynchronous insertion burn to circularize the spacecraft's orbit at geosynchronous altitude. This is followed by separation of the satellite from the now spent IUS. At this point, the TDRSS team at White Sands commands deployment of the solar arrays. The two 4.9-meter (16-foot) diameter SA antennas are then unfurled and pointed toward Earth for the spacecraft to begin its checkout. This testing will take place over the next three to five months. During this time, the ground will also command small thruster firings to slowly move the craft and position it at its desired operating location.

Joining Weitz on this mission were Pilot Karol J. Bobko and Mission Specialists Donald H. Peterson and F. Story Musgrave, both of whom would deploy the satellite from controls inside the aft flight deck. After *Challenger* was successfully inserted into a 286-kilometer (178-mile) circular orbit, the payload bay doors were opened and the TDRS-1/IUS stack was raised. Ten hours after launch Peterson flipped the switches which allowed the giant satellite to be released and gently pushed away from the Shuttle. The first engine burn went perfectly. However, the second did not; the motor shutdown prematurely.

For almost three hours, America's first TDRS appeared to be lost, deaf to all commands. At 9 a.m. EST the following morning—as Goddard engineers were busy with contingency procedures—the Goldstone tracking station received a faint indication that it had indeed separated from the spent IUS. However, its orbit was far from what was needed. Instead of a nice, circular 35,900 kilometer orbit, the incomplete engine burn had stranded TDRS-1 in a useless 35,325 by 21,790-kilometer (21,950 by 13,540- mile), elliptical orbit. Furthermore, instead of zero inclination (orbit parallel to the

Equator), it was crossing the Equator at an angle of 2.4°. As if that was not enough, the spacecraft was spinning out of control at an alarming rate of 30 revolutions per minute, or once every two seconds.

From the ground, the situation looked bleak. There was hope, however: Use the onboard ACS (designed only for station keeping maneuvers and for adjusting the satellite's location) to actually finish boosting it into geosynchronous orbit. Over the next two months, engineers at Goddard, TRW and Contel worked out a series of burns using the small (one pound thrust) ACS thrusters to carefully nudge the spacecraft into the proper orbit. Since the thrusters are so small, this orbit transfer could not be done with one maneuver. It, in fact, took 39 separate commands and consumed some 400 kilograms (900 pounds) of the usable 635 kilograms (1,400 pounds) of fuel onboard TDRS-1. The maneuvers began on 6 June 1983 and took a total of three weeks. During this time, overheating caused the total loss of one of the two sets of 12 thrusters plus one thruster from the other set.

But the patience paid off. On 29 June 1983, TDRS-1 reached its destination, parking itself over the Equator in a "figure-8" loop at 41° west longitude, just off the northeast coast of Brazil. There was much to celebrate at Goddard. As one flight controller put it, "It was a cliff hanger."[47]

A week later, TDRS-1 was turned on for testing. All went well until October when the spacecraft began to be plagued by a series of component failures. First, one of the Ku-band SA diplexers used to combine RF signals failed. Shortly thereafter, one of the Ku-band TWT amplifiers on the same antenna failed crippling the forward link relay service that it could provide. The failures continued. On 19 November 1983, one of the two TWT amplifiers serving the other SA antenna also failed. This meant that TDRS-1 had lost one of its primary capabilities, the Single Access, Ku-band, forward link relay.

One of the consequences of losing this link was that it prohibited the use of the Text And Graphics System (TAGS) onboard the Shuttle. TAGS was a high-resolution facsimile system that scanned text or graphics and converted the analog scan into a digital bit-stream. Basically, a fancy fax machine that operated via telemetry, it provided an on-orbit capability to transmit text, maps, high resolution schematics and photographs between the astronauts and Houston. In lieu of TAGS, Mission Control—not until 1989 as it turned out—had to resort to using the old S-band, Apollo-era teletype system to relay text-only instructions up to the crew (for example, procedures, weather, crew activity plan changes, etc.).[48]

Despite these annoying setbacks, Goddard continued testing over the next 12 months. The fact that the craft had lost a major link capability notwithstanding, NASA declared TDRS-1 operational in December 1984, saying "Working solo, TDRS-1 provided more communication coverage . . . than the entire network of NASA tracking stations had provided in all previous Shuttle missions."[49]

It had been a long 20 months since TDRS-1 left Pad 39A.

The ensuing years have born witness to this declaration. Besides serving as one of the two primary satellites in the early Space Network, TDRS-1, over the years, accumulated a number of firsts to its credit. It was the first satellite used to support KSC launches in the early 1990s, returning real-time telemetry and video. It also helped close the Zone of Exclusion over the Indian Ocean (explained later in the Chapter), providing 100 percent coverage for the ISS, the Space Shuttle and low Earth orbit satellites. In March 1992, Goddard called on TDRS-1 to quickly aid its Compton Gamma Ray Observatory (CGRO) when data recorders onboard the spacecraft failed.

Since the satellite was precessing (that is, changing its orbital inclination or tilt with respect to the Equator) in its orbit almost 1° per year since its deployment, it was used serendipitously in ways never expected. Due to its changing orbit, TDRS-1 was the first satellite able to see both Poles. In cooperation with the National Science Foundation (NSF), NASA put a ground station for TDRS-1 in January of 1998 at the exact location of the (true) South Pole. The terminal has since given scientists at the Amundsen-Scott Base in Antarctica the year-round ability to return high volumes of science data to the continental United States. With it, the first connection to the Internet—and the first live Web cast—from the North Pole was done as was the first Pole-to-Pole telephone call connecting the North Pole to the South Pole. The event was even recorded in 'Ripley's Believe It Or Not' and the Guinness World Records in April 1999.[50]

NASA considered retiring the aging satellite in 1998, but instead allowed the NSF and others to use it for scientific, humanitarian and educational purposes. For example, TDRS-1 was used in 1998 for a medical emergency at McMurdo Station in Antarctica. Its high-speed connectivity allowed scientists to conduct a telemedicine conference, allowing doctors in the U.S. to teleconference a welder through an operation on a woman diagnosed with breast cancer.

A second working satellite placed into orbit in January 1986 would have meant an operational TDRSS and attendant closure of most ground stations shortly thereafter. Those plans were, however, suddenly dashed when the Space Shuttle *Challenger* met with a horrific demise 73 seconds into its mission on 28 January 1986 (STS-51L). At 11:38 EST that morning, it was launched atop Pad 39B at the KSC in the 36°F chill of the south Florida winter, the coldest ever for a Shuttle mission. The mission was the most publicized NASA flight since Sally K. Ride became the first American woman in space on STS-7 two and a half years earlier. *Challenger*'s crew of seven was commanded by Shuttle veteran Francis R. "Dick" Scobee; joining him were Pilot Michael J. Smith; Mission Specialists Ellison S. Onizuka, Judith A. Resnik, and Ronald E. McNair; and Payload Specialists S. Christa McAuliffe, a high school social studies teacher from Concord, New Hampshire and Gregory B. Jarvis, an engineer with TRW. The primary mission of the planned week-long flight was to deploy and checkout TDRS-2. However, the fact that McAuliffe was going into space garnered the flight more national attention than usual from the media, much more so than on any of the previous 17 missions since STS-7.

From liftoff until telemetry was lost, no flight controller observed any indication of a problem, although post-flight analysis showed telemetry had uncovered some anomalies regarding pressures inside the starboard SRB motor shortly after liftoff. The last voice transmission was received via Ponce de Leon as Scobee acknowledged a routine main engine throttle up call from the Capcom with simply a "Roger, go at throttle up." Three seconds later, a horrified crowd—including many in the crewmember's families and students who had made the trip from New Hampshire to cheer on McAuliffe—watched, stunned, as *Challenger* erupted into a giant ball of flames.

Many unfamiliar with the Space Shuttle at first thought this was the routine separation of the SRBs. However, the onlookers soon realized that something was happening that was anything but routine when they saw the SRBs emerging from the cloud without any sign of *Challenger*. The crew

The *Challenger* crewmember remains are transferred from seven hearse vehicles to a C-141 at the Kennedy Space Center's Shuttle Landing Facility for transport to Dover Air Force Base, Delaware. The accident that claimed the lives of the five NASA astronauts and two Payload Specialists also set back construction of the TDRSS Space Network by 32 months. (NASA Image Number GPN-2000-001480)

apparently had no indication of any problems before the Orbiter rapidly broke apart.[51] No alarms ever sounded on the flight deck. The first evidence of the accident came from live video coverage on the ground and when radars at the Cape began picking up multiple objects.

A Presidential Commission (the Rogers Commission, named after Commission Chairman William P. Rogers, a former Secretary of State in the Nixon administration) was formed by President Reagan on 3 February 1986 under Executive Order 12546 to investigate the accident, which by then had assumed national tragedy proportions. Four months later, the Commission issued its report which included the following conclusion on the cause of the accident:

> The consensus of the Commission and participating investigative agencies is that the loss of the Space Shuttle *Challenger* was caused by a failure in the joint between the two lower segments of the right Solid Rocket Motor. The specific failure was the destruction of the seals that are intended to prevent hot gases from leaking through the joint during the propellant burn of the rocket motor. The evidence assembled by the Commission indicates that no other element of the Space Shuttle system contributed to this failure.[52]

Besides the tremendous shock of having lost a flight crew for the first time on an actual mission—other astronauts and astronaut candidates had been killed before during training and on ground tests—the space agency had to deal with the ramifications of a nearly three-year wait as the Shuttle would not fly again until September of 1988. The launch manifest had to be rearranged. Foremost among the considerations was to resume deployment of the TDRSs as soon as possible. Getting TDRSS operational had an extremely high priority at NASA as its capabilities were needed by so many science application satellite missions and of course, the Space Shuttle itself. The SN simply had to be established as quickly as possible after Shuttle flights resumed. In what is somewhat of a bittersweet irony, the TDRSS program in a way benefited from the *Challenger* disaster in that the hiatus allowed TDRS-1 to be shaken down as a prototype. The added time before the launch of the next spacecraft, TDRS-3, allowed problems with TDRS-1 to be fixed. This probably led to longer useful life of the succeeding spacecraft.

After the accident, Shuttle launches were put on hold indefinitely. Since *Challenger* was to have launched all the early TDRS, NASA used this down time to begin modifying the payload bay of *Discovery* for it to assume this duty. Following the Rogers investigation and an extensive redesign to the SRBs, Return-to-Flight processing finally began in earnest in September of 1987.

On 16 May 1988, TDRS-3 arrived at the KSC from California followed by its IUS eight days later. By the end of May, mechanical mating of the two was complete. The pace picked up from there. On the morning of July 4th, in a symbolic gesture befitting the moment, the entire STS-26 stack was rolled-out of the Vehicle Assembly Building to take the Shuttle's first steps back into space by making its three-mile journey to Pad 39B. Countdown tests were conducted over the next few weeks which revealed some leaks with the Main Propulsion System as well as the Orbital Maneuvering System. However, repairs were successfully done on the pad and on August 29, technicians installed the satellite into *Discovery*'s payload bay. One month later, NASA managers gave the final go-ahead for launch.[53]

At 11:37 a.m. EDT on the morning of 29 September 1988, STS-26, with NASA's most experienced crew to date, took to the skies of eastern Florida. After 32 long months, the Shuttle was back in space, this time flying with redesigned SRB field joints along with other safety and performance upgrades, including for the first time since STS-4, a (limited) crew escape capability. This time, the launch was flawless.[54]

Twelve minutes later, *Discovery* was in orbit. Onboard was TDRS-3. Six hours after reaching orbit, the crew successfully sent it on its way to its geosynchronous destination over the Pacific. NASA had for some time considered not putting the TDRS-3 payload on STS-26 since it was going to be the first mission following *Challenger*. Risk analysis showed, however, that it would have made little difference in terms of probability to mission success whether the payload was launched then or on a later mission since launch risk did not vary significantly from mission to mission. More importantly, getting TDRS-3 deployed was critical for the success of missions down the line.

With TDRS-3 (and TDRS-1) firmly in orbit, NASA finally had its long-awaited, dual-satellite SN capability. The two were referred to as TDRS-West and a TDRS-East, respectively. But the constellation was far from complete. The network called for even more satellites, including on-orbit spares plus a replacement for the one that was lost on *Challenger*. In fact, the original, first generation constellation called for six satellites total. Today, there are nine TDRSS spacecraft on orbit all together.

In the years since, NASA has been criticized (mostly from opposition in Congress) as to why there are so many satellites "up there"? After all, only two are needed to provide 85 percent coverage while three can provide 100 percent. The answer lies in something called "availability of the system." As a communications network, TDRSS, from the beginning, was designed with a very high probability that it would be there when needed. Thus, a very high mark or "figure-of-merit" was put on the system—an assurance that it was going to be available. Former Associate Administrator Charles Force explained what that meant in terms of the number of satellites required:

Table 7-1: First Generation TDRSS Constellation[55]

Satellite	Launch Date Shuttle Mission	Geosynchronous Longitude	Location
TDRS-1 (F1*)	April 4, 1983 STS-6 Challenger	49°W	Off the northeast coast of Brazil
TDRS-2	January 28, 1986 STS-51L Challenger	—	—
TDRS-3 (F3)	September 29, 1988 STS-26 Discovery	85°E	Indian Ocean
TDRS-4 (F4)	March 13, 1989 STS-29 Discovery	41°W	Atlantic Ocean east of Brazil
TDRS-5 (F5)	August 2, 1991 STS-43 Atlantis	174°W	Pacific Ocean over the Phoenix Islands
TDRS-6 (F6)	January 13, 1993 STS-54 Endeavour	47°W	Off the northeast coast of Brazil
TDRS-7 (F7)	July 13, 1995 STS-70 Discovery	171°W	Pacific Ocean over the Phoenix Islands

*GSFC designation F1 through F7 represents TDRS-1 through TDRS-7

There are more satellites and more capacity than you need because you are shooting at that mark. So that mark is what drove the number of TDRSs which were ordered and the replenishment satellites.... The reason goes right back to the criticality of it and the need to make sure that the capacity was there when needed. My analogy to a light switch: You turn on a switch and there is a satellite up there to do the job.[56]

Then the issue came up. What should NASA do with all these extra satellites—most of which were not needed yet because of the success of those already in orbit? The answer, in the eyes of the space agency, was quite simple: warehouse (store) them in orbit. Said Force:

There were some studies done, primarily by TRW ... which said ... there's nothing on the satellite that really wears out with use except the solar cells degrade slightly with time, [so] there was plenty of capacity there. The riskiest thing about a TDRS is the launch phase, as demonstrated by the fact that we lost one on *Challenger* and the first one halfway to geosync because the IUS failed. So the decision at that point was, we're better off storing them in orbit because then you get by the infant mortality—the launch failures and all that sort of stuff. So that's basically why there are so many TDRSs up there. If you look at the requirement, having 96 percent probability that you're going to have TDRSS capacity that is needed,…then you have to have x-number of TDRSs. And once you've got them, you might as well launch them and store them on-orbit.[57]

In fact, the operational availability of the TDRSS is not 96 percent but has exceeded 99 percent. This is thus a clear case where the requirement—and not the cost—drove the program.[58]

On 13 March 1989, TDRS-4 was launched on STS-29 again aboard *Discovery*. After successfully attaining orbit, it was slowly positioned as TDRS-East off the coast of Brazil. After that, TDRS-1 was slowly moved to the spare position where it has served ever since on a limited basis under the inauspicious name of WART (White Sands Complex Alternative Resource Terminal), used by the NSF in their research activities at the South Pole.

TDRS-5 followed on STS-43 on 2 August 1991, this time aboard the Shuttle *Atlantis*. Seventeen months later, on 13 January 1993, TDRS-6 was launched on STS-54 aboard *Endeavour*. The last of the first generation satellites, TDRS-7, (included with NASA's *Challenger* replacement fund) went into orbit 13 July 1995 aboard *Discovery* on STS-70. It was the replacement for the one lost on *Challenger*. With it, NASA's first generation TDRSS was completed.

Table 7-1 is a summary of the SN as it appeared during the 1990s. Since the satellites are capable of being repositioned and NASA at times changes their locations so as to maximize network efficiency or to meet specific mission demands, a good way to look at the table is that it shows the locations for a baseline TDRSS constellation.

If one were to take a close look at the satellite locations making up the TDRSS constellation, it can be seen that they are clustered in groups of roughly 130° apart in longitude around the Equator. This spacing is not by chance and has to do with where NASA wanted to put its central network ground terminal.

Take, for example, a case where two satellites are spaced 180° apart in geosynchronous orbit, one over the Eastern Hemisphere and the other over the West. In this arrangement, they would be able to provide complete global coverage. But due to curvature of Earth, however, *two* ground terminals would be required to communicate with them. If this spacing were to be reduced, however, from 180° to 130°, then only a single ground terminal would be needed.

Goddard network planners understood this well and very early on in the program, decided to take advantage of this by locating a single terminal at White Sands in southern New Mexico. The White Sands Ground Terminal (WSGT) provides a perfect line-of-sight vantage point from the western United States where communications with both TDRS-East and TDRS-West could be maintained. To protect physical security, NASA also wanted a location in the continental United States. Finally, like the Mojave Desert of California, White Sands is relatively dry in terms of annual rainfall, which is important since rain can interfere with Ku-band transmission—one of its few disadvantages.

In addition to meeting these requirements, White Sands had also continuously served NASA since 1961. Taken together, the decision to put the TDRSS ground terminal there was really quite logical. It is interesting to note that when the TDRSS Source Evaluation Board (SEB) was deciding between Western Union and RCA as to which would be awarded the contract, it gave the option for both bidders to propose putting the central ground terminal elsewhere, as long as it was within the continental United States. Neither bidder chose to do that, both opting instead to use the government-furnished land on White Sands, the birthplace of America's missile testing activities 30 years earlier.[59]

Located 25 kilometers (16 miles) northeast of the city of Las Cruces, New Mexico, the WSGT is one of the largest and most complex communication terminals ever built. Run by the Space Network Project Office at the GSFC, the WSGT provides the acquisition and relay hardware and software necessary to ensure uninterrupted communications between customer spacecraft in orbit and the NASA Integrated Services Network (NISN) that interfaces to the various spacecraft control centers. In other words, it is the critical hub on the ground that links a user spacecraft to its control center. Without it, data from the TDRSS cannot reach its user and commands cannot be sent up to the satellite.

The NISN provides the critical ground circuits which make the system a true network; without it, TDRSS would just be a collection of satellites and antennas. The ground terminal maintains each TDRS spacecraft in a nominal communication mode (Ku-band) at all times and ensures that all systems aboard the spacecraft are properly configured and functioning properly. It transmits the so-called "forward" link traffic to each TDRS spacecraft for relay to the designated user satellite. Conversely, the ground terminal receives and processes customer spacecraft "return" link, formats and then transmits the data to the NISN interface which carries the data to the rest of the user community.

In addition to providing data services, the health and status of each TDRS spacecraft must be monitored. This is done by flight controllers at White Sands who also track "the birds" in space. As with any large space project, testing and simulation are done on a regular basis so as to evaluate the performance of all the elements that make up the system. For example, "mission sims" are conducted with White Sands sending commands via the tracking and data relay satellites to the user spacecraft, ordering it to perform certain functions and self-test diagnostics. If the tests involve the Shuttle or the ISS, these commands would originate from the JSC in Houston. Otherwise, they

The White Sands Ground Terminal (WSGT) is the central hub of NASA's Tracking and Data Relay Satellite System (TDRSS). It continues the space agency's tracking and data network presence on the south New Mexico Range, a legacy that dates back to 1961. (NASA Image Number HQTC83-907)

would come from the Project Control Centers at the GSFC or the Network Control Center (NCC) at White Sands.[60]

From the outside, the complex is dominated by three 60-foot Ku-band dish antennas. Designated "North," "South," and "Central," they are the link from the ground to the TDRSS spacecraft in geosynchronous orbit 35,900 kilometers in the sky. They handle every aspect of TDRSS transmissions, from voice to television to data. Satellite commands received from various NASA sources are also modulated onto Ku-band frequencies and transmitted to orbit via the system. Because of the extremely short wavelength of Ku-band signals, there is very little room for error. These antennas are extremely precise. Surfaces of these antenna dishes cannot deviate by more than 0.5 millimeter (0.02 inches) from norm (about the width of 20 human hairs) under the extremes of the Southwest desert climate, such as temperatures and winds, plus the loading variations introduced by gravity at various pointing angles. In addition to tolerance, they also have very narrow beamwidths operating in the Ku-band. As a result, Harris had to build them to very fine specifications so that they can be pointed at anytime to within 0.03 degree and track within 0.01 degree accuracy.[61]

The complexity of the system can be illustrated by looking at what goes on inside the TDRSS Operations Control Center, the large building next to the antennas. Satellite command and control functions ordinarily found in the space segment of a traditional communication system are, for TDRSS, performed by the ground terminal. At the heart of the WSGT are the three redundant Space-to-Ground Link Terminals (SGLTs) each of which is supported by one of the Ku-band antennas to transmit and receive user traffic. Here resides over 300 racks of state-of-the-art electronics equipment that handle everything from data routing to precise timing synchronized to the United States Naval Observatory cesium clock to nanosecond—one-billionth of a second—accuracy.

The three SGLTs operate autonomously and are, for the most part, fully redundant. This means that if one of the SGLTs were to fail, then only the TDRS and services supported by that SGLT would be impacted. Breaking down the system even further, each SGLT is capable of providing four, Single Access, forward and return services for customers. In addition to SA services, two of the three SGLTs can support up to five MA return services along with one forward MA service.

From this control center, NASA can schedule TDRSS support for users and distribute the data from White Sands. Also at the ground terminal are several smaller S-band Tracking, Telemetry & Command System (STTCS) antennas. These are used to provide contingency communications to a TDRSS spacecraft in the event of a SGLT failure. They are also used to communicate with the other on-orbit spare satellites. As an everyday analogy, the STTCS is somewhat like the "service elevator" in the back of a five-star hotel that is used for maintenance, whereas the SGLTs are like the main "guest elevators" that go directly from the guest floors to the front lobby. The White Sands ground terminal and satellites are all automatic and receive their operational inputs from the NCC at the GSFC. The NCC is critical to the operation of the system and is in many ways the brain of the system. Several functions are carried out by the NCC: 1) It serves as the user interface and command center of the system. 2) It provides overall management and monitoring of the system. 3) It sets up conflict free schedules and establishes the user unique configuration details (satellite assignment, start and stop times, antenna assignment, pointing information) required for the satellites. To this end, over 40 unique configurable items can be provided by the NCC.[62]

As technically challenging as the whole process seemed, the Agency had good evidence that it was all going to work out. A data relay satellite, ATS-

6, had been used with success in 1975 on the Apollo-Soyuz Test Project. Two years later, on 6 December 1977, the Seasat Program provided for data transmission via satellite from Alaska simultaneously to the GSFC and the Naval Fleet Numerical Weather Center in Monterey California. Even though the data rate was a low 1.544 Mbps, the transmissions served as a feasibility demonstration for the WSGT which would end up using the same types of circuits.[63]

On 17 August 1981, four years after ground break on the project, Ed Smylie, Associate Administrator for Tracking and Data Acquisition, presided over the acceptance ceremony of the White Sands TDRSS Ground Terminal. Other NASA dignitaries included Jesse C. Jones, the new Facility Manager and his Deputy Louis Gomez. The opening of this new communications terminal—the largest of its kind anywhere in the world—was a much needed infusion to the south New Mexico economy which has been tied so closely to the DOD. In 1981, for instance, the value of NASA's contracts and grants to institutions in New Mexico, and White Sands in particular, provided between $20 to $30 million per year and accounted for 600 jobs and 66 contracts in the private sector and universities such as New Mexico State University in nearby Las Cruces and the New Mexico Institute of Technology in Socorro.[64]

With Holloman Air Force Base—operating right outside the gates of White Sands Missile Range—soon to be designated as the home for the Air Force's then most advanced and stealthy aircraft, the F-117A Nighthawk (better known as the Stealth Fighter), the flatlands of Otero County soon boasted some of the most advanced technology found anywhere. Added to this was the diverse work NASA was doing with the Department of the Interior in the use of remote sensing for diverse applications such as timber management, land cover classification, grasslands range management, and deer habitat identification.

On the cultural realm, satellite remote sensing technology supported by TDRSS was used by the National Parks Service to uncover features of prehistoric ruins not visible by conventional aerial photography. As an example, Smylie pointed out in his dedication speech the new insight into the society of the Anasazi Indians that had been gained by remote sensing. TDRSS continues to support Earth science research today.[65]

★ ★ ★

With six TDRSS spacecraft now in orbit, the question of reliability and the need to support more than just three operational satellites (TDRS-East, TDRS-West and the spare) became an issue. The WSGT had three antenna systems, perfect for supporting these three operational spacecraft. But now six TDRSs were in orbit all needing support from the ground. This, combined with a host of data-intensive missions that NASA was planning—missions such as the Great Observatories, Spacelab and Spacehab (orbital workshops attached to the Shuttle payload bay), Space Station *Freedom* (the canceled, U.S.-only forerunner to the ISS) and the Cosmic Background Explorer—and the huge amounts of data returned to Earth all pointed to the need for a second TDRSS ground terminal. In August of 1987, NASA approved Project 9717 to construct a Second TDRSS Ground Terminal at White Sands.

The STGT, as it would be called, is identical to the first terminal and is in fact located just five kilometers (three miles) to the north. Its purpose is really twofold: In addition to keeping up with America's spaceflight communication requirements in the 1990s and beyond, it would serve as a backup to the WSGT, eliminating it as a single point of failure in the event of a breakdown or during planned outages for system upgrades and repairs.[66] This point was driven home on 1 September 1983. On that day, controllers were busy checking out the TDRS-1 spacecraft after it had finally made it to its duty station in geosynchronous orbit, when a sudden failure at the WSGT caused a three-hour communication outage with the Shuttle (STS-8). Flight controllers did not wake the crew, however, since all indications through other communication links (transponders were in place which could operate in either TDRS or ground network mode) showed that everything was otherwise normal onboard the vehicle and that this was strictly a communications problem.[67] Nevertheless, it was a good lesson that a backup was needed. In fact, this second ground terminal was considered so important that design specifications called for it to have greater than 0.9999 reliability, or less than one hour per year of down time.[68]

In 1987, the TDRSS program office initiated competitive definition phase studies for the development of a STGT. A year later, General Electric's Military and Data Systems Operations of Valley Forge, Pennsylvania, received the prime contract to build the second terminal. This included all the design, development, installation, and testing of the $245 million worth of communication and computer hardware along with all the software. The $14 million building construction contract was awarded to Argee Corporation, a civil and mining construction company of Denver, Colorado.[69]

As massive as the original, this new terminal also boasted a 7,430-square meter (80,000-square foot) operations building, a 2,320-square meter (25,000-square foot) technical support building and an 830-square meter (9,000-square foot) power plant. Coming on the heels of the *Challenger* accident, and with the rather significant windfall to southern New Mexico economy, ground breaking for the new terminal on 9 September 1987 was quite the public affairs event. Speakers included Robert O. Aller, NASA Associate Administrator for Space Tracking and Data Systems; GSFC Deputy Director John J. Quann; and Captain Frederick H. "Rick" Hauck, Commander of the first mission following *Challenger*. In addition to representatives from State and U.S. Congress, dignitaries included Major General Joseph S. Owens, Commander of the White Sands Missile Range; John P. Stapp and Gregory P. Kennedy from Alamogordo's

Photograph of the Second TDRSS Ground Terminal (STGT) at the White Sands Missile Range in Southern New Mexico. Towering over the main Operations Building are the three 18.3-meter (60-foot) Ku-band antennas. The San Andres Mountains are in the background. (Photograph courtesy of NASA)

own International Space Hall of Fame (one of New Mexico's top tourist attractions); and even archaeologists from nearby Las Cruces.[70]

Before any concrete could be poured, though, NASA had an obligation, this one regarding the environment. In keeping with its federal mandate to protect cultural and natural resources, test excavations had to be conducted near the site of the terminal to see if construction would adversely impact any significant archaeological or historical sites. To this end, the space agency hired the firm of Batcho & Kauffman from Las Cruces to serve as archaeological consultants for this new Space Age project.

Sure enough, excavations soon uncovered Native American artifacts on the site. Further digs revealed that NASA had in fact stumbled onto quite the archaeological find. In their report, the archaeologists noted that: " . . . it soon became apparent that one of the sites contained the undisturbed remains of a pithouse settlement, while the other—located a few miles farther south—contained the remains of a temporary camp, probably once used to gather and process wild foods."[71]

Further research showed these pithouses to be a common type of dwelling used by prehistoric Indians in the Southwest United States. Charred roofing material was also found which carbon dated to some time between 650 and 750 A.D., meaning the site was more than 1,300 years old! In addition to the pithouse settlements themselves, a broad area around the dig was also excavated in what archaeologists call the "activity areas." The completeness of the find was confirmed as the activity areas contained the remains of outdoor camp and cooking fires, as well as large quantities of debris including pieces of broken pottery, several arrowheads and discarded or broken stone tools and the chips of stones leftover from making them. Also found was a large amount of burnt and unburnt animal bones—the last remains of many meals.

Because of the find, NASA had to move to a second, nearby site. It too was excavated. Though not as robust as the first site, a well-preserved roasting pit, about 1,000 years old, was found. Based on information from early settlers in the area, archaeologists were able to trace the find back to the original Mescalero Apaches of the Southwest.

Construction of the terminal eventually embarked on a plot of land near the archaeological find. As serendipity would have it, what started out as NASA simply fulfilling a legal obligation unexpectedly turned into a portal to the past. As one of the archaeologists on the project put it: "While construction is about to begin on this new, high technology facility—to give us another window into space—archaeologists have, likewise, been able to open a small, yet intimate, window into the dim past."[72]

Two and a half years after the ground breaking ceremony, Agency officials once again returned to White Sands, this time to hold a formal ribbon cutting ceremony dedicating the new and second White Sands terminal. Present at the February 1990 ceremony were NASA Administrator Richard H. Truly and his wife; Goddard Center Director John W. Townsend, Jr.; a contingent of New Mexico officials from Albuquerque and Las Cruces; and astronauts John E. Blaha and James F. Buchli, crewmembers of STS-29 that deployed TDRS-4.[73]

With civil construction finished and the new terminal set to open, the Agency wanted something special to tie the White Sands Complex (note the new name) to the Native American and Southwestern roots of New Mexico. After considering several options, the Office of Space Communications, along with the nonprofit New Mexico Space Grant Consortium and New Mexico State University, decided to sponsor a "Name the Ground Terminals" contest.

In keeping with the spirit of the "Land of Enchantment" and the Agency's charter, entries had to 1) Relate to Native American, Hispanic or African American local culture; 2) Be appropriate for space communications and America's involvement in space; 3) Limited to one to two words in length; and 4) Show relationship between the two names. Teams from elementary, middle and high schools in qualifying school districts of southern New Mexico competed. These teams had to abide by some simple rules, such as four students per team along with a team coordinator. Teachers were responsible for guiding their team's activities and for submitting their entry. And each team could submit only two names, one for each ground terminal.[74]

Just as NASA had hoped, the contest proved to be popular, especially among elementary and middle schoolers. More than 100 entries were received. From these, two names—submitted by a team of four girls from Zia Middle School in Las Cruces—were selected: Cacique (kah-see-keh) which means "leader" and Danzante (dahn-zahn-teh) which means "dancer". Roots of the winning names can be traced back to the Tortugas Indians who preserve their culture through traditional dance. In reaching the names, "the students compared the TDRSS to the Tortugas dancers. The dancers com-

The Compton Gamma-Ray Observatory (CGRO) is deployed by the Remote Manipulator System aboard the Space Shuttle Atlantis during STS-37 in April 1991. For nearly nine years, the observatory studied gamma-rays from objects like black holes, pulsars, quasars, neutron stars, and other celestial objects. The information returned have provided scientists clues to the birth, evolution and death of stars, galaxies, and the universe. It reentered Earth's atmosphere and ended its very successful mission in June 2000. (NASA Image Number MSFC-0003356)

municate through complex maneuvers as do the TDRSS satellites, [and] the ground terminals are the leaders of this orbital dance," said Wilson T. Lundy, Manager of the White Sands Complex, in an interview after the winning entrants were selected.[75]

NASA was elated. As Charles Force put it, "To those familiar with the culture of the Southwest, these names will give meaning to the purpose of the stations. To those who understand the role of the stations, the names will convey appreciation for the culture of the area."[76] Although the names of the stations were never really embraced by the technical community, the contest was politically successful and had more than fulfilled its purpose.

As for the four girls from Zia Middle School, they received a two-day, all expenses paid trip to tour the JSC in neighboring Texas. In a ceremony on 17 May 1993, the names for the White Sands terminals were officially announced, with presentation of awards to the students by retired Apollo 8 Commander and Las Cruces businessman Frank Borman. A year later, the Danzante terminal was accepted by NASA and declared a fully operational part of the TDRSS.

* * *

On the morning of 5 April 1991, the Shuttle *Atlantis* took off on a six-day mission, the highlight of which was deployment of the CGRO. Named after Ohio Nobel Prize laureate Arthur Holly Compton for his research demonstrating the particle behavior of electromagnetic radiation, the second of NASA's "Great Observatories" to be launched into space, the CGRO, at 17 metric tons (37,500 pounds), was the heaviest astrophysical payload ever flown into space.

The Great Observatories of NASA were four of the largest and most powerful space-based telescopes ever put into orbit. Each was similar in terms of its size, cost and scope of the program, and all have since made a substantial contribution to our understanding of the deep space environment, greatly expanding our knowledge of the known universe. Each of these four space-based observatories was designed to investigate a specific region of the electromagnetic spectrum.

Undoubtedly the best known of the four is the first one to be put into space: the $1.5 billion Hubble Space Telescope (HST). Launched aboard STS-31 on 24 April 1990, the HST primarily observes the visible spectrum. Besides the incredible photographs that have since come from the telescope, it also received a lot of media scrutiny early-on over its "blurred vision," a manifestation of a manufacturing imperfection in which the objective mirror was ground too flat by 2.2-microns, or 1/50th the width of a human hair. Demonstrating the irreplaceable value of human spaceflight, this error was corrected when the crew of STS-61, over the course of four spacewalks, installed and checked out corrective optics to the telescope in December 1993.

Besides the HST and CGRO, there is the Chandra X-ray Observatory, launched on 23 July 1999 aboard STS-93 and the Space Infrared Telescope Facility (SIRTF) whose primary mission, as its name implies, is observation of the infrared spectrum. (The SIRTF, launched on 25 August 2003 aboard a Delta II rocket, was later renamed the Spitzer Space Telescope.)[77] Aside from performing each telescope's own mission, most of which cannot be replicated by ground observatories, the Great Observatories program allows the four to synergistically interact with each other for greater combined scientific returns. Each astronomical object in the sky radiates in different wavelengths. But by training two or more observatories on an object, combined data can be returned to paint a much more comprehensive picture than is possible with just a single instrument.[78]

After its deployment from STS-37, the CGRO operated as advertised for almost a year, returning more data on that portion of the electromagnetic spectrum than the previous six decades put together. But in March 1992, it suffered a failure of its two onboard tape recorders which restricted downlinks of scientific data to real time only. With the tape recorders gone, CGRO was able to relay only slightly more than half of the science data it collected, because it could not point at a TDRS all the time.

While TDRSS coverage had been about 65 percent of each orbit, scientists could not even collect that percentage of data anymore because Compton's instruments had to be turned off during the part of each orbit when it passed through the elevated background radiation of the South Atlantic Anomaly—a region of significantly increased space radiation experienced by satellites passing over the South Atlantic Ocean.[79] This reduction in data return presented an obstacle to the Goddard science team. NASA, understandably, wanted to get back to the point where all of the data could be retrieved. Furthermore, real-time data dumps could only be done at the very slow rate of 32 kilobits-per-second whereas the playback rate was 512 kilobits-per-second.[80]

Considering all these factors, in March 1992, Goddard's Mission Operations and Data Systems Division was tasked to study approaches to solve this problem utilizing any combination of ground or space resources available. Analysis quickly ruled out an independent, Compton-only, ground station as a solution due to potential high cost with a relatively small increase in additional coverage. An on-orbit Shuttle repair was also looked at but proved too costly, even if just one time. But the same study showed that a TDRSS solution could produce (up to) full, 100 percent coverage for the Compton observatory.[81]

The solution was this. One of the existing TDRS spacecraft had to be moved and located somewhere over the Indian Ocean. Despite the fact that TDRS-1 was near the end of its 10 year design life, it was apparent that its remaining functionality—fuel, health, and condition of onboard instruments—was still meeting the requirements needed for an Indian Ocean satellite. Since this location could not be viewed by the White Sands Complex

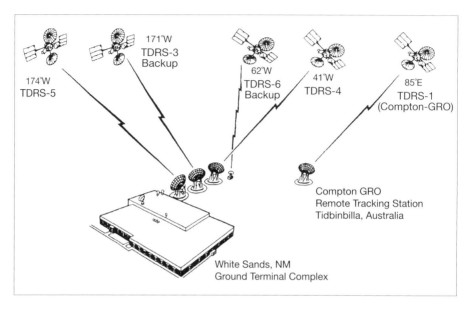

The first generation Tracking and Data Relay Satellite System (TDRSS) constellation as it appeared in 1994, with five orbiting satellites—two operational and three backups—in communication with the White Sands Ground Terminal (WSGT) and the GRO Remote Terminal System. (Space Shuttle Mission STS-54 Press Kit, January 1993, NASA Headquarters)

(being inside the so-called Zone of Exclusion from North America), the solution was to consider a ground terminal which could see a TDRS spacecraft if placed over the Indian Ocean.

The existing DSN sites at Madrid and Canberra could observe the TDRS and were thus (initially) the prime candidates. Of the two, Canberra had a slightly better line-of-sight. In addition, it had the advantage of being located in an English speaking country and had a NASA-like culture in its operating infrastructure, the Australian Space Office (ASO). Other ground locations were examined too, for example eastern Africa, but were disqualified mainly because they were not under direct NASA control.[82]

Following completion of the study that summer, Goddard sent out a site survey team, which along with members of the ASO, visited five sites throughout the commonwealth. Reminiscent of the old STADAN days, the team considered such factors as existing hardware, accessibility to long distance communications, transportation and overall logistical support requirements. Based on this survey, the NASA site at the Canberra Deep Space Communication Complex (CDSCC) was selected. The pace of establishing the site was of a high priority since a period of "best science" solar activity was then fast approaching.

The $12 million, GRO Remote Terminal System (GRTS) project was started without delay on 1 September 1992. Scheduled for completion in 13 months, the station was built leveraging maximum use of existing equipment. Essentially all of the TT&C equipment was transferred from existing resources at other Goddard facilities. Redundancy in design was exploited, to the extent feasible and practical, so as to attain good mission assurance. In addition, the TT&C equipment used was purposely identical to the existing equipment on the CDSCC Deep Space side so that any additional training, repair and logistics would be minimized. All of the remaining hardware, such as the 9-meter (30-foot) S-band and 5-meter (16.4-foot) Ku-band antennas, were bought using existing, commercial-off-the-shelf (COTS) designs. In fact, the overall design of the GRTS was based on that recently used to complete the STGT at the White Sands Complex.

With the help of Raytheon Service Company as the procurement agent and Allied Signal Technical Services Corporation providing technical support, Goddard was able to complete the entire procurement process—specifications, solicitations, and negotiation—by January 1993. The Australian contribution was significant too, as all of the construction was done in four months. This included two antennas, two new S- and Ku-band transmitter buildings and a two-kilometer fiber optic cable-run to a remote calibration site—an amazing feat in that amount of time.[83]

On 29 November 1993, White Sands sent a series of commands to begin drifting TDRS-1 from its location over the Phoenix Islands in the Pacific to the Indian Ocean. The trip took 73 days. A week later, the nearly completed GRTS at Tidbinbilla made first contact with the satellite. Then on 9 February 1994, commands were sent to stabilize the spacecraft at its duty station 85° East longitude over the middle of the Indian Ocean. TDRS-1 was now perched atop the Eastern Hemisphere and NASA finally had a truly global SN. Data from the Compton was received by TDRS-1, downlinked to Tidbinbilla, relayed up to an Intelsat commercial satellite where it was downlinked to a commercial terminal on the West Coast and then routed to White Sands. From there, the data was distributed to scientists around the world. Control of TDRS-1 and the Tidbinbilla ground terminal remained at White Sands, marking the first time NASA controlled an out-of-view TDRS from that location.

On 14 March 1994, the Agency officially announced the opening of the new, remote ground station in Tidbinbilla, Australia. "With activation of this ground facility, the TDRS System can, for the first time, provide global coverage," said Charles Force in declaring the new TDRSS station operational. "While the new ground station is devoted to Compton at this time, it has the potential for use by other Earth-orbiting spacecraft."[84]

Compton scientists were elated. Frank J. Stocklin, a mission manager at Goddard compared the added capability to the repair of the Hubble Space Telescope:

> We're very pleased that this project came in on budget and on time and that we are able to collect additional, significant data from Compton in a cost-effective manner. It's difficult to place a dollar value on the additional science data obtained in this effort, but the restoration of data recovery capability is similar to that done for the HST, and marks the second successful recovery of a major NASA observatory.[85]

Almost immediately, Compton scientists saw a 30 percent jump in data returned from their observatory. As useful as the Tidbinbilla station at the CDSCC was, though, it still had its fair share of drawbacks. First, the location resulted in a lower than desirable elevation look angle to the TDRS in orbit. Another problem was related to the inability of TDRS-1 to point its Space Ground Link (SGL) antenna far enough south to Canberra to maximize the coverage duration for users besides just the Compton observatory. While not a serious problem, another location could be better. Then there was the cost factor. The ongoing grip of a fiscal mandate to reduce annual maintenance and operating costs required some form of ground station automation. And finally, there was the geopolitical factor. While the British Commonwealth is among the strongest of America's allies—cooperation of Australia with NASA had been impeccable since the days of Minitrack—the United States wanted something as important as an overseas TDRSS ground terminal on American soil, if at all possible. TDRSS had become a national resource. Although not a military asset, the missions and programs it supported had national security implications. A "U.S. territory-based solution" was highly desirable.

Guam, once again, stood out. In addition to being a longtime U.S. territory, there is the stability offered by virtue of having key DOD presence on the island. From its location in the Mariana Islands, Guam is closer to the Equator and longitude to a TDRS spacecraft over the Indian Ocean, allowing it to accommodate much higher antenna elevation angles than is possible from Australia. Besides, the Agency had only just left the island in 1989, finally closing down the Guam STDN station after 24 years as one of the most successful stations in the history of the Agency's networks. Thomas A. Gitlin, Goddard's former ground terminal Project Manager summarized it concisely: "NASA built the Guam ground station to significantly expand the quantity and quality of services we provide to all our customers."[86]

In 1995, NASA was ready to begin funding for a new Guam Remote Ground Terminal, or GRGT. With the acceptance of the second terminal at White Sands the previous year and the launch of TDRS-7 to complete the

Chapter 7 \ A Network in Space

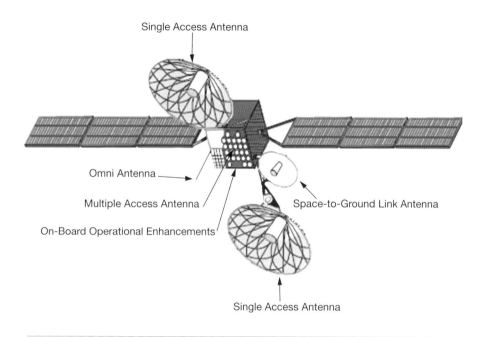

Multiple Access Antenna
- 32 receive antenna elements
- 15 transmit antenna elements
- S-band communications:
  - 2.1064 GHz (Forward)
  - 2.2875 GHz (Return)
- LHC polarization
- ±13° conical Field-of-view

Single Access Antenna
- Tri-frequency communications:
  - S-band: 2.025–2.120 GHz (Forward), 2.200–2.300 GHz (Return)
  - Ku-band: 13.775 GHz (Forward), 15.0034 GHz (Return)
  - Ka-band: 22.55–23.55 GHz (Forward), 25.25–27.50 GHz (Return)
- Circular polarization (LHC or RHC)
- Field of View:
  - Primary: ±22° E-W, ±28° N-S rectangular
  - Outboard: 76.8° E-W
  - Inboard: 24° E-W
  - Extended: ±30.5° N-S elliptical

Space-Ground Link Antenna
- 2.4m Ku-band antenna:
  - 14.6–15.25 GHz (uplink)
  - 13.4–14.05 GHz (downlink)
- WSC/GRGT-TDRS uplink/downlink
- Orthogonal, linear polarization
- Modified frequency plan allows collocation

Enhancements
- On-board SA antenna control
- Autonomous recovery from anomalies
- Improved monitoring

Omni Antenna
- S-band TT&C
- LHC polarization

The second generation Tracking and Data Relay Satellite. (TDRS-H, I, J). (Adapted from Space Network User's Guide, SNUG-450, Revision 8, NASA Goddard Space Flight Center)

satellite constellation, $9 million of SN funds became available for this project. The remaining $12.4 million needed (for a total of $21.4 million) also came from within the SN program office, but in two parts—from phase-down of the Compton GRTS in Tidbinbilla and from an unexpected source: greater-than-anticipated reimbursements by the Columbia Communications Corporation for revenues from their agreement with NASA for the lease of excess C-band services on the TDRSS. This last point could be called the "remnant" of the Western Union debacle, albeit a positive one. Under the original TDRSS contract, a reimbursable, long-term plan for using the commercial capability built into the original Western Union TDRS design was negotiated. Despite a long, drawn-out, legal process to recover the expected commercial reimbursement (involving the Small Business Administration and the courts), the nightmarish process did eventually return funds to the Agency and was a good use for a C-band system that was otherwise totally superfluous.[87]

The GRGT was designed from the beginning to be a fully automated, remote station, identical in most respects to its White Sands counterpart in the United States. Situated on the secure grounds of the Computer and Telecommunications Area, Master Station Receiver Site of the U.S. Navy base, the station is distinguished by two large radomes which enclose the 5-meter (16.4-foot) Ku-band and 9-meter (30-foot) S-band antennas, protecting them from the typhoons of the central Pacific. Equal in performance with the terminals in New Mexico, the Guam terminal provides relay services in the form of two S-band and two Ku-band forward and return links. High rate, forward service to customer satellites is done at 25 million bits-per-second (Mb/s) while the return service rate is double that, at 50 Mb/s.

Three years after getting the go-ahead, the GRGT was officially opened in a ribbon-cutting ceremony held on 15 July 1998. Although Governor Guerrero was not present this time, the legacy he helped set in bringing NASA's first tracking station to the island three decades earlier had, in a way, come full circle. With the Guam terminal operational, the SN's Zone of Exclusion was closed and TDRSS could now provide 100 percent coverage regardless of where a satellite is in low-Earth orbit. With the project completed, the original Compton remote terminal in Tidbinbella was shut down as planned.[88]

* * *

Even before the first TDRS was deployed by the crew of STS-6 in April 1983, NASA was already planning for the day when the original TDRSS spacecraft would need to be replaced or replenished after their projected 10-year service life expired. The space agency (and space communications in general) could take advantage of an increase in capabilities brought on by a more advanced, second generation of TDRSS spacecraft. A big factor was that with TDRS-1 through 7, communication links for the Space Shuttle, the HST and its Great Observatory companions, and other Earth-orbiting space missions are limited to the S- and Ku-bands.

In 1981 and 1982, the Office of Space Tracking and Data Systems (OSTDA) conducted a "Prephase A Advanced Study" to look at an advanced TDRS System in which communications would utilize the even more efficient Ka-band of the radio frequency spectrum. Even at that time, the increasing number of users in the S-band was starting to crowd that part of the RF spectrum. It was obvious that the congestion was only going to get worse as the number of satellite users increased in the coming years. With the second generation—or TDRS-II—spacecraft, users would be able to take advantage of Ka-band links to transmit at higher data rates. Along with the higher frequency, smaller antennas could be used than those required at Ku-band—just like smaller antennas are required for Ku-band compared to S-band (and VHF before that).[89]

Following the cessation of all Shuttle flights that ensued after *Challenger*, a Phase A Preliminary Analysis for the TDRS-II was conducted, even as the initial satellite constellation was still being completed. A Phase B Definition Study followed in August of 1990. With this year-long study, specific requirements of a TDRS-II spacecraft were defined, along with specifications and a roadmap of the potential migration of services to the Ka-band. Issues which would affect this migration of services to Ka-band were addressed, such as availability of commercial off-the-shelf space-qualified antennas and equipment with acceptable performance, weight, size, power consumption, and cost. On the user end, the study looked at the development and qualification of customer antennas which would be needed.

The replenishment program would have three TDRS-II spacecraft—designated TDRS-H, I and J—that would support customer services currently provided by TDRS-1 through 7. The three new satellites would be functionally equivalent to the original spacecraft with the exception of the added Ka-band communications capability and an improved MA capability. But there was a major difference, one primarily philosophical. The original TDRS spacecraft—not including TDRS-7—hosted a Ku-band commercial payload which was to have been used by Western Union but was never activated, and a commercial C-band antenna and payload package, two of which are operated by a commercial service provider. To stay far away from the "shared system" approach this time, TDRS-H, I and J were dedicated from the beginning to NASA missions and did not include a commercial Ku or C-band payload. To minimize impact to the user community, the spacecraft was designed such that Ka-band used the same SGL design that the original Ku-band used. In this way, transmissions at the new frequency were essentially transparent to the ground station.[90]

Looking like a high-end version of the original spacecraft, the second generation TDRSS spacecraft was still dominated by two 4.5-meter

(14.8-foot) diameter steerable SA antennas and a pair of wing-like, solar arrays spanning almost 21 meters (68 feet) from one end to the other. But with a fully-fueled launch weight of 3,175 kilograms (7,000 pounds), it was nearly 900 kilograms (2,000 pounds) heavier than the original.[91] Based on the then newly developed Hughes Spacecraft 601 bus structure, the electrical power,

TDRS-H, I, and J could provide over two and a half times the data relay capability of its predecessors by using Ka-band and other new features. (Photograph courtesy of NASA, *www.gsfc.nasa.gov/topstory/20021127tdrs_j.html*, accessed October 2, 2005)

attitude determination and control system, and the TT&C units were all mounted on the central bus structure, as were the solar arrays.

While the original TDRS used hydrazine monopropellant, the new spacecraft now used the higher performing bi-propellant combination of monomethyl hydrazine fuel and nitrogen tetroxide oxidizer for attitude control and main propulsion. This was a proven propellant combination that has been used in the Apollo spacecraft and the Space Shuttle. The RCS used this propulsion system to feed a 110-pound thrust (490-newton) liquid apogee kick motor (used for orbit insertion), along with four 2-pound thrusters (9-newton) and eight 5-pound thrusters (22-newton) mounted around the periphery of the main spacecraft bus to support on-orbit operations over its 15-year service life.[92]

In addition to the RCS jets, attitude control was passively maintained using a gimbaled momentum wheel for three-axis torquing and angular momentum "storage." Continuously operating gyros—updated by Earth and Sun sensors on the spacecraft—provided highly accurate, three-axis attitude sensing to point the spacecraft and its antennas in the proper attitude

Unlike the first generation of Tracking and Data Relay Satellites, TDRS-H, I and J were launched using expendable launch vehicles. Here, TDRS-H rises from PAD36A, Cape Canaveral Air Force Station at 8:56 a.m. EDT on 30 June 2000 atop an Atlas IIA/Centaur launch vehicle. The new satellites augmented TDRSS's existing S- and Ku-band capabilities by adding a Ka-band capability. (NASA Image Number KSC-00PP-0825)

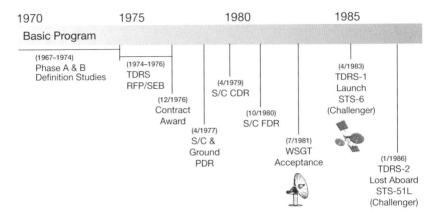

Acronyms:
CDR–Critical Design Review
FDR–Final Design Review
GRGT–Guam Remote Ground Terminal
NAR–Non-Advocate Review
PDR–Preliminary Design Review
PER–Pre-Environmental Review
RFP–Request for Proposal
SEB–Source Evaluation Board
STGT–Second TDRSS Ground Terminal
STS–Space Transportation System
TDRS–Tracking and Data Relay System
TDRSS–Tracking and Data Relay Satellite System
WSGT–White Sands Ground Terminal

Chronology of the NASA Tracking and Data Relay Satellite System (TDRSS), from concept to reality. (NASA Goddard Space Flight Center)

Chapter 7 \ A Network in Space 297

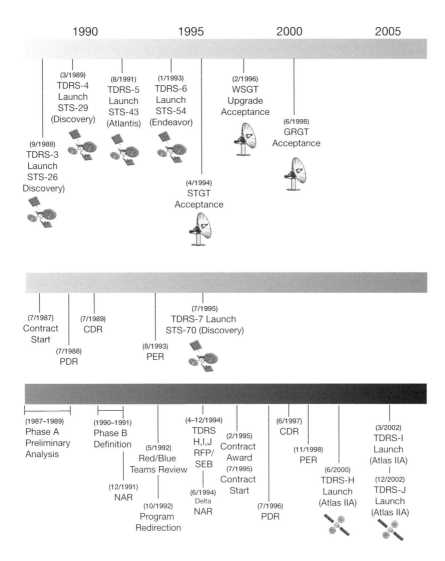

in the weightlessness of space. Since the nature of Ka-band transmissions required a narrower beam and thus tighter pointing accuracy than Ku or S-band, the rate gyros used on TDRSS-II were much more robust and had significantly fewer moving parts (that can wear out) than those on the first generation satellites.[93]

Integrated into the main bus structure was a system of heat pipes, multi-layer insulation, radiators and thermostatic heater controls that provided thermal control to the spacecraft—a necessity in the harsh environment of geosynchronous orbit. Then there were the two wing-like power arrays covered with silicon solar cells designed to last 15 years. They provided approximately 2,300 watts of power, enough to light some 30 common household light bulbs. Besides providing electrical power to the spacecraft, they also charged four nickel-hydrogen battery packs which supplied power when the spacecraft was in darkness.[94]

Just like the first generation TDRS, the most prominent part of spacecrafts H, I, and J were the two, mechanically steerable SA antennas. Made of a flexible, graphite reinforced, epoxy mesh, the antennas were furled into a taco-like shape and stored for launch. Once deployed, they unfurled and with an innovative "spring-back" design, fine adjustments could be made to compensate for on-orbit changes in the dish contour from things like heating and cooling in the vacuum of space. The SA antennas used a tri-band electronic feed—the device at the focus of the antenna which receives and transmits signals—to accommodate frequencies in S-, Ku- and Ka-bands. With S-band, user satellites with lower gain (less sensitive) antennas, or MA users temporarily requiring an increased data rate, could be accommodated.[95]

It was used, for instance, to support human missions, science data missions such as the HST, and satellite data dumps. With Ka-band, higher bandwidth items such as high-resolution digital television—including all Space Shuttle video—could be relayed. Also, more transmission traffic and higher volumes of data could be dumped to the ground. Finally, with the significant increase in transmission performance (so called "figure-of-merit" increase) afforded by Ka-band, transmission rates approaching the realm of a billion bits-per-second (1 Gbps) were possible. At the TDRS-II specification of 800 million bits-per-second, it was over two and half times faster than what was possible with the original TDRS operating at Ku-band.[96] Again using the encyclopedia analogy, that was somewhat akin to downloading ten 20-volume encyclopedias each second.

Request for Proposals to build the three next generation satellites were issued in April of 1994. After a six-month evaluation, the SEB consisting of members from GSFC and NASA Headquarters presented its recommendation to the Source Selection Official, Charles Force. Force and the SEB were convinced that Hughes Space and Communications of Los Angeles (now Boeing Satellite Systems after its acquisition in 2000) was the best contractor for the job. Not wanting a repeat of the whole Western Union affair, the agency this time went with a well-established satellite manufacturer and the producer of the commercially proven 601 spacecraft bus design. (Founded by billionaire aviator Howard R. Hughes, Jr., the company was in fact the world's largest supplier of commercial satellites in 1995.)

Work started in July, five months after the official announcement. This time, the progress was smooth. Over the next two years, the contractor worked with GSFC to move TDRS-II from a set of requirements onto the drawing table and finally into a design which would fly. After passing the Critical Design Review in June 1997, the pace picked up as manufacturing and testing on the first new satellite, TDRS-H, entered final production.[97]

One major difference between the second generation TDRS spacecraft and their predecessors was in the way they went into space. The original TDRS were launched by the Space Shuttle exclusively. In a move that can still be traced back to the *Challenger* accident, TDRS-H, I, and J were launched by an intermediate class of expendable launch vehicles, the Lockheed Martin Atlas II-A. Developed to fulfill an expendable launch vehicle requirement to supplement United States launch capability following the accident, the Atlas II-A, along with its variants, was a two-and-a half stage liquid propellant rocket. (The Centaur upper-stage was a so-called "half stage" since it was used to position the payload into a separation orbit after booster burnout.) Following separation from the Atlas, the TDRS spacecraft was injected into its final orbit using its own apogee kick motor.

Launch services using the Atlas II-A were finalized as early as 1997. Nearly three years later, the new TDRS-II spacecraft was ready. On 30 June 2000, TDRS-H successfully lifted off from Launch Complex 36 at Cape Canaveral. Since the satellite was not launched by NASA, Boeing had the overall responsibility to make sure that it got onto orbit as advertised prior to the Agency taking control. It attained orbit without any problems. Its acceptance by NASA was delayed, though, due to lower than expected performance of the new MA phased array antenna. As a result, ground controllers discovered that 5 of the 18 communications services provided by TDRS-H performed at less than full capability.

This degradation puzzled both Boeing and NASA since the spacecraft had checked out perfectly on the ground. After a month of troubleshooting, the culprit was found. Randy H. Brinkley, President of Boeing Satellite Systems explained at the time that the hidden problem was traced back to a material defect. "We identified the cause of the problem to be rooted in one specific material used in the assembly of the antenna and implemented straightforward corrective measures for TDRS-I and TDRS-J. We are certain that a repeat of this performance shortfall will not occur."[98]

Manufacturing changes were implemented and 18 months later, on 8 March 2002, the next spacecraft was launched. This time, it performed flawlessly. TDRS-I was followed into orbit 9 months later with the launch of the final satellite, TDRS-J, on 4 December 2002.

The three satellites were initially launched into their on-orbit "storage locations" over the Phoenix Islands in the mid-Pacific; off the west coast of South America near Ecuador; and off the Brazilian coast over the Atlantic. There, the satellites stayed, almost in a "garaged" fashion, until they were needed. The advantage of having these spacecraft in orbit was that the Explorations, Operations, Communications and Navigation Systems Division of Goddard, who manages the SN, may change the geosynchronous location of any TDRS to any other geosynchronous location assigned to NASA. This allowed collocation of two spacecraft in one longitudinal setting. Two second generation spacecraft could be located together, if needed, or one first generation with one second generation. This was quite useful and allowed the use of two partially failed spacecraft to be collocated to conserve the limited slots available at geosynchronous altitude and to pool together their capabilities.[99]

With TDRS-II available, the SN was much more flexible and more options could be exercised to optimize the TDRSS network for all users. A case in point was a high data rate user such as a remote sensing mission with large amounts of imaging data. This satellite, which could have onboard several bandwidth-intensive instruments, may generate up to three-*terabits* (3,000,000,000,000) per day of science data. On top of that was the required "overhead" information such as data for link protocols and error correction coding, adding another 16 percent or more to the raw science data.[100] With the first generation TDRS, it would have taken over three hours each day just to transfer this data from the spacecraft to the ground. A Ka-band TDRS-II SA link at a rate of 800 Mb/s reduced this to about 72 minutes per day. NASA may schedule this data transfer in a number of ways. For instance, 5 minutes of TDRSS service for every orbit of a satellite or 10 minutes of service on every other orbit. If the data was time sensitive, Ka-band service allowed for near instantaneous availability of the data to its users, much more so than with the original system. While S- and Ku-band capabilities also provided near instantaneous services, they required significantly longer transfer times.[101]

Now consider the case of a small, low-data rate user such as a single instrument satellite. It too wanted to use the TDRSS to get data to the ground. Although low-data rate users did not require the wide bandwidth channels available at Ka-band, they *could still benefit* from Ka-band services in terms of antenna requirements. Take a small low-Earth orbiting spacecraft that had only one or two low data rate instruments. It may generate only 20 gigabits (20,000,000,000) of science data per day. Ka-band MA service at 4 Mb/s could transfer all that data in less than 7 minutes on each orbit or 13 minutes on every other orbit. In this case, the user only needed a very small 10-centimeter (4-inch) diameter parabolic dish or a phased array antenna on his satellite. Although small, the parabolic dish would still require a tracking mechanism to keep it pointed at the TDRS spacecraft. On the other hand, a phased array antenna would have been especially beneficial since the absence of a steearable antenna greatly simplified attitude control of the user satellite and minimized moving parts on the spacecraft.[102]

NASA now had its long awaited TDRSS. With it, the expansive network of worldwide ground stations seemed to be a thing of the past. Gary A. Morse, former Network Director at Goddard's Network Control Center, reflected on the change TDRSS brought to those who worked on NASA's spaceflight tracking networks:

> The concept of the SN was culture shock. Here, instead of a worldwide net of ground stations, we had two satellites looking down and providing 85 percent orbit coverage, continuous command and telemetry. We were no longer confined to six-minute passes over stationary ground equipment. We had to learn an entirely new technology and apply it. With the old ground net, we had to rely on redundancy. This switch fails, the backup is activated by an operator reaching over and flipping another switch. The new SN was less real-time redundant than the GN had been. We had relied heavily on that redundancy to remain transparent. Any mission was about the spacecraft that was flying, not on what might be going on inside the tracking network. It's our job to focus on the mission . . . The network was there to serve the user, to serve the guy that's flying his spacecraft . . . We might be launching and flying fewer spacecraft now, but those in orbit and the ones planned for launch were more complex. Data rates were higher. The stakes were higher.[103]

★ ★ ★

As passé as the ground network may have seemed at the time, the advent of TDRSS did not eliminate ground stations all together but merely transformed them into a different role. The GN was supposed to have been shut down with TDRSS and the SN was supposed to have taken on the load for near-Earth activities. But it did not quite happen that way. One reason was that a satellite with fairly high data demands still had to have a steerable dish in order to communicate with a TDRS. That was an expensive thing to put on a satellite, even today. To get around this, the satellite could instead downlink to a ground station using only a fixed, much cheaper antenna since a ground station was much closer—only some 1,000 kilometers or 600 miles away—rather than the TDRS orbiting 35,900 kilometers (22,300 miles) overhead.

Users, including NASA itself, understood this. It was inherently less expensive for many cost-constrained, particularly Earth science missions, to build small ground stations dedicated specifically to support their own missions. Many of these places were either unattended or minimally attended stations (for safety) to further reduce cost. Since the TT&C service *was* the MA service and TDRSS could provide that at anytime and anywhere, an interesting synergy developed. Satellites quite often downlinked their high rate science data to their ground station but still used the TDRSS for the lower data, lower cost MA capability to monitor its health and status.[104]

No longer would a STDN ground station be used to track a spacecraft orbiting around Earth or to talk to astronauts in space. Ground stations now had a new mission, and that mission could be summed up in one word: Science. With TDRSS operational, there was no longer the need for ground stations to assume the role of the traditional "tracking station." Emphasis of a GN was now on data acquisition at remote outposts and rocket ranges to support range safety, Earth science and space research. This paradigm shift in the role of the ground stations soon made NASA a key player in what became the commercialization of space, taking the Agency to ever more remote regions of the globe, even to the North and South Pole.

CHAPTER 8

THE NEW LANDSCAPE

An operational TDRSS did not mean that a GN was not needed. It still was, just not in the same way as before. Phase out from the STDN organization did not put an end to NASA's ground station activities. Many sites operated like they did before TDRSS came along, only now they did so for different reasons. No longer called the STDN, the GN played a different role to support a different mission.

In the Pacific, the Kauai Station supported the University of Hawaii specifically, and the Earth science community at large, operating as the Kokee Park Geophysical Observatory (KPGO). As far back as 1981, operations at Kauai had reduced from 24/7 around-the-clock operations to a standard eight-hour, five-days-a-week schedule. Like many of the stations in the network, Hawaii had seen its fair share of "close calls" when it came to closing. Originally scheduled for complete phase out in April of 1984, it kept getting postponed while NASA awaited TDRSS to come online.[1]

The original plan was to transfer the equipment and tracking responsibilities of the station to the Navy's Pacific Missile Range Facility (PMRF) on nearby Barking Sands. The memorandum of agreement for this transfer had in fact been signed-off by both NASA and the DOD when the *Challenger* accident happened in January 1986. After STS-26, the station was officially closed on 30 September 1989 after TDRS-4 was checked out and declared operational, and the transfer to PMRF took take place at that time. The Navy took possession of most of the equipment with the exception of one key asset which NASA retained, the 9-meter (30-foot) S-band antenna system used by KPGO.

One of KPGO's first assignment in its new Earth observation role was to support Goddard's Crustal Dynamics Project. It joined several other observatories in the continental United States, Japan, Chile, and Australia to make ultra-precise position determination of the crust using the tracking measurement technique called Very Long Baseline Interferometry, or VLBI. Kokee Park participated in these NASA and Naval Observatory sponsored experiments and because of its location in the mid-Pacific tectonic plate, was among the most active of the more than 30 observatories around the world.[2]

In this application of VLBI, several radio telescopes (observatories) simultaneously received signals from extra-galactic quasar radio wave sources. Using lasers and the most precise clocks in the world, the difference in time of arrival of the signals due to the slightly different path lengths from the quasar to each VLBI observatory around the world could be determined to an accuracy of 1×10^{-11} seconds (10-trillionth of a second) and their relative positions measured to better than 1 centimeter (0.4 inches).[3] These ultra-precise position measurements, when made repeatedly over several years to decades, allowed scientists to plot the contemporary motion of the tectonic plates—the enormous pieces of Earth's crust—as they moved slowly with respect to each other. The observatories were also able to monitor other geodynamic parameters such as the very complex variation of Earth's spin rate with the minute wobble of the spin axis.

All these gave insight, unavailable before this time, into global geophysics and the underlying forces that led to earthquakes, for example. VLBI measurements which have been made at KPGO since 1984 and which continue today, showed clearly that the Hawaiian Islands (located on the Pacific plate) move at a rate of some 9 centimeters (3.5 inches) a year with respect to the North American plate.

In this new mission, Kokee Park was a principal ground station that diversified to support science application satellites from across several U.S. government agencies. One was the Department of Commerce's PEACESAT (Pan-Pacific Education and Communication Experiments by Satellite) program, which provided medical, educational, and cultural satellite communications between Hawaii and the remote islands in the Pacific basin. It also supported GSFC's Interplanetary Monitoring Platform-8 (IMP-8) that monitored Earth's magnetic field and solar wind activities. In addition, KPGO supported the GOES (Geostationary Observational Environmental Satellite) program for the state of Hawaii.

In South America, the station at Santiago—one of the original Minitrack sites established in 1957 and which had been mostly operated by the Chileans—was completely turned over in 1988 to the University of Chile, who operates it to this day. NASA, though, still has a stake in the station, but now strictly as a customer. Because of its optimal location in the Southern Hemisphere, the United States pays Chile about half a million dollars a year to support a finite number of satellite passes. The number averages out to about two or three passes a day depending on what missions need support. Bill Watson at Headquarters explained how the Agency uses the station today to meet its data pass requirements.

> Some days we take none, some days we take a lot depending on what's going on. Sometimes Santiago is one of a few Southern Hemisphere stations that we have so when something is happening in the Southern Hemisphere and we need coverage, that is a convenient place. Sometimes there are planetary flybys, JPL satellites whizzing by the Earth. They are going so fast near the Earth that their big antennas can't slew fast enough to track, so stations like Santiago will support it.[4]

In effect, a station that the United States in cooperation with the government of Chile started nearly 50 years ago continues in its legacy today.

The surest sign that the era of NASA's world-wide network of spaceflight tracking stations have come and gone was when Bermuda was finally phased out in 1997. Since 1962, when it first gave John Glenn the "go for orbit" call, Bermuda had supported every human spaceflight that NASA had flown, making the critical go/no-go call on all of them—an impressive resume of 118 missions. On 19 November 1997, *Columbia* took to the air on the 88th flight of the Shuttle program. In a Space Shuttle first, the entire stack was rolled from its usual belly-up to a belly-down position in a 40-second Roll-To-Heads-Up (RTHU) maneuver six-minutes after liftoff. Prior to this flight, such a maneuver would have been used only if a Trans-Atlantic Abort emergency landing were declared by Mission Control due to a failed main engine or the loss of cabin pressure during the crew's ascent into orbit.[5]

The RTHU maneuver was added to eliminate the Shuttle's large External Tank from obstructing the communication line-of-sight between the vehicle's antennas and the TDRS-East spacecraft. By doing so, a smooth handover from Merritt Island to TDRSS could be made with only a momentary gap in coverage. Up until that mission, the Shuttle switched over to the space-based tracking satellites only after reaching orbit some eight and a half minutes after launch. The RTHU maneuver—used ever since on all low inclination, easterly launches—allowed the Orbiter to communicate with TDRS about two and a half minutes sooner. (Higher inclination launches towards the northeast for flights to the ISS did not have to perform the roll maneuver due to the availability of DOD tracking stations along the East Coast.) Although tricky, the roll maneuver did not unduly stress the vehicle since it was done well after the SRBs had jettisoned and the Shuttle itself had passed through the thickest part of the atmosphere so that aerodynamic stresses were not a problem.[6]

Bermuda was needed no more. Ultimately though, the decision to close the site came down to cost. The closing saved NASA $5 million a year, which coincidently, was the same amount it cost to build the station in 1961.[7]

With Bermuda closed, Merritt Island/Ponce de Leon (MILA/PDL) became the only source of tracking data for the first seven minutes of each Space Shuttle launch. Despite the phase out of all the original ground stations in the STDN, MILA still remains. In fact, it is as essential today after over 100 Space Shuttle launches as it was for STS-1 back in 1981.

Located adjacent to Launch Complex 39 at the KSC, MILA (acronym for "Merritt Island Launch Annex to Cape Canaveral," the early name of the area that was eventually renamed the John F. Kennedy Space Center) was greatly expanded in 1972 right after Apollo 17. The site was used to get Shuttle data to the Launch Control Center at Kennedy during prelaunch testing and terminal countdown. Once the vehicle cleared the tower, MILA transmitted data to Mission Control in Houston. The GSFC first established MILA in 1966 as a primary MSFN station to provide Earth orbit support for Apollo. The station received the first television signals using Unified S-Band during the Apollo Saturn 203 mission on 5 July 1966 on a flight first testing the performance of the liquid hydrogen fuel in the S-IVB third stage to verify its on-orbit restart capability.

Shortly thereafter, GSFC worked with JSC and equipped the station with a complete set of flight control consoles in order to train Mission Control engineers during prelaunch testing of the CSM and LM. The consoles were used until the end of the program in December 1972. In the mid-1970s when S-band transmitters were added to NASA's Delta and Atlas-Centaur expendable launch vehicles, MILA became really busy, supporting those programs as well as Skylab and Apollo-Soyuz. When the STADAN station at Fort Myers was shut down in 1972, its VHF telemetry and communication equipment were relocated to MILA, greatly enhancing the station's capability to also support application satellites programs.[8] With 13 antennas, including a 9-meter (30-foot) USB system, C-band radar, full TT&C capabilities and a UHF air-to-ground voice link for backup, MILA was (and is) NASA's primary launch area tracking station.[9]

As development of the large SRBs of the Space Shuttle neared completion in the mid to late 1970s, GSFC, working with the MSFC, predicted a potential "plume attenuation" problem in which the high temperature, highly reflective plasma in the rockets' exhaust interfered with MILA's reception of signals from the Shuttle early in its ascent. The phenomenon would have been something akin to trying to follow the flight of a bird with a pair of binoculars while looking through a cloud. To solve this problem, a site with a different

Prior to STS-87, Shuttle flights on easterly trajectories went all the way into orbit on their backs. The Shuttle now performs a Roll-to-Heads-Up (RTHU) maneuver prior to main engine cutoff so that communication with TDRSS can be established some two and a half minutes sooner. This allowed the Bermuda Station to be closed down in 1997. (NASA Image Number GPN-2000-000736)

look-angle had to be found. What followed was the Ponce de Leon Station (PDL) "wing-site" that was set up in 1979. Located 64 kilometers (40 miles) north of MILA on 1.4 acres of U.S. Coast Guard property, it was just south of the Ponce de Leon Inlet at New Smyrna Beach. PDL provided a different viewing angle, putting it outside of the "plume shadow." A 4.3-meter (14-foot) USB system was setup specifically to circumvent this problem.

Upon loss-of-signal at MILA, PDL took over as the primary station during a launch, communicating with the Shuttle during its second minute of flight. PDL, however, could not directly communicate with Mission Control at the JSC; MILA still had to do this. Therefore, a three-hop, microwave system with towers at Shiloh and North Wilson were built to relay data from the wing-site to the main location (again, not unlike relaying of cell phone calls). Strictly a supplement to Merritt Island, Ponce de Leon was normally not even staffed, with two or three technicians dispatched to the station to support flight readiness, countdown activities and the actual launch.[10]

With PDL tagged to cover this 60-second gap, according to Shuttle flight rules, a backup to the site itself had to be identified. This dual-redundant requirement harbored back to the early days of NASA human flight operations where a back up was required for any system designated as primary. To this end, a search was conducted in the southern Florida area to find a location suitable to back up Ponce de Leon. Communication link analysis showed that the Air Force's Jonathan Dickinson Missile Tracking Annex (JDMTA) some 150 kilometers (95 miles) south of the Cape near Jupiter, Florida, could back up PDL for S-band downlink. The DOD had constructed this facility in 1985 and 1986 on 11 acres of land in the state park to provide launch support for their launches and missile testing activities. This allowed the Air Force to permanently shut down its more expensive Grand Bahama tracking station. (The latter had provided launch support for over three decades, from 1954 to 1987, first for the Air Force and then for NASA.) Jonathan Dickinson already had everything that Goddard engineers were looking for, including radar, telemetry, a microwave relay to the Cape, and a command destruct system that could be remotely activated from the Cape if it were ever necessary to protect life and property should a launch go awry.

Since MILA was so crucial, the site continually evolved and was upgraded. The most dramatic change was its transition from a mostly human-operated site to autonomous operation, which has, not surprisingly, significantly reduced costs. While not a switchover to purely unattended (or "lights-out") operations, the change brought on by the ever increasing reliance on automation and computer processing has been beneficial, significantly reducing the station's staffing requirements. During the height of Apollo and for STS-1, for instance, the station employed upwards of 140 workers. That number has dropped dramatically to where less than 40 people are now required.[11]

Even as staffing was being reduced, modernization of technology increased. In 1995, the station went to an "all fiber" system, with fiber optics replacing all the communication lines between MILA, PDL and the control facilities at the KSC. A year later, a UHF voice system with a powerful, state-of-the-art quad-helix antenna was installed to support the Shuttle in the event of a Return to Launch Site (RTLS) abort. Today, Merritt Island has become a full-service spaceport communications facility, boasting a suite of 15 antennas that support all phases of a Shuttle flight—from prelaunch checkout to launch, on-orbit (via TDRSS) and landing. Leveraging each other's assets has enabled the DOD and NASA (and more recently the commercial launch industry) to support a wide range of space launches from Florida. As former Station Director Tony Ippolito put it, "All of this has allowed us a more business oriented approach in the operation of MILA."[12]

With near-Earth space communications now well covered by the TDRSS and the SN (with the Jet Propulsion Laboratory's DSN handling planetary work), the emphasis for NASA to support suborbital science missions has, in turn, made Goddard's Wallops Flight Facility home to the GN's most

Merritt Island, MILA (top) and Ponce de Leon, PDL (bottom) provided uninterrupted launch vehicle tracking out of the Kennedy Space Center (KSC). Shown are the MILA operations building along with the station's two 9-meter (30-foot) Unified S-band (USB) antennas used for tracking, telemetry, command (TT&C), and voice. The less complex PDL "wing station" had a 4.3-meter (14.1-foot) antenna used to cover loss-of-signal at MILA from the exhaust of the Shuttle's Solid Rocket Boosters (SRB). (Un-numbered Kennedy Space Center images, *science.ksc.nasa.gov/facilities/mila/milstor.html*, *science.ksc.nasa.gov/facilities/mila/pdl.html*, accessed 21 November 2005))

extensively equipped facility. Located on Wallops Island off the Delmarva Peninsula coast of Virginia, Wallops is NASA's lead facility for implementing its suborbital and special low-orbit research projects. Established by the NACA in 1945, the 6,200-acre facility is today staffed by 1,000 full-time government personnel and contractors who support everything from sounding rocket and balloon launches to conducting unpiloted aerial vehicle research.[13]

The beginnings of Wallops date back to the end of the Second World War. In 1945, NACA authorized the LRC to develop the small offshore island into an aeronautical range where rocket propelled models can be launched to conduct studies of the upper atmosphere. In this way, Wallops became the oldest civilian launch site in the United States. The facility allowed Langley scientists to have many more options in conducting their research, like overcoming the limited capabilities of the wind tunnels of the day, for example. With the establishment of NASA in 1958, the creation of the "manned-satellite" (Mercury) program and Wallops's close association with Langley and its STG, much of the activities there quickly turned to developing the components needed for putting a human in space. This included designing capsule escape techniques, pressure testing of the early blunt-body aerodynamic designs and flight test support of heat shield development and ocean recovery techniques.[14]

In addition to the emphasis put on Mercury, research in the aviation arena continued. The facility's airport, for instance, was used to develop and test runway surface designs for aircraft noise reduction. And it was at Wallops that the Scout launch vehicle solidified its place in history as the premier rocket for launching small payloads for the scientific community, with a remarkable 100 percent success rate since 1976. It was here that the Scout became the first solid fuel rocket to place a satellite into orbit when, on 16 February 1961, it successfully launched a 44-kilogram (96-pound) NASA atmospheric research payload into orbit.[15]

On 19 October 1981, the Wallops Flight Center, as it was then called, was consolidated under GSFC management and redesignated the Suborbital Projects and Operations Directorate, otherwise known as the Wallops Flight Facility. Less than five years later, in April 1986, the tracking station that was part of the NTTF located on the grounds of GSFC, was transferred to Wallops. The flight facility now had the added responsibility for capturing small satellite telemetry, tracking, and command. Many of the first satellites supported from the facility would go on to become some of NASA's most successful orbital science platforms. Among them were the IUE, the Inter-planetary Monitoring Platform (IMP-7), and the Cosmic Background Explorer (COBE). To better handle the additional workload, the facility soon underwent a one-year modification where existing hardware was supplemented with equipment from former STDN stations around the world that were then being phased out. A new communications system was added as part of the upgrade to transmit data from Wallops to the Project Control Centers located back at Goddard.[16]

In the late 1990's, the facility began developing ways to really expand its sphere of operations so as to more effectively support launches at locations away from Wallops Island and the immediate Virginia coast area. Mission operations at Wallops took on a new dimension when it began operating the Mobile Range Control System, or MRCS. Developed by the Center's Electrical Systems Branch, the MRCS is a self-contained, transportable launch system that can be loaded into a military cargo aircraft such as the C-130 and flown around the world to conduct satellite launches at remote locations as needed. It in fact acts somewhat like a transportable range, equipped with an Uninterruptible Power Supply, a range safety display and redundant command destruct transmitters for flight termination along with all the necessary computers and communication equipment needed to support a launch in a "turnkey" fashion.[17]

Before there was the MRCS, setting up a mobile range was much more cumbersome and logistically demanding, translating into higher cost. Equipment in several vans and trailers had to be transported either by air or by sea and put together upon arrival at the remote location. One former MRCS Project Manager noted the tremendous advantage this new system offered, saying "In comparison with the older collection of subsystems in separate trailers, the fully integrated MRCS can be completely tested prior to shipment. This helps reduce mission support and cost."[18]

True to its calling, the Wallops's mobile range has been well traveled since 1997, supporting launches from the nearby Coquina Outer Banks of North Carolina, to the Canary Islands in the East Atlantic and even as far north as Kodiak Island, Alaska. To support the commercial launch market, the MRCS was in 1999, granted a license by the FAA's Office of the Associate Administrator for Commercial Space Transportation (FAA/AST), which allows private paying customers from the U.S. commercial launch industry to use the system to launch their payload into space.[19]

All these developments have made Wallops Island (also known to the commercial launch sector as the Mid-Atlantic Regional Spaceport) America's preeminent small rocket facility. As a controlled range, it has the authority to clear airspace and reroute planes in times of need. Since Wallops's mission is so diverse, the ground station there is somewhat unique in that it has a combination of some very old antennas alongside state-of-the-art equipment. It still operates, for example, an original VHF antenna for ISS and Russian Soyuz voice support. A VHF Satellite Automatic Tracking Antenna/Satellite Command Antenna on Medium Pedestal— SATAN/SCAMP telemetry/command system—from the 1960s can also be found still operating there. Although rendered obsolete when stations began using microwaves to transmit data over long distances, this old system was kept to support the facility's suborbital and short-range data needs. Also, there are the original

The remote barrier-island location of Wallops Island on Virginia's Eastern Shore makes it ideal for testing aircraft models and launching small rockets. As the space program evolved, it became one of the Agency's mainstays for launching sounding rockets carrying scientific experiments into the upper atmosphere. In the 1980s, however, a proposal emerged to close Wallops as a way of reducing NASA's operating costs. Instead, officials decided to incorporate the facility into the Goddard Space Flight Center (GSFC) as it relied on the facility for satellite launch, tracking and data support. In this way, Wallops Island Station became the Wallops Flight Facility managed under the Suborbital Projects and Operations Directorate at GSFC. (NASA Image Number GPN-2000-001323)

9-meter (30-foot) S-band antennas that tracked Apollo astronauts to the Moon and back. These "antiques" can be found still being used everyday alongside the station's state-of-the-art 11-meter (36-foot) S-/X-band dual-feed antenna. As one NASA manager puts it, "They have practically one of every kind of antenna out there," which, in some ways, makes Wallops the perfect setting as a nostalgic rocket range.[20] It bridges the gap between an old fashioned test range nestled along the Atlantic coast and the modern twenty-first century spaceport.

* * *

Satellites and spacecraft circling Earth today rely on both the SN and the GN in different ways. The GN of today is used primarily to support aeronautical and atmospheric research, range safety, and high inclination (high latitude) orbital communications. It is in this setting that the new era of NASA's communications network is found, the hallmark of which are *technology expansion* and *commercialization*.

First, the rapid expansion—indeed evolution—in digital telecommunications technology over the last quarter-century has made NASA's space communications a truly global amalgamation that connects every corner of the world. This same technological evolution has also greatly improved the ability of today's stations to perform TT&C functions compared to the previous STDN generation. Station autonomy has greatly reduced the requirements for human staffing. The objective is not to eliminate human-in-the-loop but let automation do what can be done in terms of scheduling, redundancy, and self-testing. Advancements in digital signaling and transmission techniques have allowed for the ever increasing demand for higher bandwidths (traffic) and lower bit-error-rates (accuracy) to be accommodated.

The other trait which can be used to describe the Agency's network operations is commercialization. This should not be surprising when one looks at the trend of space communication in which NASA has historically set the precedence but is now heavily influenced by the commercial sector. Just like the demand for better technology is always a driver, as space moves from the realm of government sponsorship to being a commercial commodity with increasing private industry participation, cost reduction,—and more importantly—profit in today's world of real-time global communications is more important than ever. It is these fundamental paradigm shifts that have taken NASA's STDN of the past to where it is today. This shift has enabled NASA in recent years to put ground stations in very remote regions of the globe where it was just simply not feasible a generation ago.

Take Antarctica, for example. The manpower that would have been needed to make a continuously operating ground station cost effective from such a location would have, in the past, been difficult at best. On top of that would have been the technical challenge of how one would get the data received at the station in a timely manner to their users who may be scattered across many continents.

In 1956, the U.S. Navy established McMurdo Station on the continent of Antarctica. At 77° 50° south latitude, McMurdo is well inside the Antarctic Circle and is the southern most harbor in the world. It is also Antarctica's largest community and the continent's center of activity. Built on the bare volcanic rock of Hut Point Peninsula on Ross Island, it is the farthest south solid ground that is accessible by ship. As early as 1901, McMurdo took on some sense of import when it became the staging point for the race to plant the first flag at the South Pole. Among the landmarks still preserved (by the New Zealand government) from that era is Hut Point, left behind by the doomed expedition of British Naval officer Robert F. Scott and his party in 1910. That year, Scott—with his team of four companions—embarked on an expedition with the aim of becoming the first man to reach the South Pole. The 2,964-kilometer (1,842-mile) trip was the longest continuous sled journey ever attempted in the polar regions. On 18 January 1912, they reached the bottom of the world only to find the tent and flag of the Norwegian explorer Roald E. G. Amundsen, who had achieved the goal only five-weeks earlier. Demoralized and short on supplies, Scott and his men never made it back to McMurdo. The return journey ended in the loss of the entire party. Scott came to within 18 kilometers (11 miles) of a supply depot when he and his remaining two teammates perished of starvation and exposure. Their remains, along with diaries left by Scott in his tent, were found by a search party almost eight months later.[21]

Since 1956, McMurdo has grown from an outpost of a few buildings to the largest community on the icy continent with more than 100 structures, an outlying airport (Williams Field) with landing strips on sea ice and shelf ice, and a helicopter pad. Despite its remote location, McMurdo is among the most ethnically diverse communities per capita anywhere to be found. During the summer months, the population can swell to over 1,000 people, attracting scientists, construction workers and polar explorers from all nations around the world. During the harsh winter months of March to October, the population usually drops down to below 250 people who, except for time of emergencies, find themselves pretty much isolated for the winter.

Like a small town, there is a freshwater system, sewer, telephone, and power lines linking the buildings. Science equipment at McMurdo include diving equipment, recompression chambers for diving accident victims, cosmic ray monitors, and facilities to study magnetosphere and ionosphere phenomena. From the runways of Williams Field 16 kilometers (10 miles) away, flights span the continent and to airbases in and out of New Zealand. While skid-equipped planes can fly in and out of the frozen landing strips year-round, it was not until 1992 when a permanent, hard-ice runway on the Ross Ice Shelf was completed that larger transporters equipped with wheeled landing gears could come and go more frequently thereby greatly increasing the availability of supplies to the delight of the personnel stationed there.[22]

It is in this unique part of the world that NASA teamed up with the National Science Foundation (NSF) to establish the southern most satellite data acquisition station in the world.[23] The McMurdo Ground Station (MGS) is today home to a 10-meter (32-foot) S- and X-dual band NASA antenna located atop the 152-meter (500-foot) Arrival Heights peak. From this vantage point, it has a fantastic view in all directions. Looking south, it can see satellites on the other side of the Pole. NASA's original requirement there was to support a joint effort by the two Agencies (along with international partners) to radar map the entire Antarctic continent by satellite.

Operational since 1996, MGS started out collecting X-band telemetry (frequencies in the 5- to 11-gigahertz range in the electromagnetic spectrum, higher than C-band but lower than Ku-band) on about 25 passes each day from ERS-1 and ERS-2 the European Earth Resource Satellites, and the Canadian synthetic aperture radar mapping satellite *RADARSAT*. The

Chapter 8 \ The New Landscape 315

McMurdo, Antarctica is the world's southern-most port and home to numerous expeditions to the South Pole since 1901. The McMurdo Ground Station is located on nearby Arrival Heights Peak. (Photograph courtesy of NASA)

station's S-band capability was put to use not long after in August of 1997 when a NASA Lewis land imaging satellite malfunctioned and began tumbling shortly after launch. Because MGS could see virtually every pass, it was a real asset in the rescue attempt. Unfortunately the spacecraft could not be stabilized and was consumed in a fiery reentry just 36 days after its launch. NASA today uses McMurdo as a data collection hub for satellites monitoring ice movements in the Southern Hemisphere. Such data is used immediately on site by scientists on the continent as well as by those planning re-supply shipping routes in and out of Antarctica.[24]

Despite the fact that few other ground stations have the capability of MGS to collect the enormous volume of data that can only be done at the Poles, communications in and out of the continent is still not so good. The only way to get data out is through something called the McMurdo TDRSS Relay System, or MTRS, which consists of two antennas (4- and 7-meter [13- and 23-foot] dishes) that actually communicate with the TDRSS. NASA uses a nearby microwave tower to relay signals to the MTRS at a place called Black Island located about 50 kilometers (30 miles) closer to the Equator. Due to the curvature of Earth and the way a satellite travels in polar orbit, near the poles, even this relatively short distance can make a big difference to provide a much better view to relay satellites. The drawback for NASA, however, is that since the link is shared with the NSF, it is only used occasionally so as not to overwhelm the NSF's ability to send data off the continent on the always busy TDRSS.

Rounding out Antarctic communications is a small system located right at the South Pole. Although not really part of NASA's GN, it allows polar scientists there to communicate with TDRS-1—the original satellite—on brief occasions when it pops above the horizon while performing its "figure-8" loop in the vicinity of the Equator. These ongoing efforts to build a good communications network on this most desolate of places has only recently culminated, allowing the inhabitants to join that most global of communities: the Internet. This accomplishment is not lost on those who run the space agency's networks. The proclamation "We brought internet to the South Pole!" sums up the Agency's legacy on Antarctica rather nicely.[25]

*　*　*

Antarctica and the South Pole are not the only places to have been "tamed." In this new era of ground stations, NASA has also been busy on the

A team from the Goddard Space Flight Center (GSFC) visit the Intelsat communications relay station on Black Island in 1999. Note the microwave tower link back to McMurdo. (Photograph courtesy of NASA)

other end of the globe. While the mid-latitude location of the continental United States makes for a good setting for launching science payloads into orbit (Wallops, Cape Canaveral), it cannot however, provide routine, low-cost, launch access to investigate interesting activities that permeate the upper atmosphere in the polar regions, activities such as Aurora Borealis, or the northern lights. To do this, one must venture near the Arctic Circle, to a place called Poker Flat some 30 miles north of Fairbanks, Alaska.

Owned and operated by the University of Alaska's Geophysical Institute since 1968, Poker Flat Research Range has been primarily dedicated to the launch of sounding rockets for the purpose of middle to upper atmospheric research. The first rocket was launched there in 1954. The rather enticing name Poker Flat is believed to have been taken from American author and poet F. Bret Harte's rags-to-riches short story, *The Outcasts of Poker Flat*, which in a way describes the inauspicious beginnings of the original ad hoc launch site that was constructed from begged and borrowed materials. But the range could have simply been named after nearby Poker Creek. In any case, Poker Flat is today the only nongovernment, university owned and operated range in the world. It is also the only high-latitude, polar region, rocket launching facility in the United States.

Because of its importance, NASA has funded the operation of the range under a cooperative agreement with the University of Alaska's Geophysical Institute since 1979, assuming funding responsibilities previously held by the National Oceanic and Atmospheric Administration (NOAA). Much bigger than the birthplace of America's missile activities at White Sands, Poker Flat is in fact the world's largest, land-based rocket range. It consists of a chain of downrange flight and observation sites spanning inland Alaska to Spitsbergen in the Arctic Ocean that are used to monitor and help recover payloads. Since it is an active rocket range, NASA and the university have to coordinate their activities with many U.S. government agencies. The FAA must approve and coordinate the air space during launches. Also, since the range is so large, permission to impact rockets and their payload on its 26 million acres of land has to be authorized by a whole host of government agencies, including: the Bureau of Land Management; the U.S. Fish and Wildlife Service; the State of Alaska Division of Lands; Doyon, Ltd. (the largest private landowner in the state of Alaska); and the Village Traditional Councils of Venetie and Arctic Village. Unlike bygone days at the dawn of the Space Age, environmental regulations mandate much of what can go on at these ranges today.[26]

As with the oil pipelines a quarter-century before, Alaska today serves as the great northern frontier. But this time, instead of energy, the commodity is information. More specifically, the information age revolution and commercialization of space. In fact, AGS, the Alaska Ground Station at Poker Flat, is not really even a NASA owned station at all. Rather, it is part of a commercial network of ground stations called DataLynx, which is owned and operated by Honeywell Technology Solutions, Inc. With U.S. and international partners such as Universal Space Network, the Australian CSIRO, and the Japanese Institute for Aerospace Technology (JAXA), just to name a few, DataLynx has today expanded to over 20 ground stations on six continents.

Mirroring in many ways NASA's STDN of the previous generation, DataLynx stations today operate in a latitude band spanning 78° north at Svalbard, Norway, down to 33° south at the Santiago Station operated by the University of Chile. In fact, locations of many of the stations harbor back to the STDN days (and even earlier), with places like Hartebeesthoek, South Africa; Perth, Australia; and as mentioned, Santiago. Other places such as Beijing, China—a location which would have been impossible to imagine during the Cold War—have become part of this new age in commercial networking designed to serve satellite-using customers from around the world. With profit openly the bottom line, DataLynx, which depends highly on automation and "lights-out" operations, advertises itself as a " rapid, proven, reliable, cost effective mission-critical . . . distributed partner network," one that offers 24/7 command, control, and communications for "broad and flexible solutions, reducing cost and risk to our clients so that they can focus their resources on their core businesses."[27]

One of these clients is none other than NASA, the one who subsidizes Poker Flat, the very range that the Alaska Ground Station sits on. With $5 to $6 million a year, the Agency literally buys a minimum number of passes each day using the station's 11-meter (36-foot) antenna. (AGS also supplements this system with a smaller 5-meter (16.4-foot) S-band system called the Low Earth Orbit Terminal as well as a somewhat larger 8-meter (26.24-foot) transportable S-band antenna called the Transportable Orbital Tracking System.) Assuming 36 passes a day, this averages out to approximately $400 per pass, a figure that is much more economical to NASA than what it would cost to otherwise engineer, build, operate, and maintain a station of its own. It is therefore a truly joint government/industry arrangement that in the end benefits both NASA and the DatyLynx stakeholders.

Program Executive Bill Watson explained the arrangement which in a way captures the business-end of how ground stations in the twenty-first century works: "In this case, Honeywell not only owns DataLynx but they also won the current contract for operating the GNs. So they consolidated the deal and said we will treat the government assets and commercial assets as a pooled resource. You give us a guaranteed annual amount of revenue and we will guarantee you a minimum number of passes per day, and then if we [NASA] go over that, we pay by the pass. So it's a quasi-government commercial activity."[28]

While the Alaska station at Poker Flat is located just outside of the Arctic Circle, there is yet another ground station which continues this joint, government-to-private enterprise theme but is situated a mere 965 kilometers (600 miles) from the North Pole itself. It is here on the Norwegian archipelago of Svalbard that the world's northern-most, permanent, satellite ground station can be found at 81° latitude near the top of the world.

The earliest written records documenting the existence of these frozen polar islands date back to the late twelfth century by the Vikings as they sojourned about the Arctic Ocean. For the next 400 years, though, Svalbard—which means "cold edge" or "cold coast"—was largely forgotten. Then in 1596, the islands were accidentally rediscovered by an expedition led by the great Dutch seaman Willem Barents while searching for a Northeast trade passage to Asia. This was followed by the English explorer Henry Hudson, who mapped the area and reported good whaling there. This spurred a bitter quarrel between English and Dutch whalers over the territory. In 1618, a compromise was reached, with the Dutch limiting their operations to the northern part, leaving the rest to the English, the French and the Hanseatic League (an alliance of trading cities that maintained a trade monopoly over northern Europe between the 13th and 17th centuries). The Danes also claimed the archipelagos as part of Greenland.

Over the next 300 years, various countries such as Norway, Russia, and Sweden laid claim to the islands, this especially after coal—the great source of energy that could empower the new steam engines—was discovered there in the late nineteenth century. Norway finally took formal possession of Svalbard in 1925 after a treaty was signed in Paris after the First World War. (Russia, who did not sign the accords, was to dispute Norway's stance on the islands well into the latter part of the century.) The islands again came to the forefront when, during the Second World War, it was the scene of some very intense naval battles between Germany and the Allies due to its rich deposit of coal.

Although Svalbard's claim to fame for years had been its geographical setting serving as the staging point for North Pole explorers, it was not until 1990 that Norway officially opened up the region to general tourism, greatly expanding its economic base and spurring on the development of new industries. In 1997, the Norwegian Space Center (NSC), along with the private space conglomerate Konnesberg Satellite Services (KSAT), began putting together the Svalbard Satellite Ground Station at Platåberget, near the town of Longyearbyen. SvalSat, or SGS, is today a truly general purpose facility, providing customers with tracking, telemetry, and data returns from a host of polar orbiting satellites.[29]

While NASA knew that the Alaska Ground Station was in a fairly good location at 65° latitude, it could, nevertheless, only observe about 14 out of every 16 passes that were actually available each 24 hours. With SGS being less than 1,000 kilometers from the North Pole, it could literally see every single polar orbit pass. Therefore, when Norway approached NASA to join their operations, it was a rather easy decision. The Agency first put up a trailer (often covered by a tent to keep the snow off) and an antenna in 1997. The site has steadily grown since then into sort of a "space park" of the Arctic. NASA now

With its extreme northern location on the Svalbard archipelago (78°13' latitude), SGS is the only ground station in the world able to provide all-orbit support of polar orbiting satellites. Six multi-mission antenna systems, along with several minor antenna systems, are used for TT&C and operations. One of the systems is dedicated to the NASA Ground Network (in shared operations with EUMETSAT) and is operated locally by SvalSat. The remaining antennas are remotely controlled and operated from Tromsø at the Tromsø Network Operations Center 1,000 kilometers (600 miles) to the south on the Norwegian mainland. (KSAT photograph, *https://www.spacecommunications.nasa.gov/spacecomm*, accessed 22 August 2007)

operates an 11-meter (36-foot) S/X-band antenna there. Here, the Americans join the Norwegian Space Center and Kongsberg, as well as EUMETSAT (the European Organization for the Exploitation of Meteorological Satellites), to make up the world's northern-most tracking station.

Surrounded by "the King of the Arctic" polar bears—personnel are required to carry a weapon when working outside—the space park continues to grow. One of its hallmarks is the ability to get large amounts of data quickly off the island with the use of fiber optics. NASA arranged in 2004 to have the NSC install redundant fiber optics all the way back to the United States, much to the delight of data users on the North American continent. With this capability, NASA's communications to Norway today is actually much more robust than that to Alaska. Taking the Agency's domestic, government-industry relationship in Alaska to an international level, the operation in Svalbard serves as a model for the way space is being treated openly as an international commercial commodity.

Unlike days bygone, the United States and the (former) Soviet Union are by no means the only torch bearers. Watson explained:

The Norwegians made us a deal. They said you're [NASA] paying $6 million *a year* now for commercial relay satellite services to haul data out of Norway. For $5 million for *five years*, we will install the fiber and then give you the next *15 years* for free. How could we turn that deal down! And what they did was they made a similar offer to NOAA and so between NASA and NOAA, they got a revenue stream of about $10 million a year. They went off to a commercial financier and got the money, . . . basically borrowed it against a promissory. The implication was that NASA and NOAA would commit to pay for five-years. . . . They brought two ships in and laid redundant fiber, two different paths so if one gets cut, we still have the other. It was a remarkable plan.[30]

Svalbard has proven to be a win-win situation for both NASA and the Norwegians. What began as a joint venture to gather a few more satellite passes has basically turned this faraway mining town into what is one of the best wired and well connected places anywhere in the world. In a way, NASA's operations at Svalbard bring the GN of the twenty-first century full circle to how it all began nearly half a century ago. Back then, the emphasis was on hiring local people and training them to "nationalize the station." This worked well at many places, from Chile to Ecuador to Guam. With Norway, this is very much a continuation of that legacy, except NASA no longer has to provide all the technology and set all the precedence. At Svalbard, the Norwegians did not need that "leg up" to turn their concept into a reality.

In a sense, while America was winning the space race, the rest of the world caught up.

✳   ✳   ✳

Throughout the 1990s, the phrase "Faster, Better, Cheaper" became somewhat of a choreographed aphorism that drove much of the way business was conducted in the high technology world, both in the private sector and in the government. This approach impacted the space program in ways ranging from economics to performance and, some would argue, safety. For critics, "Faster, Better, Cheaper" was usually followed by "Two out of three aren't bad!"

In this era of commercialization, the approach as to how ground stations were to be built and how they were to be used began to change. "Lights out" operations and station autonomy entered the scene. Companies like USN and DataLynx entered the playing field. These multimission network terminals offered users the advantage of low cost services based on the philosophy of "pay only for what you use." Like Santiago, they provided services on a retainer basis with added "per-pass" cost on actual usage, targeting not only commercial users of satellite services but also government users like NASA.

It was in this renewed atmosphere of cost awareness and infrastructure reduction that NASA tried implementing an across-the-board streamlining of its organizations. This included, in particular, its tracking and space communications operations. The goal was actually rather sweeping but straight to the point: back the government out of day-to-day operations. To do this, all management of space operations was consolidated under a single office at the JSC in Houston. The office was named the Space Operations Management Office. Known as SOMO, the name would seemingly take on a life of its own in the coming years, the mere mention of which, even today, conjures up strong feelings on the part of those who were involved.

The decision to establish SOMO was prompted by an Agency-wide examination of space operations requirements as part of a so-called Zero Base Review, completed in 1995. In this review, representatives from NASA Field Centers evaluated opportunities for consolidation, privatization, and commercialization of existing government functions across NASA. The review team recommended several initiatives to achieve cost savings while ensuring a continued, high quality of operations services. It was decided that a single but consolidated, management structure could be implemented that would best accommodate these goals. SOMO would be that management organization. Theoretically, it could be a centralized office that could quickly respond to service requirements as identified by specific NASA programs and projects, and even to external (non-Agency) customer requests for similar services—as long as NASA operations were not interfered with. Whether this was a good idea or not, at least on paper it seemed like it could work.[31]

Under Administrator Daniel S. Goldin's direction, the SOMO was established the following year. Its central objective was to ensure that existing NASA assets were used as efficiently as possible and that duplication was avoided. This objective unavoidably resulted in shuffling of responsibilities (and power) between various organizations within NASA. Not only that, it would also go one step farther by *eliminating* certain offices.[32]

SOMO was purposely designed to be small, with key positions held by about three-dozen individuals from NASA Field Centers who, for all intents and purposes, made the decisions in carrying out all the Agency's space operations. Specifically, the Data Services Manager was from the JPL; the Missions Services Manager was from Goddard; and the Commercialization and NISN Manager was represented by the MSFC.[33]

Thus from JSC, the office soon ended up basically managing all of NASA's space operations, including its vital communication networks and tracking systems. The TDRSS Space Network became a part of it as well as the GN, the NISN and JPL's DSN. In other words, management of the networks suddenly—and for those involved, unbelievably—now came under the auspices of the JSC. This gave JSC an inordinate amount of power. While the move did not take away the GSFC's day-to-day responsibility of operating the network (nor JPL for its Deep Space activities), it did significantly erode the role that Headquarters had in Washington. The Office of Space Communications (OSC, organizational designation "Code O") responded unreservedly to the sweeping mandate to, in effect, do more with less. Staffing at OSC was first reduced by 35 percent, then by 60 percent, and finally, 85 percent. This included elimination of 16 senior level GS-15 and higher positions. With its role greatly reduced, the job of Headquarters was relegated to conducting external interface, determining program requirements and strategic planning.[34] In short, Code O was one of those, which for all intents and purposes, eliminated. The numbers back this up. Prior to consolidation, Code O had in excess of 50 people at Headquarters working the Agency's tracking and communications needs. After the reorganization, scarcely eight remained.[35]

To put it bluntly, space communications—at Headquarters as well as Goddard—was being gutted.

The idea of consolidation was radical but seemed noble enough at the time. NASA established the SOMO (Code M) to oversee its space operations activities and to implement a single Consolidated Space Operations Contract (CSOC) as the initial step to reduce the cost of operations. In this "integrated operations architecture", many of the trends which were permeating space communications in general—trends such as the aforementioned automation and privatization—could take effect. The thinking was that a single, large contract would naturally be more efficient than a plethora of smaller ones. On 25 September 1998, NASA awarded the enormous $3.4 billion operations contract—five year base with an option for an additional five years—to a team led by Lockheed Martin. It was one of the most valuable outsourcing programs ever undertaken by a civilian agency. Under CSOC, five contracts which had up until then operated independently were consolidated on the first day that the contract took effect. (Ten more separate contracts transitioned to CSOC from 1999 to 2004.)[36]

Even NASA was fully aware, however, that the envisioned cost reduction could not take place overnight. While some reduction could be gained initially, the contractor work force supporting space operations at five NASA Centers would have to be reduced gradually over the 10-year period at a rate of slightly less than 100 jobs per year on the average. The idea was that, by implementing the reductions over a decade, essentially all of the attrition could be absorbed through planned retirement, personal job changes and reassignment of contractor employees to other programs.

Even as CSOC was being awarded to industry in 1998, SOMO began exploring several additional commercialization initiatives with the aim of realizing some further, longer term cost savings. For example, services with USN were established. Another component was to provide, using the CSOC contract-vehicle with private industry, opportunities to offset some of NASA's operating cost by marketing unused capacity on the TDRSS. One such initiative—originally conceived by Code O—was to provide "TDRSS time" to commercial oil exploration vessels at sea, as there was virtually no commercially available communication satellite that could support the transfer of the large amounts of data for such application. Through these and other initiatives, SOMO at the JSC projected that a $1.4 billion cost saving could potentially be realized over the next 10 years.[37]

As optimistic as JSC was on the outside, however, it could simply not shed a barrier that soon (and clearly) manifested itself as a growing cancer in the whole SOMO idea. When the office was formed in 1996, program responsibilities moved from Washington to Houston. This had huge repercussions. While the various Field Centers still had their programs to work, this shift in responsibility to Houston naturally did not sit well with them. According to Robert Spearing, who witnessed the whole thing and was one of those actually recruited back from private industry by NASA to help fix the problem, it was putting in charge "an organization that they [the Field Centers] felt was ill-equipped to deal with their issues." Said Spearing,

> What was not understood well was what happens to the work that has to go on at Headquarters is very difficult for a Field Center to perform because they don't have the skills for that. A lot of our work here at Headquarters relates to working with the other agencies of the federal government, both civilian and military, and also working with the Congress to advocate our programs. So you ended up having a very limited capability to perform that function here at Headquarters.[38]

Essentially, SOMO had grossly underestimated the importance of having a team in Washington to take care of business with the rest of the federal government (and international partners). Said Spearing:

> You have to have the right talent here in town for our relationships internationally. . . . When you work with international organizations like the European Space Agency or the Italian Space Agency or the Japanese Space Agencies, all of these organizations look to NASA Headquarters. When someone from a Field Center goes and represents NASA to these organizations, they see that as somewhat of a mismatch. So there were some lessons learned.[39]

By 1998, just two years after SOMO was set up, it was already clear to most that in the process of trying to work issues between the JSC, Headquarters, and the other Field Centers like Goddard, Marshall, and JPL, that it just was not working. Space Communications had been relegated to an office under Code M. There were open and often ugly struggles over who was

Chapter 8 \ The New Landscape

NASA's management of its tracking and space communications activities began at Headquarters with the establishment of the Office of Tracking and Data Acquisition (OTDA) on 1 November 1961. This photograph shows its staff in the early 1980s. OTDA was reorganized into the Office of Space Tracking and Data Acquisition (OSTDA) in 1983 and then into the Office of Space Operations (OSO) in 1987. It became the Office of Space Communications (OSC) in 1992. Edmond C. Buckley was OTDA's first Associate Administrator (1961–1968). He was succeeded in that position by Gerald M. Truszynski (1968–1978), William C. Schneider (1978–1980), Robert E. Smylie (1980-1983), Robert O. Aller (1983–1989), Charles T. Force (1989–1996), Wilson T. Lundy (interim Deputy AA 1996), David W. Harris (interim Deputy AA 1997–1998), Robert E. Spearing (1998–present). (Photograph courtesy of Charles Force)

in charge and who had control. Beset by quagmire, something had to be done. The SOMO in Houston eventually came around to this realization. A rather laborious process followed in which the OSC was slowly restored. First, the position of Deputy Associate Administrator for Space Communications under Code M was abolished. It was then placed under the Office of the Deputy Associate Administrator for the Space Shuttle. Later in the year, it was again transferred, this time under the Director for Resources Management. By 2001, Stan C. Newberry, who by then was heading up the SOMO, was holding talks about the future of the office with top agency officials at Headquarters. Not soon thereafter, and with little fanfare, Administrator Dan Goldin dissolved the controversial office and restored program management of space operations back to Washington where they remain today. With the restoration of Code O, management for tracking, data acquisition, and space communications at NASA were reconsolidated. Ten years after the controversial reorganization began, the OSC was finally restored to the structure Buckley had setup 45 years earlier.

In April 2004, at the end of the five-year base period, the CSOC industry contract to the Lockheed-Martin team was terminated and not exercised into the option period.[40] OSC at Headquarters not only survived the SOMO fiasco, it broke apart CSOC. The separate set of five contracts (now collectively called Space Mission Communications and Data Services) divided CSOC's functions fairly equally across three NASA centers—Goddard, Kennedy, and Marshall. The new contracts were collectively worth about $400 million a year, roughly equal to the annual average of the old CSOC. But now Headquarters intended to award, manage and determine the specifics of each contract separately.[41]

Regarding the whole affair, James Costrell of the OSC said, "It was a new concept to NASA. The theory was that spacecom (space communications) is spacecom, that it's all the same. So NASA charged ahead with the idea that a consolidated contract would result in efficiencies."[42]

Of course, it didn't quite work out that way. With the contract managed by SOMO from Houston, other Field Centers found that Lockheed-Martin and its subcontractors could not respond to their needs very effectively. The contract had other weaknesses too. Among the worst was that NASA had to renegotiate prices with the contractors whenever conditions changed. For instance, if the Agency ended a mission, closed a tracking station or took any action that altered the work needed from the industry team, Agency and company managers had to agree on cost revisions. Through the lessons of SOMO and consolidation, NASA learned (the hard way) that one size does not always fit; divide and conquer sometimes works better. "What the agency finally came to grips with was that there are some fundamental differences between the various communications activities," Costrell said.[43]

Although the controversy and subsequent fallout of what is now simply referred to as "consolidation" inside NASA may have in many ways been a failed experiment and a bitter pill to swallow (many were reassigned from GSFC and Headquarters or left the Agency), the ensuing years have shown the resiliency of America's space agency as an organization to overcome and move forward from its setbacks.[44] The revolutionary, new kind of network ushered in by TDRSS and the new mission of today's network of ground stations have helped set the stage for the coming decade. On 14 January 2004, President George W. Bush announced his "New Vision for Space Exploration" to return astronauts to the Moon by the year 2020 to be followed by mankind's first journeys to a neighboring planet.

Space communications have indeed come a long way since engineers first tried to keep track on a little sphere called Sputnik as it beeped it way around the globe. Today, that same technology—which allowed the world to watch live broadcasts of the 1964 Tokyo Games to the "space handshake" of Apollo-Soyuz and webcasting of telemedicine from Antarctica—continues to provide people around the world with sharing of everyday technologies not possible before. Such is the diversity that has allowed live communications from virtually anywhere in the world. Families who have not seen each other in years can stay in touch using cellular telephone networks and video networking on their personal computers. Brilliant HDTV via satellite can be enjoyed, for example, by the outdoorsman on a camping excursion hundreds of kilometers from the nearest city.

As NASA prepares to send humans back to the Moon and launch evermore ambitious space missions into Earth orbit and beyond, the tracking and communications network which Jack Mengel, Ozzie Covington, Ed Buckley and so many others began not so long ago will be there to meet the challenge.

CHAPTER 9

A LEGACY

NASA's STDN supported every U.S. space mission since 1958. Its desire was to stay inconspicuous, the more invisible the better. Much like an offensive lineman in the game of football or a player in an orchestra, they did not want their name called because it usually meant there was a penalty or a wrong note. Yet, the team could not win without the lineman nor could the orchestra music make without every member playing the right note.

How will space communications progress in the decades ahead and how will it best be able to build on the accomplishments of the past?

This is the question the NASA finds itself asking at the dawn of the twenty-first century. But this is not new.

> Man has made remarkable strides in penetrating the atmosphere surrounding his planet and even venturing into space. Scores of objects have been launched into space, many to roam the solar system forever. We stand now on the threshold of a new era of discovery.[1]

This was not the reflection of a modern day philosopher, but rather, something that Ed Buckley said in 1966 even before man had first left the confines of Earth. Bill Wood echoed this sentiment when he was asked in 1983 even before the first TDRS was launched just what the future holds for spaceflight tracking and communications.

> Today, the problems are hardly any different: how to handle reliably and economically a data flow now grown to 50 million bits-per-second and how to make the best use of these data in our search for new knowledge for the benefit of all mankind. We may have new and more advanced technology, but the challenge remains.... Our motto was a rather basic one: 'Close is good enough when you are playing horseshoes, but that is not good enough for manned spaceflight!'[2]

The challenge remains. Instead of megabytes (millions of bytes of digital data), today space communications work in gigabytes and terabytes (billions of bytes and higher). Just like it took the pioneers of aviation nearly half a century to go from 10 miles per hour to the speed of sound, the last 50 years of the twentieth century has seen tracking and space communications go from the picket line of Minitrack to the near-instantaneous, on-demand services offered by an invisible network that sees an entire hemisphere of Earth from 36,000 kilometers away.

In an age when global weather forecasts, spectacular images of celestial objects never before seen from space, and live images of astronauts living and working in space are taken for granted, it is difficult to imagine a time when America was struggling to put satellites into space. Engineers were not even sure whether or not they could be reliably tracked let alone prove useful on an everyday basis. From optical and radar tracking to radio interferometry, and from the large, automatic tracking antennas of the 1960s to the SN of today, advancement in communications technology has undergone many evolutions during this time. In fact, entirely new industries have been spawned. Though it may sound a bit trite, the continuing legacy of the spaceflight networks through its many incarnations has not only produced America's success in space but has, in its own way, contributed to such spin-offs as calculators, personal computers, digital watches, cellular telephones, and internet links to remote corners of the world. From VHF and UHF to S-band and Ka-band, the exploitation of higher and higher frequencies across the radio spectrum has enabled the spaceflight networks to meet and plan for the ever-increasing demand for higher bandwidths that are needed both today and for visions on the horizon.

But more than the technology itself, the history of America's STDN is a testament to the people behind the scenes who made it work, both as an organization and as a technological marvel. Leading the way has been the organization at NASA's GSFC, the Agency's focal point for all near-Earth communications and scientific satellite work, a role which it continues to enjoy today. Henry Clements, who was the first Network Controller during Project Mercury, reflected on the role of this Center.

> As for the support provided by the Goddard tracking and communications team, it was outstanding—though it has been largely unrecognized even within NASA's own family. Management, outside the JSC, all too often failed to recognize the very important contributions made by these men and women, engineers and computer experts who were stationed around the world. We could not have done it without them. They got the data to us.[3]

Vern Stelter, who ran the Center's Communications Division from 1962 to 1973, summarized what he and his people were all about. "We were there, but determined to be invisible. Ours was a service on which the programs could depend. We were there when needed."[4]

This unpretentious mindset that the spaceflight tracking network be an "invisible network"—or as former Associate Administrator Charles Force put it like a light switch, always there when you turn it on—was something that Ozzie Covington always stood by. Said Covington unpretentiously years ago:

> I must confess, I had never been able to actually pinpoint what my contributions were. Granted, we pushed the state of the art in the areas of tracking, communications, and computer applications. Maybe it was the assembling of a first rate team of men and women—both in NASA and private industry—who in fact deserve the credit. We were only in the background, providing the links between the astronauts and the Earth.[5]

This first rate team was in fact a testimony to what can be accomplished when industry and the government develop a high level of trust. From Bendix field technicians like Gary Schulz to Senior Managers like Glenn Smith and Cliff Benson or Program Manager Larry Jochen, working the networks became a way of life. Like the early days of Mercury when the Agency had "Go Fever", many had "Island Fever". Some ended up spending their entire careers with the contractor teams, hopping from one locale to another. Some brought their families, others found new ones. As one former supervisor put it, "We didn't make a lot of money, but we had a lot of fun!"[6]

Murray Weingarten, President and Chairman of the Board for Bendix from 1973 to 1989 and who was perhaps the one most influential in establishing this esprit de corps of contractors, once summarized this legacy:

During the peak of the U.S. space program, some 2,300 Bendix people were committed to this effort. It was a good marriage, based on professional relationships and a dedication that would be difficult for some to understand. It was a productive partnership for both the government and private industry. There is no doubt that it has been effective, and we take great pride that the Congress of the United States referred to these people—both government and industry—as the 'unsung heroes of the space program.'[7]

Author Alfred Rosenthal in 1982 interviewed Gerald M. Truszynski, NASA's top official for Tracking and Data Acquisition from 1968 to 1978, and asked him to describe how the Agency (in particular, the importance of Headquarters charting the course and delegating the responsibilities to the Field Centers) and its contractors were able to meet the unique challenges of the time. The fabric of Truszynski's remark is as true today as when he first spoke compellingly of it 25 years ago:

One of the major reasons for the outstanding success of the NASA tracking and data acquisition networks lies in the organizational and management approach taken by NASA in this vital area of flight program support. While the variety of these programs was quite broad—ranging from research sounding rockets through scientific satellites, manned missions of great complexity and far ranging planetary missions—all needed the very necessary common denominator of reliable, and in most cases, worldwide tracking and data acquisition support for their accomplishment.

In the beginning days of the space program, there was a tendency to look upon tracking and data support as an associated part of major flight program functions, or as a necessary part of launch vehicle operations. However, early in the 1960s, we were successful in making the point to NASA management that there was a need to organize the tracking and data acquisition and communications function as a single, centralized entity, responsible for the development, implementation and operation of these facilities for support of all of NASA's flight programs. This resulted in a highly efficient structure and gave us the necessary resident technical expertise—in one office—to plan, develop, budget and defend before Congress the requirements for this key activity in an integrated fashion.

The office was able to become an integral part of the overall program planning function at NASA Headquarters and was involved, early on, in the evolution of every major program and thus able to translate mission requirements into network requirements in a timely manner. We now could plan our own destiny. We were given control over our own financial resources along with the technical expertise in the NASA Field Centers—primarily the Goddard Space Flight Center and the Jet Propulsion Laboratory—where major elements of these centers were directly associated with the tracking and data acquisition function.

The dedication of the people at these Centers was a major factor in the success of our program, and deserving of particular mention. We involved them directly in our planning and, with the splendid cooperation of the Center Directors, had the ability to deal quite directly with the appropriate technical groups to handle our problems with a minimum of administrative delay. Over the years, we evolved a network capability which was extremely reliable by requiring that systems committed for implementation into the network were within the state of the art and had the necessary developmental and test lead times to assure their operational integrity. This, despite the fact that we were working under too stringent, fixed time constraints. Because of this record, we were able to earn the confidence of the flight programs, the support of our management, and the Congress, which gave us the financial resources to get the job done.

Another important element in the success of our operations was the good international cooperation we enjoyed where we were required to establish tracking stations in foreign countries. We, at the outset, always approached each country involved as partners, never attempting to or even suggesting that we establish 'Little Americas.' We encouraged the active participation of the host country in the planning, construction, and subsequent operation of the tracking stations. As a result, we were never refused permission to establish our facilities. Zanzibar [and Havana] was the only facility we had to vacate on short notice when a coup toppled the government.

In the final analysis, the success of any activity usually can be traced to the individual efforts of the personnel who were highly skilled and dedicated in their efforts to provide the highest quality and roost reliable support possible to the space flight programs. The late Congressman Olin Teague referred to these individuals as 'the unsung heroes of the space program.' I certainly share his sentiments and thank each one for a job well done.[8]

As the space agency builds on these accomplishments moving into the future, some challenges remain the same. But some are quite different. Take the ISS and the Space Shuttle—which NASA plans to retire in 2010 after completion of the ISS. To support a Shuttle launch, it is not just the Agency's Space and Ground Networks that are involved. It is a collaborative effort

Official crew portrait of STS-107 which broke apart during reentry on 1 February 2003. From left to right are David M. Brown, Rick D. Husband, Laurel Clark, Kalpana Chawla, Michael P. Anderson, William C. McCool, Ilan Ramon. (NASA Image Number KSC-01PP-1639)

between NASA and the DOD, with international partners also having a stake. In a sense, NASA integrates a network from different organizations in order to meet a particular mission need. And when that mission is over, the network is broken up to allow the different stakeholders to return to their primary functions. This kind of "virtual networks" allows the Agency the flexibility it needs to accommodate many different types of missions.

  A case in point is the new communications requirement stipulated for the Space Shuttle after *Columbia* broke apart on reentry during the final minutes of STS-107 on 1 February 2003. Foam and ice debris from the Shuttle's giant External Tank during launch punched a hole in the leading edge of the left wing which led to thermal protection breakdown in the 1,650°C (3,000°F) searing heat of reentry, killing the crew of seven. (The crew member on that fateful day were Commander Rick D. Husband; Pilot William C. McCool; Mission Specialists Kalpana Chawla, David M. Brown and Laurel Clark; Payload Commander Michael P. Anderson; and Payload Specialist Ilan Ramon of the Israeli Air Force.)

Flight rules now require continuous, live, high-resolution video of the External Tank during the Shuttle's ascent into orbit. To this end, Enhanced Launch Vehicle Imaging System, or ELVIS, cameras are mounted on the Orbiter, SRB and the tank itself. The goal is to provide the ground with engineering and visual data to assess the vehicle condition and tracking of debris during launch and ascent. To meet this flight-critical ("Crit 1") safety requirement, the integrated network stations of Merritt Island/Ponce de Leon, Wallops Island and the Jonathan Dickinson Annex provide the necessary and seamless link needed for ELVIS to work.[9]

With respect to communications with the ISS, several upgrades have been implemented in recent years or will be in the near future. One such modification goes by the catchy acronym of IDEA: ISS Downlink Enhancement Architecture. IDEA is in essence a modified ground system infrastructure that provides the space station with the ability to increase its science data return rate three-fold, from 50 megabits-per-second to 150 over the station's Ku-band downlink. A fiber optic ground network began in 2004 enabling JSC and MSFC to receive this high-rate data. It became operational in 2005.[10]

For TDRSS and the SN, GSFC, and JSC have been working with the ESA since 1998 to make the system compatible and ready to support the latter's much anticipated Automated Transfer Vehicle, or ATV. ESA's ATV will be an automated, resupply ship designed to dock to the ISS and provide the crew with dry cargo, oxygen, water, and propellant. After cargo is unloaded, it will be reloaded with waste products, undocked, and set on a course for destructive reentry.

The first craft—to be named Jules Verne after the nineteenth century French science fiction writer—is considered by ESA as the most sophisticated space vehicle ever to be built in Europe. To support these partner objectives, Goddard completed a series of communications compatibility tests in Bremen, Germany in 2004.[11] Parallel with this effort, NASA SN engineers are working with the Japanese Space Agency NASDA to develop a tracking and communication solution for their H-II Transfer Vehicle (HTV), the Japanese version of Europe's ATV. Coding, data rate, and modulation upgrades to the TDRSS are anticipated to be complete no later than 2007.[12]

Finally, there is the familiar matter of television. What started humbly on Apollo 7 has come full circle, with HDTV. Just like consumer demands for better and better pictures from sporting events to big-screen IMAX pictures, images from space are no different. Starting with STS-114 (the first flight after the *Columbia* disaster), real-time HDTV was downlinked. Future HDTV sponsors include the Japanese along with the American cable television's Discovery Channel. Using customer-furnished hardware, HDTV

Backdropped against water and clouds, the International Space Station (ISS) was photographed by the crew of STS-102 on 1 March 2001 as they headed home in the Space Shuttle *Discovery*. In the foreground is a Russian Soyuz still docked to the station. Major construction of the ISS is scheduled for completion in 2010, at which time, NASA will transition American human space transportation from the Space Shuttle to the Crew Exploration Vehicle. (NASA Image Number MSFC-0102549)

signals will be downlinked from the ISS via TDRSS and distributed to users at the NASA Field Centers and to domestic and foreign customers.

The trend is clear. Space communications will remain an international activity, just like it was when it all started back in the 1950s. As it did then, NASA Headquarters will play a leading role to establish partnerships with the international space community. In the late 1980s, the Space Networks Interoperability Panel (SNIP) was informally created at an international conference under the direction of then Associate Administrator Robert Aller. At that time, differing space data relay systems were in various stages of planning and development by NASA, ESA, and NASDA. It was the first forum specifically designed to discuss, anticipate and try to resolve differences in the design and operation of the different systems with their stakeholders. The routine meetings among the agencies identified, for example, frequency differences, on-orbit locations, user operation limitations, and even emergency backup support scenarios in the event of total space communication system failures.[13]

This panel was followed in the June of 1999 by the formal establishment of the Interagency Operations Advisory Group, or IOAG. Here, top officials from NASA, Italy's Agenzia Spatiale Italiana (ASI), the French Centre National D'Etudes Spatiales (CNES), Germany's Deutsches Zentrum für Luftund Raumfahrt (DLR), the European Space Agency (ESA), and the Japan Aerospace Exploration Agency (JAXA), meet annually (at rotating host countries) to coordinate space communications policies, procedures, technical interfaces, and many other matters related to interoperability. The group itself does not do—for the most part—the technical work. It relies primarily on technical work already completed by other organizations, for example, that develop standards for space systems. However, when a deficiency or inconsistency is discovered, the IOAG may recommend to such organizations that they address the missing areas in their work. By doing so, a common framework is laid that enables synergy and cooperative efforts among all the international partners.[14]

In the same vein, in 1995 and 1996, NASA was an active participant in the National Facilities Study—a joint effort by the DOD, NOAA, NASA, and other U.S. government participants—the basic premise of which was to reduce duplication and identify areas of national need. One of the sub-panels was devoted to aerospace tracking and communications capabilities. It was in this panel that issues regarding frequency usage, allocation, and station locations were addressed. There was also extensive discussion regarding synergies between NASA and NOAA in places like Alaska for mutual cost savings and improved coverage. Like the IOAG and its predecessor the SNIP, the National Facilities Study addressed and tried to resolve the spectrum differences and expedite cooperative operations amongst the agencies.[15]

With respect to ground station operations, since Earth is round and the United States obviously does not own territory everywhere, it has been and always will be an international activity. In other words, if NASA has a requirement on foreign soil, it has no choice but to go to the other country. This actually has inherent advantages. Said Bill Watson:

> I think the really neat thing about the early ground network was that many of these countries welcomed us in because they wanted to join the space program. But they also welcomed us in on the condition that we train their people, hire them to become technically literate and competent. We did a lot of that. We had a NTTF at Greenbelt that ran classes for years and put thousands

of people through and trained them up in receivers and recorders and computers and how to solder and all kinds of techniques. They then took that back home and became technically competent in the countries that they came from. I think we still see that desire to engage, even with our peers today.[16]

While Earth science will always be there serving to anchor near-Earth space activities, the future for NASA—as it was in the past—is exploration. Here, synergies exist between science and sending astronauts back to the Moon and onto Mars, neither of which can be done without defining the space communication requirements. NASA Headquarters has set up, for this purpose, a Communication and Navigation Architecture Working Group to define and lay the foundation for its communication and navigation infrastructure for the next 25 years. One concept is a plan for an Integrated Near-Earth Network (INEN).

Although only a proposal, the elegance of an INEN is attractive since it would involve building a network one mission at a time. One of the first ingredients is Goddard's Solar Dynamics Observatory (SDO), a geosynchronous mission that will monitor solar storms and send back information which will be a benefit to astronauts in space in case of storms. It will do this, plus provide warnings of commercial communications disruptions here on Earth.

In conjunction with development of the SDO is the LRO, or Lunar Reconnaissance Orbiter. To support these projects, NASA's next expansion in ground stations may again be at White Sands, New Mexico where three 60-foot Ka-band antennas could be built to support these missions. Both the SDO and LRO will utilize the Ka-band frequency for communications where ever higher bandwidths and data rates can be accommodated. (Today most of NASA's communications is at X-band which is fine for data rates of around 150 megabits-per-second.) With the advancement to Ka-band, though, data rates can be increased by over a factor of three, to 500 megabits-per-second.[17]

With Ka-band capability, the kernel—or seeds—for an exploration network at White Sands could be established. The idea is that once SDO and LRO are over, this resource could then become part of a Ka-band network for lunar exploration. As the need arises, S-band commercial sites—for instance, in Australia and South Africa—can be used to supplement White Sands. As it does today, NASA can buy these lower rate services commercially. As the Exploration Program matures and as the need for high rate data requirements expands, the Agency might then consider putting additional Ka-band dishes in places like South Africa, Australia, or Madrid to complete its mid latitude, high data rate network for exploration.[18]

The idea is that although the DSN has traditionally been responsible for planetary communications, it will help support near-Earth work as well since GSFC has closed down its large tracking antennas in places like

Chapter 9 \ A Legacy   339

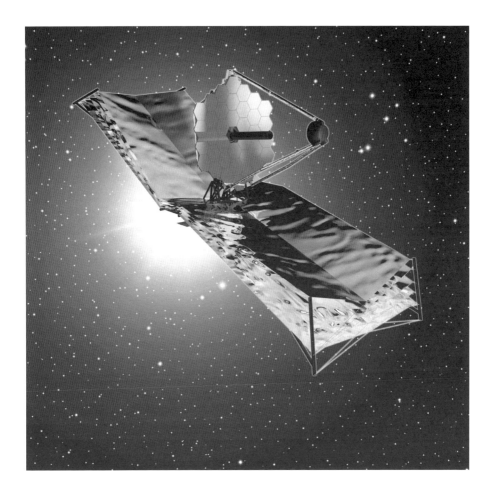

Pictured is the chosen artist's rendering of NASA's next generation space telescope. A successor to the Hubble Space Telescope (HST), the futuristic James Webb Space Telescope (JWST) is named in honor of NASA's second administrator, James E. Webb. To further our understanding of the way our universe formed, NASA is developing the JWST to observe the first stars and galaxies in the universe. The new telescope will carry a near-infrared camera, a multi-object spectrometer and a mid-infrared camera-spectrometer. The JWST is scheduled for launch in 2010 aboard an expendable launch vehicle. It will take the spacecraft three months to reach its destination, an orbit of 1,513,000 kilometers (940,000 miles) in space. The Marshall Space Flight Center (MSFC) is supporting Goddard in developing the JWST by creating an ultra-lightweight mirror for the telescope at Marshall's Space Optics Manufacturing Technology Center. The program has a number of industry, academic, and government partners, as well as the European Space Agency and the Canadian Space Agency. (NASA Image Number MSFC-0202886)

Rosman and Fairbanks. Since the TDRSS coverage zone stops at around 12,000 kilometers (7,400 miles)—the point at which TDRSS can no longer provide continuous communications with a spacecraft—new 18-meter (60-foot) Ka-band systems will be added to cover the gap between near-Earth and deep space. It in effect pushes the boundary of the NEN out to somewhere around two million kilometers (1,240,000 miles), which is where near-Earth transitions to deep space from a spaceflight point of view.[19]

Of particular interest in recent years are spacecraft that can be located at five distinct points in space where the gravitational pull of Earth, Sun, and Moon all balance out. A craft positioned at one of these "Lagrange Points" (named after Italian-French mathematician Joseph Louis Lagrange) can "hover" there in a so-called lissajou orbit—somewhat akin to the "figure-8" loop that a geosynchronous satellite does over Earth except the orbit would not be stable. One point in particular, called "L2," is approximately 1.5 million kilometers (932,000 miles) away from Earth beyond the Moon. In the coming decade, NASA plans to put several astronomical observatories there, including the much anticipated James Webb Space Telescope—the follow up to the HST. Space will be a busy place.[20]

Under the envisioned Integrated Near-Earth Network, the Agency could use a combination of 18-meter Ka-band antennas plus the TDRSS to provide seamless coverage on a space mission. Take a Mars exploration mission for instance. As it is launched out of the KSC, it would be supported from a Merritt Island/Ponce de Leon-like station, but equipped with Ka-band. Just like the Shuttle, it would then transition to TDRSS soon after launch. But as the vehicle leaves Earth's orbit and out of TDRSS coverage, the new Ka-band antennas at mid-latitude locations such as New Mexico, South Africa, or Madrid and Australia would pick up its signals much like the Apollo stations did with their Unified S-Band antennas in the 1960s and 1970s. Finally, as it goes beyond two million kilometers towards Mars, coverage will transition over to the DSN.[21]

NASA's tracking and data network will have then come full-circle, albeit this time with much faster data rates, much higher bandwidths and much more autonomy than before. This is really but a reflection of the cyclical nature of space exploration. The first satellites went into orbit to explore Earth. This was followed by space probes to the Moon and the planets. After America sent 12 astronauts to the Moon and won the space race, the Agency once again concentrated on Earth science, both un-crewed and with human presence (the ISS). In the coming decades, this cycle will likely shift once again to emphasize human exploration, not only of the Moon but this time out to Mars and beyond.

To this end, NASA's Exploration Systems Mission Directorate plans to start flying the Crew Exploration Vehicle soon after the three Space Shuttle Orbiters are retired in 2010. To ensure continuity, the TDRSS will have to deploy new satellites in the 2012 to 2015 time frame. As astronauts return to the Moon, the Agency will likely need some type of lunar relay system to be able to communicate with the spacecraft when it is on the back side of the Moon. Much has happened in the three-decades since Apollo last went to the Moon. Tragedy, unfortunately, has played a major role. To enter lunar orbit, the Apollo Service Module had to fire its engine on the backside of the Moon out of communication with Mission Control. After *Challenger* and *Columbia*, the Agency will likely not want to fire another rocket without having a communication link and know what happened. Thus, there is expectation at NASA that a TDRSS-like system on the backside of the Moon may have to be built before returning astronauts there.

The challenges don't stop there. Farther down the road, as mankind places our first steps on Mars, high-definition television will have already been a fixture for many years. The world is going to want to see pictures *a lot* clearer than what it saw with Armstrong and Aldrin. To do this, an optimal array of antennas and frequencies will be needed. The magnitude of the challenge should not be understated. For example, architectural studies have shown that in order to receive 100-megabits of low bit-error data from Mars, 100 to 300 12-meter (40-foot) antennas will need to be arrayed together in one location![22] Not impossible but clearly a challenge. Then there is the issue of relay and perhaps more importantly, delay. It takes light and radio signals anywhere from 3 to 18 minutes to reach Earth from Mars—the exact time depending on where the two planets are with respect to each other. One might ask, does the information we receive have to be real-time or is it good enough to just have the information tell what happened?

Delay-tolerant networking is even today a major issue for NASA (and for information technology at large). A simple example would be the case of a person using wireless internet access on a bus or a train. As he travels through a tunnel and loses his session, he would not want to have to start all over again when he resumes the session on the other end. The issue is how to get internet standards to evolve to the point where they can reliably cope with delays.

NASA will be busy.

With President Bush recommitting America to human space exploration, the National Aeronautics and Space Administration has been offered an opportunity that was not really all that well formed prior to 2004. With that vision comes the opportunity for NASA to take a new look at where it is going as the nation's space agency, and in particular, where it is going from a space communications point of view. The two are inextricably tied together. In sort of a twist on the popular American Express advertising slogan, those who work the networks at the Agency have adopted as their unofficial tagline "Space Communications: Don't Leave Earth Without It" when describing the indispensable role that the tracking and communications networks have played over the years.[23]

Because space communications is not easy, those who build the network are really building an enabling capability for the future. Just as the MSFN enabled the United States to safely go to the Moon and back, today the TDRSS enables the ISS and other data-rich spacecraft to return the amount of data that they were designed for. In many ways, as the previous generation did for today, those who now work the Space and Ground Networks are setting up for what is to come for the next generation.

The story of America's global spaceflight tracking network is ultimately the story of the men and women who made space communications a reality before it became the neat thing that it is today, something that we cannot live without. When Apollo 11 landed on the Moon in 1969, 56-kilobit connections to ground stations was a big deal. Today, the space agency hauls 4.5-*terabytes* (that is, 4.5-trillion bytes of digital data) a day back to Earth at an average rate of 100-megabits (100,000,000) per second.[24]

As one walks the hallways of NASA, whether it be the GSFC, the JSC or Headquarters, there are televisions all around showing videos of astronauts working in the ISS, spectacular images of celestial bodies from across the galaxy Telescope or live pictures of storm patterns developing on the other side of the planet. As Bob Spearing, NASA's former top official for space communications, put it:

> None of that—*none of it*—would be there without space communications. So when I walk by and somebody is looking at the television screen, I ask 'Do you know how that picture got here?' Most of them say no. Most people don't have any idea. I'll then go through a little talk about how the picture got here—and there is a real appreciation then. From a legacy point of view, there are a couple of things. One thing is that communications is ubiquitous and it is a capability that we all assume and just move on. We don't give it a second thought until it doesn't happen. If you look back over time, I believe you'll be hard pressed to find a mission that was compromised in any way by lack of communications. So our legacy is we deliver the goods and we always have. Every mission requires it and we have always delivered. . . . We made it work![25]

As the NASA enters the second half of its first century, a new generation of space probes and human explorers will lead the way back to the Moon, eventually venturing to Mars and beyond. As they make these journeys, men and women here on Earth will track them across that vast ocean of space.

After all, someone will have to stock the ships for the new Columbus and the new Magellan!

ENDNOTES

Preface

1   Much of the above description on the DSN was adapted from JPL Publication 400-326 "Goldstone Deep Space Communication Complex."

Foreword

1   NASA Management Instruction NMI 1162.1W
2   "Review of the Space Communications Program of NASA's Space Operations Mission Directorate" by the National Research Council of The National Academies, The National Academies Press, 2007.

Introduction

1   Robert Godwin, *Friendship 7: The First Flight of John Glenn. The NASA Mission Reports,* (Apogee Books, Ontario, Canada, 1999), Appendix B: Air-Ground Communications of the MA-6 Flight.
2   Alfred Rosenthal, *Vital Links* (NASA Goddard Space Flight Center, 1983), p. 196.
3   Ibid, p. 60.

4   Schneider interview from *Vital Links*.
5   Mengel interview from *Vital Links*.
6   Kraft interview from *Vital Links*.
7   Carolynne White, Goddard Space Flight Center News Release 89-2, 13 January 1989, Folder Number 8813, NASA Historical Reference Collection, NASA History Division, NASA Headquarters, Washington, DC.
8   See Note 6.
9   Force interview.
10 Prior to the Early Bird, satellites like Echo I and Echo II were passive, giant, shiny and highly reflective balloons that could only bounce, but not control, signals from one point on Earth to another.
11 See Note 6.
12 Ibid.
13 Force interview.

Chapter 1

1   The Oxford English Dictionary references the first use of the term "missile gap" from The Economist article dated 13 June 1959: "The Air Force gets an additional $170 million to help close the 'missile gap.'" John F. Kennedy is particularly connected with the phrase as he used it frequently during the 1960 presidential election campaign to attack the Republicans for their supposed complacency on the subject of Soviet Intercontinental Ballistic Missiles. Warnings and calls to address imbalances between the armed capabilities of the two countries were not new. A "bomber gap" had exercised political concerns a few years previously. What was different about the missile gap was the fear that a distant country could now strike without warning from far away. History has shown that this so-called missile gap was in fact very real. Within three years after the death of Joseph Stalin, the Soviet Union would announce that it had an effective Intercontinental Ballistic Missile force capable of striking North America and the United States. At that time, the U.S. could only field the Army's Redstone rocket, a tactical system with a range of 200 miles.
2   Mengel interview from *Vital Links*.
3   David West Reynolds, "Apollo: The Epic Journey to the Moon" (Tehabi Books, Inc., 2002), pp. 24–25.
4   Origin of the term "WAC" is somewhat unclear and may have stood for "Without Attitude Control," referring to the fact that the simple rocket was purely ballistic, with no stabilization and guidance system. The label was probably not associated with the more well-known Women's Army Corps.
5   The first monkey launched by the United States was Albert, a rhesus monkey, who on 11 June 1948 flew to nearly 40 miles altitude on a V-2 rocket

at White Sands. However, Albert died of suffocation during the flight. It was followed by Albert II who survived the V-2 flight but died on impact on 14 June 1949. However, Albert II became the first monkey in space as his flight exceeded 400,000 feet, the theoretical boundary of the detectable atmosphere. Albert III died at 35,000 feet in an explosion of his V-2 during launch on 16 September 1949. On the last V-2 monkey flight, Albert IV died on impact on 8 December 8 1949. Albert II and Albert IV were rhesus monkeys while Albert III was a cynomolgus monkey.

6   Dave Harris email, 17 April 2006.
7   K. G. Henize, "The Baker-Nunn Satellite Tracking Camera," *Sky and Telescope*, 16, 1957: pp. 108–111.
8   A 16th magnitude object is 25 times dimmer than the faintest star visible to the naked eye. Baker-Nunn cameras were able to observe the 6-inch diameter Vanguard 1 satellite at an altitude of 2,400 miles.
9   Kathleen M. Mogan and Frank P. Mintz, *Keeping Track* (NASA Goddard Space Flight Center, 1992), p. 9.
10 Alfred Rosenthal, *Vital Links* (NASA Goddard Space Flight Center, 1983), pp. 22–24.
11 Cristine Russell, "Watch Ends for Sputnik Spotters," *Washington Star*, A3, 26 June 1975.
12 Optical systems suffered from limitations imposed by weather and lighting conditions. Nevertheless, they proved very useful for tracking satellites with expired or failed beacons. Also, by obtaining highly accurate readings of satellite positions, the optical sites could provide information on changing global elevations caused by alterations in Earth's crust. In fact, one of the final functions of the SAO network was to serve as a progenitor to the NASA/Goddard Laser Crustal Dynamics program.
13 Perhaps Whites Sands biggest claim to fame in the early days of the range was the detonation of the world's first atomic bomb at Trinity Site, 16 July 1945, on the north part of the range.
14 "1947 History," White Sands Missile Range Fact Sheet, *http://www.wsmr.army.mil/pao/FactSheets/1947his.htm* (accessed 16 November 2005).
15 Barnes interview from *Vital Links*.
16 Covington interview from *Vital Links*.
17 "SingleObject Tracking Radars," *http://www.wsmr.army.mil/capabilities/nr/testing/range_inst/radar/sinobj.html* (accessed 30 August 2005).
18 Constance McLaughlin Green and Milton Lomask, *Vanguard: A History* (NASA SP-4202, 1970).
19 Rosenthal, pp. 17–18.
20 Ibid.
21 Mengel interview from *Vital Links*.
22 Ibid.

23 Zale, "Fast Summary: Minitrack System," 1958, Folder Number 8800, NASA Historical Reference Collection, NASA History Division, NASA Headquarters, Washington, DC.
24 Looney interview from *Vital Links*.
25 Rosenthal, pp. 19-20.
26 John T. Mengel and Paul Herget, "Tracking Satellites by Radio," *Scientific American,* January 1958, 198, no.1, Folder Number 8800, NASA Historical Reference Collection, NASA History Division, NASA Headquarters, Washington, DC.
27 MOTS cameras were all implemented with a solenoid assembly that would very briefly move its 8x10 inch photographic plate against a spring load. The station timing system would pulse the solenoid in a time-coded fashion, thus allowing the satellite to be displaced against the stellar background, enabling a position versus time solution. (Dave Harris email, 4/17/06)
28 See Note 26 and Ibid. In addition to performing system calibrations, MOTS was eventually used as an optical tracking system for satellites in low-Earth orbit. When NASA prepared to launch the passive Echo 1 into orbit on 12 August 1960, it needed to modify the MOTS cameras so that they could successfully photograph it and define the orbit. While inflation of the Echo balloon took place over South Africa, the first tracking images were taken by the San Diego Minitrack station. A couple of years later, on 31 October 1962, ANNA 1B was launched. The first orbital spacecraft designed to support the combined geodetic surveying requirements of the Air Force, NASA, Navy and Army, it carried a flashing light triggered by a time code. Photographing the satellite's flashes against the stellar background allowed for long distance baselines to be measured with great precision. For this effort, the Goddard Space Flight Center worked with domestic and international universities to establish a ground network of cameras in addition to those at the existing Minitrack stations. Stations were established in Jamaica, Puerto Rico, Edinburg, Texas (at Pan American University of Texas) and Sudbury, Ontario (at Laurentian University). In addition, Kodak developed a new emulsion to effectively increase film speed for the MOTS cameras.
29 Ibid.
30 See Note 21.
31 Rosenthal, p. 21.
32 Rosenthal, p. 22.
33 See Note 24.
34 The San Diego station was actually located on Brown Field Naval Auxiliary Air Station in nearby Chula Vista. When Brown Field closed in 1960, the station was moved to Mojave (soon known as Goldstone).
35 Mogan and Mintz, pp. 6-8.

36 McKeehan interview from *Vital Links*.
37 Ibid.
38 Ibid.
39 Rosenthal, p. 29.
40 Constance McLaughlin Green and Milton Lomask, *Vanguard: A History* (National Aeronautics and Space Administration Special Publication-4202, 1969).
41 Throughout most of Project Vanguard, another high-speed computer was used. An IBM 709 was installed at Patrick Air Force Base near Cape Canaveral. During launch and ascent, the vehicle was tracked using radar. After third stage burnout, the radar track data was processed by the IBM 709 and sent via teletype to the NRL Computing Center in Washington, DC. This preliminary data provided a best estimate of position and velocity at booster burnout from which the preliminary orbit determination was sent by teletype to the Minitrack stations enabling them to initialize their observation of the satellite passes.
42 See Note 26.
43 See Note 24.
44 Ibid.
45 Organization for Civil Space Programs, Memorandum for the President, 5 March 1958, Executive Office of the President, President's Advisory Committee on Government Organization, Eisenhower Library.
46 The NACA Director was selected by its 17-member governing committee.
47 See Note 45.
48 Ibid.
49 Ibid.
50 "The National Aeronautics and Space Administration: U.S. Centennial of Flight Commission," http://www.centennialofflight.gov/essay/Evolution_of_Technology/NASA/Tech2.htm (accessed 01 September 2005).

Chapter 2

1 JPL was transferred to NASA in 1958. It was staffed and operated by the California Institute of Technology under contract to NASA, an arrangement that continues to this day.
2 Jane Van Nimmen, Leonard C. Bruno and Robert L. Rosholt, *NASA Historical Data Book Volume I: NASA Resources 1958-1968* (National Aeronautics and Space Administration Special Publication-4012, 1988), pp. 4–5.
3 Weingarten interview from *Vital Links*.
4 Ibid.

5   Bodin interview from *Keeping Track.*
6   Nimmen, Bruno and Rosholt, pp. 309–311.
7   Dave Harris email, 16 March 2006.
8   Mengel interview from *Vital Links.*
9   Spearing interview.
10 Dunseith interview *Vital Links.*
11 At 59 missions, the Explorer series of satellites would go on to become the most successful series of science satellites ever, finally ending with Explorer 59 (the Solar Mesosphere Explorer) launched on 6 October 1981.
12 Looney interview from *Vital Links.*
13 Ibid.
14 Dave Harris email.
15 STADAN performed one other major function not covered by the acronym: satellite command, where vehicle control instructions were uplinked to the satellite from the ground.
16 Kathleen M. Mogan and Frank P. Mintz, *Keeping Track* (NASA Goddard Space Flight Center, 1992), p. 15.
17 Alfred Rosenthal, *Vital Links* (NASA Goddard Space Flight Center, 1983), p. 54.
18 "Carnarvon Tracking Station," Technical Secretariat Group, Weapons Research Establishment, Department of Supply, 1963, p. 6.
19 Kronmiller interview from *Keeping Track.*
20 See Note 18.
21 Rosenthal, p. 55.
22 William M. Hocking, "The Evolution of the STDN," *International Telemetering Conference,* 16, 80-01-03.
23 Nimmen, Bruno and Rosholt, pp. 57–59.
24 Edmond C. Buckley "Requirement for Establishment and Operation of 40-foot Parabolic Antenna at the Existing Minitrack Station in Santiago, Chile," Memorandum for the Office of International Programs, Folder Number 8799, NASA Historical Reference Collection, NASA History Division, NASA Headquarters, Washington, DC.
25 See Note 5.
26 Chet Matthes was later the Station Director at Fort Myers, Florida. He passed away in 1988.
27 Clotaire Wood, "Morning Report to the Administrator," 2 January 1959, Folder Number 8807, NASA Historical Reference Collection, NASA History Division, NASA Headquarters, Washington, DC.
28 "Fort Stewart History," *http://www.stewart.army.mil/ima/sites/about/history.asp* (accessed 5 September 2005).
29 Mogan and Mintz, p. 21.
30 The STADAN station at Fairbanks actually consisted of sites at nearby College and Gilmore Creek.

31 "Budget Estimates: FY 1969," IV, AO 2-34, National Aeronautics and Space Administration.
32 Gerald M. Truszynski, "Fatal Accident at NASA's Alaska Tracking Station," Memorandum to the Administrator, 7 May 1969, Folder Number 8805, NASA Historical Reference Collection, NASA History Division, NASA Headquarters, Washington, DC.
33 Glen Nagle interview.
34 See Note 18, p. 1.
35 Robert A. Leslie to The Minister of Science, "An Annual Contribution by Australia to the Cost of the U.S. NASA Tracking Stations in Australia," 74/2967, 4 February 1976, Historical Records Box 18, NASA Canberra Office, Yarralumla, Australian Capital Territory.
36 "STADAN Facility, Orroral Valley, ACT: Information Brochure," The Department of Supply, Melbourne, Victoria, p. 1.
37 Ibid.
38 Ibid.
39 R. C. Davey to The Secretary of the Department of Science, "Australian Contribution to NASA Activities in Australia," 70/703, 17 October 1974, Historical Records Box 18, NASA Canberra Office, Yarralumla, Australian Capital Territory.
40 NASA News Release 63-240, 24 October 1963. Folder 8820, NASA Historical Reference Collection, NASA Headquarters, Washington, DC. Goddard's Applications Technology Satellite (ATS) project would become one of the most scientifically productive satellite programs ever devised by NASA. For example, using the Hawaii Station, the University of Hawaii utilized ATS-3 to communicate with South Pacific islands. Medical data and problems were relayed to the hospital at the university from these remote islands and in return, the university transmitted back the remedies.
41 Mogan and Mintz, pp. 18–19.
42 The $5 million tag broke down as follows: Site development and utilities $477K; Facility construction $1.1M; Tracking and data equipment $3.1M; Design and engineering services $337K. ("Satellite Tracking and Data Acquisition Facilities Fiscal Year 1962 Construction and Facilities Estimates, Rosman Data Acquisition Facility, Project No. 3379," 14 Feb 1962, Folder 8820, NASA Historical Reference Collection, NASA History Division, NASA Headquarters, Washington, DC.)
43 "Data Acquisition Facility, Rosman, NC," NASA Goddard Space Flight Center Brochure, Folder Number 8820, NASA Historical Reference Collection, NASA History Division, NASA Headquarters, Washington, DC.
44 Ibid.
45 Ibid.

46 NASA News Release 63-279, Folder Number 8815, NASA Historical Reference Collection, NASA History Division, NASA Headquarters, Washington, DC.
47 McKeehan interview from *Vital Links*.
48 Transportable equipment was also used in Brasilia, Brazil to obtain telemetry data on the Van Allen Radiation Belt on Explorer 15 in 1962. Much of this equipment was also sent to Majunga in early 1964.
49 Letter from the Director, Office of Tracking and Data Acquisition to the Office of International Programs, "Establishment of a Station in South Africa to Provide Supplementary Coverage to a Station Located in Zanzibar," 03 March 1964, Folder Number 8815, NASA Historical Reference Collection, NASA History Division, NASA Headquarters, Washington, DC.
50 See Note 47.
51 In addition to an Assistant Station Director at some sites, another Civil Servant was occasionally assigned to a station, such as a Training Supervisor or a NASCOM Switching Center Supervisor. (Charles Force email 21 March 2006 and review notes 14 September 2006)
52 These were not the same aerospace primes that Goddard contracted to design, develop, and fabricate the satellites themselves. This list included aerospace giants such as Grumman Aircraft Engineering Corporation on the Orbiting Astronomical Observatories (OAO), Hughes Aircraft Company for the Applications Technology Satellites (ATS) and General Electric for the Nimbus spacecraft.
53 The other six companies ahead of Bendix were all prime contractors working directly on the Apollo spacecraft or its Saturn V launch vehicle: North American Rockwell, Grumman, Boeing, McDonnell, General Electric, and IBM.
54 Nimmen, Bruno and Rosholt, pp. 203–226.
55 Hunsicker interview from *Keeping Track*.
56 Blossom Point operated as part of STADAN until September 1966, when NASA ceased joint operations with the Navy. The site is still used today by the NRL as a primary Navy satellite control facility. ["Blossom Point," *http://www.globalsecurity.org/military/facility/blossom-point.htm* (accessed 02 September 2005)]
57 Mogan and Mintz, pp. 22–23, 42–43.
58 Stelter interview from *Vital Links*.
59 Rosenthal, pp. 181–182.
60 For example, Orroral Valley used the switching center located at Deakin, in the Canberra suburbs. In fact, NASA was Telstra's (Australian telephone company) biggest customer in the 1960s and 1970s. (Mike Dinn email, 16 March 2006)
61 See note 58.

62 Rosenthal, pp. 182–183.
63 Ibid., p. 59.
64 Ibid.
65 Ibid., pp. 185–186.
66 Ibid., pp. 187–188.

Chapter 3

1   This includes 302 paid employees stationed at Wallops.
2   *Orbiting Solar Observatory Satellite OSO-1* (National Aeronautics and Space Administration Special Publication-57, 1965).
3   Gagarin was promoted to the rank of Flight Major immediately upon his return.
4   Kenneth Gatland, "Manned Spacecraft," (Macmillan Publishing Company, Inc., New York, 1976), pp. 104–111.
5   From the minutes, Panel for Manned Space Flight, September 24, 30 and October 1, 1958, NASA Historical Reference Collection, NASA History Division, NASA Headquarters, Washington, DC.
6   William R. Corliss, *History of the Goddard Networks* (NASA Goddard Space Flight Center, 1969), pp. 92–101.
7   Bill Wood interview from *Vital Links*.
8   Kathleen M. Mogan and Frank P. Mintz, *Keeping Track* (NASA Goddard Space Flight Center, 1992), p. 29.
9   Alfred Rosenthal, *Vital Links* (NASA Goddard Space Flight Center, 1983), pp. 69–70.
10 See Note 7.
11 Although planned, live television from the spacecraft never became a reality on Mercury. It had to wait until Apollo 7.
12 See Note 7.
13 Rosenthal, p. 75.
14 Mogan and Mintz, pp. 30–31.
15 "The Manned Space Flight Tracking Network (MSFTN)," description brochure, NASA Goddard Space Flight Center, 1965.
16 Rosenthal, p. 83.
17 Bill Wood email, 25 April 2006.
18 Bill Wood thinks the Super Constellation operated by NASA may in fact have been none other than President Eisenhower's old airplane, the *Columbine*.
19 John Saxon, who coordinated operations at the Honeysuckle Creek station, tells of how reports of UFO sightings would rise whenever testing with the Super Constellation, with all its blinking lights, was done.
20 Dave Harris email, 21 March 2006.
21 See Note 15.

22 Covington interview from *Vital Links*.
23 Henry Thompson and his wife tragically perished when Air New Zealand Flight 901 crashed into Mount Erebus in Antarctica on 28 November 1979 killing all 257 onboard. (Charles T. Force interview)
24 See Note 22.
25 Bill Wood email.
26 See Note 22.
27 Ibid. Former Goddard Center Director Harry Goett called Ozzie Covington 'Mr. Manned Net' and some of his colleagues called him "Mr. Radar." They all credit him with a unique understanding of the technical and management requirements that made the network succeed.
28 Prior to the formation of the Office of Tracking and Data Acquisition, there was considerable debate at NASA about the most effective way to organize Headquarters management. Some preferred a functional management structure (propulsion, space research, etc.) while others wanted project type management with special offices for human spaceflight, scientific satellites, and so on. The final decision favored a project management structure.
29 Western Electric, Inc., "Final Project Report to NASA: Project Mercury," NAS1-430, pp. 42–45, June 1961.
30 Ibid.
31 See Note 22.
32 Lynn Dunseith interview from *Vital Links*.
33 The first two human Mercury flights (Alan B. Shepard and Virgil I. "Gus" Grissom) were suborbital, using the Army's Redstone rocket as launch vehicles. Because of their short range, Bermuda was not needed. It was not until beginning with the next flight when John Glenn was launched atop an Air Force Atlas ICBM for an attempt at orbital flight did the Bermuda Station come into play.
34 *The Royal Gazette*, 11 April 1961, p. 1, Folder Number 8808, NASA Historical Reference Collection, NASA History Division, NASA Headquarters, Washington, DC.
35 Robert E. Spearing interview.
36 See Note 34.
37 See Note 29.
38 E. J. Kerrigan, Memorandum of Record, undated. Folder Number 8810, NASA Historical Reference Collection, NASA History Division, NASA Headquarters, Washington, DC.
39 "The History of Bendix," *http://www.bfec.us/bfectxt6.htm* (accessed 18 June 2005).
40 Virg True email to Charles Force, 24 April 2006.
41 Ibid.
42 Ibid.
43 Ibid.

44 Ibid.
45 Russell Howe, *The Washington Post*, 03 December 1961, Folder Number 8819, NASA Historical Reference Collection, NASA History Division, NASA Headquarters, Washington, DC.
46 Henry Tanner, "Nigeria Tracking Station Set Up in Desolate Area Near Aged City," *The New York Times*, 21 February 1962, Folder Number 8819, NASA Historical Reference Collection, NASA History Division, NASA Headquarters, Washington, DC.
47 Hugh L. Dryden to Keith Glennan, "Mercury Tracking Station in Mexico," 17 September 1959, Folder Number 8817, NASA Historical Reference Collection, NASA History Division, NASA Headquarters, Washington, DC.
48 Ibid.
49 Edmond C. Buckley to Director, Office of Space Flight Programs, "Review of Some of the Inter-governmental Agreements for the Mexican Station," 23 April 1961, Folder Number 8817, NASA Historical Reference Collection, NASA History Division, NASA Headquarters, Washington, DC.
50 Ibid.
51 Ibid.
52 Ibid.
53 Ray W. Hooker, Assistant Chief, Engineering Service Division, "Tracking and Ground Instrumentation Systems for Project Mercury, Special Report on African Sites," Memorandum for File, 20 October 1959, Folder Number 8819, NASA Historical Reference Collection, NASA History Division, NASA Headquarters, Washington, DC.
54 Ibid.
55 Ibid.
56 Henry Clements interview from *Vital Links*.
57 Swenson, Grimwood, Alexander, *This New Ocean: A History of Project Mercury* (National Aeronautics and Space Administration Special Publication-4201, 1998), pp. 213–220, 647–648.
58 N. Pozinsky to Ruech, Naval Facilities Engineering Command, "Grand Canary Island: Replacement Housing for Permanent and Migrant Farm Workers," Folder Number 8809, NASA Historical Reference Collection, NASA History Division, NASA Headquarters, Washington, DC.
59 Ibid.
60 Corliss, pp. 150–155.
61 Mogan and Mintz, pp. 35–36.
62 Ibid.
63 It is well known within NASA circles that Donald K. "Deke" Slayton, who for many years ran Flight Crew Operations in Houston, would assign members of the Astronaut Office to "nice locations" such as Australia or Hawaii as reward (rest & recreation) for his people.

64 "Carnarvon Tracking Station," Technical Secretariat Group, Weapons Research Establishment, Department of Supply, 1963, p. 7.
65 See Note 15.
66 Rosenthal, pp. 89–91.
67 Mogan and Mintz, p. 35.
68 Rosenthal, pp. 83–84.

Chapter 4

1 William R. Corliss, History of the Goddard Networks (NASA Goddard Space Flight Center, 1969), pp. 156–157.
2 David West Reynolds, "Apollo: The Epic Journey to the Moon" (Tehabi Books, Inc., 2002), p. 41.
3 Memorandum for the Vice President, John F. Kennedy, 20 April 1961 and Robert C. Seamans, Jr., *Project Apollo: The Tough Decisions* (National Aeronautics and Space Administration Special Publication-2005-4537, 2005), p. 13.
4 Robert C. Seamans, Jr., *Project Apollo: The Tough Decisions* (Monographs in Aerospace History No. 37, SP-2005-4537), pp. 11–21.
5 Alfred Rosenthal, *Vital Links* (NASA Goddard Space Flight Center, 1983), pp. 111–113.
6 Barton C. Hacker and James M. Grimwood, *On the Shoulders of Titans* (National Aeronautics and Space Administration Special Publication-4203, 1977).
7 See Note 1.
8 James R. Hansen, "First Man: The Life of Neil Armstrong," p. 236, Simon and Shuster, 2005.
9 Lyn Dunseith interview from *Vital Links*.
10 George Gray, "The UNIVAC 418 Computer," *Unisys History Newsletter*, 4:2, August 2000.
11 Hacker and Grimwood, p. 592.
12 "Carnarvon Tracking Station," Technical Secretariat Group, Weapons Research Establishment, Department of Supply, 1963, p. 4.
13 Rosenthal, pp. 114–116.
14 "The Manned Space Flight Tracking Network (MSFTN)," description brochure, NASA Goddard Space Flight Center, 1965.
15 Bill Schneider interview from *Vital Links*.
16 In November 1963, Secretary of Defense Robert S. McNamara reorganized the DOD national ranges, transferring much of the Pacific Missile Range from the Navy to the Air Force. McNamara also entrusted the latter with the overall management of all DOD ground tracking facilities. Following those actions, the Air Force renamed the Atlantic Missile

Range and the Pacific Missile Range to Eastern Test Range (ETR) and Western Test Range (WTR), respectively.
17 R. Owen, "Evolution of Telemetry, Command and Voice on the MSFN," memorandum to OWU, Goddard Space Flight Center, 29 January 1991.
18 Hacker and Grimwood, pp. 590–592.
19 Pan American World Airways, Inc., Aerospace Services Division, Public Relations MU517, Folder Number 8807, NASA Historical Reference Collection, NASA History Division , NASA Headquarters, Washington, DC.
20 Robert C. Seamans, Jr., Harold Brown, "Memorandum of Agreement on Operation and Support of NASA Instrumentation Facilities on Antigua and Ascension," 22 May 1965, Folder Number 8807, NASA Historical Reference Collection, NASA History Division, NASA Headquarters, Washington, DC.
21 Ibid.
22 Proceedings of the Apollo Unified S-Band Technical Conference, NASA-SP-87, 1965.
23 "Tracking and Data Acquisition Facility Ascension Island," 25 June 1965, Folder Number 8807, NASA Historical Reference Collection, NASA History Division, NASA Headquarters, Washington, DC.
24 Benson interview.
25 See Note 22.
26 Gerald M. Truszynski, "Construction of Apollo Unified S-Band Facilities on Antigua B.W.I. and Grand Bahama Island," 16 June 1965, Folder Number 8806, NASA Historical Reference Collection, NASA History Division, NASA Headquarters, Washington, DC.
27 Norman Pozinsky, "Memorandum of Communication: Agreement to Establish the Apollo Station on Antigua, BWI," 20 July 1966, Folder Number 8806, NASA Historical Reference Collection, NASA History Division, NASA Headquarters, Washington, DC.
28 H. R. Brockett to Carl Jones "Antigua Apollo Station," 28 July 1966, Folder Number 8806, NASA Historical Reference Collection, NASA History Division, NASA Headquarters, Washington, DC.
29 Ref: Edmund C. Buckley to Dr. Harry J. Goett, Technical Note, 22 July 1964, Folder Number 8810, NASA Historical Reference Collection, NASA History Division, NASA Headquarters, Washington, DC.
30 "Interagency Meeting to Discuss U.S. Requirements for Canton Island," Office of Tracking and Data Acquisition, Memorandum for File, 05 August 1964, Folder Number 8810, NASA Historical Reference Collection, NASA History Division, NASA Headquarters, Washington, DC.
31 Bendix was also the contractor for the Federal Aviation Administration. The FAA was still providing onsite support on Canton Island to the Department of Defense even though its own requirement there had ceased nearly a year ago.

32 E. C. Buckley, " Ascension Island Power Plant," Technical Note Director of the Office of Tracking and Data Acquisition to AFETR, Patrick Air Force Base, 18 June 1965, Folder Number 8807, NASA Historical Reference Collection, NASA History Division, NASA Headquarters, Washington, DC.
33 James C. Bavely, "Ascension Island Operations," Technical Note to H. R. Brockett, 26 April 1967, Folder Number 8807, NASA Historical Reference Collection, NASA History Division, NASA Headquarters, Washington, DC.
34 Letter from NASA Headquarters to Dr. H. J. Goett, "Zanzibar Station," 23 May 1963, Folder Number 8824, NASA Historical Reference Collection, NASA History Division, NASA Headquarters, Washington, DC.
35 Letter from L. F. Griffin to Edmund C. Buckley, Bendix Field Engineering Corporation, 11 March 1964, Folder Number 8824, NASA Historical Reference Collection, NASA History Division, NASA Headquarters, Washington, DC.
36 Administrator's Briefing Memorandum, 22 January 1964, Folder Number 8824, NASA Historical Reference Collection, NASA History Division, NASA Headquarters, Washington, DC.
37 Letter from NASA Administrator to the Secretary of State, 19 March 1964, Folder Number 8824, NASA Historical Reference Collection, NASA History Division, NASA Headquarters, Washington, DC.
38 N. R. Brockett, "Status of Zanzibar," Memorandum for Mr. Buckley, 13 April 1964, Folder Number 8824, NASA Historical Reference Collection, NASA History Division, NASA Headquarters, Washington, DC.
39 Howard Simons, "LBJ Orders Space Station Out of Zanzibar," The Washington Post, 08 April 1964, Folder Number 8824, NASA Historical Reference Collection, NASA History Division, NASA Headquarters, Washington, DC.
40 NASA News Release 63-279, Folder Number 8815, NASA Historical Reference Collection, NASA History Division, NASA Headquarters, Washington, DC.
41 Letter from the Director, Office of Tracking and Data Acquisition to the Office of International Programs, "Establishment of a Station in South Africa to Provide Supplementary Coverage to a Station Located in Zanzibar," 03 March 1964, Folder Number 8815, NASA Historical Reference Collection, NASA History Division, NASA Headquarters, Washington, DC.
42 James C. Bavely, "Staffing at Malagasy for Satellite and Gemini Programs," 13 October 1964, Folder Number 8815, NASA Historical Reference Collection, NASA History Division, NASA Headquarters, Washington, DC.

43 Former Deputy Associate Administrator David Harris recalls that land on Guam became an issue later on when the landlord decided to significantly increase the leasing cost.
44 Edmund C. Buckley to Dr. Hugh L. Dryden, "Visit of Governor Manuel F. L. Guerrero Territory of Guam," 10 June 1964, Folder Number 8813, NASA Historical Reference Collection, NASA History Division, NASA Headquarters, Washington, DC.
45 Force interview.
46 The section of the Operations Building labeled Diner was a cafeteria added after the station was operational. The only eating facility initially provided was a small kitchen, sized and equipped as one in an average home—and totally inadequate for the station staff. There were no restaurants within reasonable driving distance. During operations, personnel had to remain on-site anyway. Station Director Force requested vending machines to remedy this deficiency. But Goddard denied his request, explaining that since the station had no on-site cafeteria (as most stations had), it clearly lacked the expertise to stock food vending machines! (Charles T. Force email 14 September 2006.)
47 See Note 44.
48 See Note 45.
49 "New Bermuda Radar Selected for Apollo Moon Mission Support," GSFC News Release No. G-9-65, 08 April 1965, Folder Number 8808, NASA Historical Reference Collection, NASA History Division, NASA Headquarters, Washington, DC.
50 Director, Office of Tracking and Data Acquisition, "Unified S-Band Facility Bermuda; Project 9831 Revised," Administrator's Briefing Memorandum, 10 March 1965, Folder Number 8808, NASA Historical Reference Collection, NASA History Division, NASA Headquarters, Washington, DC.
51 Lorne M. Robinson to T. V. Lucas, "Chronology of Requirements for Unified S-band Systems at Grand Bahama Island (GBI) and Grand Turk Island (GTI)," 09 February 1966, Folder Number 8811, NASA Historical Reference Collection, NASA History Division, NASA Headquarters, Washington, DC.
52 Ibid.
53 "Mexico, U.S. Extend Agreement for Operation of Guaymas Tracking Station," NASA News Release 65-76, 4 March 1965, Folder Number 8817, NASA Historical Reference Collection, NASA History Division, NASA Headquarters, Washington, DC.
54 N. Pozinsky, Chief, Facilities and Station Implementation to Director, Office of Tracking and Data Acquisition, "Guaymas Station Land Requirements for Apollo USB," 10 May 1965, Folder Number 8817,

NASA Historical Reference Collection, NASA History Division, NASA Headquarters, Washington, DC.
55 Edmond C. Buckley to Morton Berndt, "Guaymas Station Importance," 6 August 1965, Folder Number 8817, NASA Historical Reference Collection, NASA Headquarters, Washington, DC.
56 Former Associate Administrator for Tracking and Data Systems, Bill Schneider, would call Gemini 8 as the network's finest moment during the Gemini program. He explained why. "During Gemini 8 as astronauts Armstrong and Scott docked with their Agena target vehicle, just as they left the line of sight of the Madrid tracking station and before they were acquired by Australia, a technical problem with the Agena caused them to spin wildly. So, after they were reacquired, the decision was made to end the flight prematurely and arrange for recovery at an alternate location in the Pacific. This was the first and the only time that NASA had to use a contingency landing site. With the help of the very excellent tracking network, we landed the spacecraft safely near a U.S. destroyer in the middle of the Pacific Ocean. It was a real test of our ability to communicate and to locate a spacecraft in a far away, strange and unexpected environment." (Schneider interview from *Keeping Track*.)
57 Kenneth Gatland, *Manned Spacecraft* (Macmillan Publishing Company, Inc., New York, 1976), pp. 185, 275–276.

Chapter 5

1 Kathleen M. Mogan and Frank P. Mintz, *Keeping Track* (NASA Goddard Space Flight Center, 1992), p. 44.
2 Ibid.
3 Alfred Rosenthal, *Vital Links* (NASA Goddard Space Flight Center, 1983), p. 136. Three locations spaced approximately 120° apart in longitude were required to provide seamless coverage of a lunar spacecraft over 24 hours.
4 Mogan and Mintz, pp. 48-49.
5 R. Owen, "Evolution of Telemetry, Command and Voice on the MSFN," memorandum to OWU, Goddard Space Flight Center, 29 January 1991.
6 Ibid.
7 Linda Neuman Ezell, *NASA Historical Data Book, Volume III: Programs and Projects 1969-1978* (National Aeronautics and Space Administration Special Publication-4012, 1994), Table 6-32.
8 "The Worldwide Deep Space Network" (NASA Jet Propulsion Laboratory JPL 400-326, May 1989).
9 William R. Corliss, *History of the Goddard Networks* (NASA Goddard Space Flight Center, 1969), p. 131.

10   Mogan and Mintz, pp. 49–51.
11   Gerald M. Truszynski to Donald Crabill, "The Need for Apollo Ships," 05 March 1964 Folder Number 8792, NASA Historical Reference Collection, NASA History Division, NASA Headquarters, Washington, DC.
12   In the letter, Truszynski referred to these stations as Pretoria and Mauritus (Mauritius), respectively.
13   See Note 11.
14   Ibid.
15   Ibid.
16   There was originally concern by NASA regarding the reliability of communications with the tracking ships. Because of this, NASA installed flight control consoles in them—a move that was distinctively counter to the MSFN trend toward centralizing all control functions at the Mission Control Center in Houston. Though installed, these control consoles were never actually used on a real mission.
17   A particular concern of Apollo engineers was reentry tracking, that is, locating the position of the Command Module during the communication blackout while the spacecraft was surrounded by a layer of super-heated plasma during reentry. Because the transponders were also blacked out during this period, the C-band radars lost this tracking aid. Reentry tracking of the CM returning from the Moon was more difficult than for vehicles returning just from Earth orbit since reentry could occur over a larger corridor at a much higher velocity. In addition, the Command Module had a center-of-gravity offset allowing the capsule to fly a lifting trajectory using roll maneuvers. In this manner, the trajectory could be controlled. For energy management, the capsule first plunged into the atmosphere and then briefly skipped out of the atmosphere before making its final descent towards the recovery area. During the early 1960s, the acquisition problem was considered potentially serious. A radio interferometer was even proposed by Mengel's group as a solution. In the end, a skin-tracking radar was designed and installed on the recovery ship *Huntsville*.
18   Apollo Instrumentation Ships Technical and Management Problems, 12 June 1967, Folder 8792, NASA Historical Reference Collection, NASA History Division, NASA Headquarters, Washington, DC.
19   Ibid.
20   Ibid.
21   Letter from T. O. Paine to Senator Clinton P. Anderson, 28 October 1969, NASA Office of the Administrator, Folder Number 8792, NASA Historical Reference Collection, NASA History Division, NASA Headquarters, Washington, DC.
22   Ibid.
23   Ibid.
24   Ibid.

25 Ibid.
26 Edmond C. Buckley to Harry J. Goett, Requirements for Instrumentation Aircraft for Support of Apollo, 8 January 1966, Folder Number 8798, NASA Historical Reference Collection, NASA History Division, NASA Headquarters, Washington, DC.
27 Ibid.
28 Letter from H. R. Brockett to Robert C. Seamans, Jr., 29 May 1967, ARIA Aircraft, Folder Number 8798, NASA Historical Reference Collection, NASA History Division, NASA Headquarters, Washington, DC.
29 "Fact Sheet: Advanced Range Instrumentation Aircraft," Office of Information, United States Air Force Systems Command, November 1976.
30 W. L. Folsom, Action Item (OTDA) Clarification of NASA Requirements for ARIA, 02 May 1966, Folder Number 8798, NASA Historical Reference Collection, NASA History Division, NASA Headquarters, Washington, DC.
31 As part of an Air Force consolidation of large test and evaluation aircraft, basing for the ARIA was transferred from the Eastern Test Range at Patrick Air Force Base in Florida to the 4950th Test Wing at Wright Patterson Air Force Base near Dayton, Ohio in December 1975. In 1994, the ARIA fleet was relocated to Edwards Air Force Base, California, as part of the 452nd Flight Test Squadron in the 412th Test Wing. The final flight of ARIA took place on 24 August 2001.
32 Edmond C. Buckley was the NASA Vice Chairman of the panel. Christopher C. Kraft represented the Manned Spacecraft Center and Ozro M. Covington the Goddard Space Flight Center. The Air Force Eastern Test Range at the time had a critical need for the ARIA in order to support testing activities for development of the Navy's Poseidon submarine launched ballistic missile.
33 Edmond C. Buckley, NASA/DOD ARIA Management Agreement, 26 January 1965, Folder Number 8798, NASA Historical Reference Collection, NASA History Division, NASA Headquarters, Washington, DC.
34 George E. Mueller to Buckley, Instrumentation Aircraft Support, 11 May 1964, Folder Number 8798, NASA Historical Reference Collection, NASA History Division, NASA Headquarters, Washington, DC.
35 Letter from H. R. Brockett to Robert C. Seamans, Jr., 29 May 1967, A/Ria Aircraft, Folder Number 8798, NASA Historical Reference Collection, NASA History Division, NASA Headquarters, Washington, DC.
36 See Note 7.
37 "ARIA 328 Memorial," *http://www.flyaria.com/memorial/1999/personal.htm* (accessed 27 December 2005).
38 "Intelsat," Jet Propulsion Laboratory Mission and Spacecraft Library, *http://msl.jpl.nasa.gov/Programs/intelsat.html* (accessed 12 October 2005).

39 At the speed of light, it takes approximately 1.3 seconds for radio signals to travel from Earth to the Moon.
40 They were: Antigua, Ascension, Bermuda, Canary Island, Canberra, Cape Canaveral, Carnarvon, Corpus Christi, Goldstone, Guam, Guaymas, Hawaii, Grand Bahamas, Madrid.
41 Corliss, pp. 244–249.
42 Ibid.
43 Force email.
44 Letter from James Webb to E. S. Terlaje, 7 April 1967, Folder Number 8813, NASA Historical Reference Collection, NASA History Division, NASA Headquarters, Washington, DC.
45 Brooks, Grimwood, Swenson, *Chariots for Apollo* (National Aeronautics and Space Administration Special Publication-4205, 1979), Chapter 9.
46 Phillips to Administrator, "Apollo 5 Mission (SA-204/LM-1) Post Launch Report #1," 12 February 1968.
47 See Note 45.
48 Ibid.
49 Ibid.
50 The first use of television on a human flight by NASA was actually on Mercury Atlas 9, where a camera was trained on astronaut L. Gordon Cooper mainly to monitor his movements and behavior during the 34-hour mission. Three Mercury ground stations were set up to receive the slow-scan television pictures. However, they could not be broadcast live to an audience.
51 Henry P. Yschek, MSC, to North American, Contract Change Authorization 95, 24 Sept. 1963; William A. Lee to Mgr., ASPO, "The case for television transmission during LEM descent and ascent," 27 April 1964.
52 Phillips to Manager, ASPO, "Apollo On-board TV," 10 April 1968.
53 Walter Cunningham, *The All-American Boys* (Simon & Shuster, Inc., 2004), p. 155.
54 "The Fresnedillas (Madrid, Spain) MSFN Station," http://www.honeysucklecreek.net/other_stations/fresnedillas/intro.html (accessed 02 November 2005).
55 Apollo 8 Mission Commentary, 24 December 1968, tapes 242-3, 277-1.
56 Robert Zimmerman, *Genesis: The Story of Apollo 8* (Random House, Inc., 1998), pp. 210–211, 240–249.
57 "American Experience: Race to the Moon," http://www.pbs.org/wgbh/amex/moon/peopleevents/e_telecasts.html (accessed 07 November 2005).
58 Brooks, Grimwood, Swenson, Chapter 12.
59 "Apollo 15 Camera Equipment," http://www.lpi.usra.edu/expmoon/Apollo15/A15_Photography_cameras.html#COLORTV (accessed 07 November 2005).
60 Bill Wood email.

61 Kerrie Dougherty, John Sarkissian, "Dishing Up the Data: The Role of Australian Space Tracking and Radio Astronomy Facilities in the Exploration of the Solar System," IAC-02-IAA.2.3.01, 2002. Also see John Sarkissian, "On Eagles Wings: The Parkes Observatory Support of Apollo 11 Mission," Astronomical Society of Australia, Volume 18, 2001.
62 Ibid.
63 Ibid.
64 Ibid.
65 Glen Nagle interview.
66 See Note 61 and Nagle interview.
67 See Note 65.
68 See Note 61.
69 Ozzie Covington interview from *Vital Links*.
70 Bill Wood interview from *Vital Links*.
71 Dinn interview.
72 Ibid.
73 Robert Owen interview from *Vital Links*.
74 "Apollo and the Dish Down Under," CSIRO Media Release 2000/266, 12 October 2000.
75 See Note 61.
76 Tom Reid was the colorful Station Director at Honeysuckle Creek during the Apollo missions. He had a long and distinguished association with NASA as was one of the longest serving Station Directors. He first became involved with NASA while working with the Weapons Research Establishment at Woomera in the 1958-1962 time period. He became the first Station Director at Orroral Valley in 1964, and in 1967 was appointed Director at Honeysuckle Creek, where he provided inspiration and leadership during the lunar landings. In 1970, he began an 18-year assignment as Director of the Tidbinbilla Deep Space Communications Complex. Reid was awarded the Member of the Order of the British Empire (MBE) in January 1970 in recognition of his contribution to Australia meeting its commitments to the Manned Space program. It was actually bestowed upon him by Her Majesty the Queen when she was in Australia during the Apollo 13 mission. She happened to ask him how the mission was going when she presented it. (Charles T. Force review comments, 14 September 2006.)
77 Lyn Dunseith interview from *Vital Links*.
78 Chris Kraft interview from *Vital Links*.
79 "Experiment Operations During Apollo EVAs," ARES Publications, *http://ares.jsc.nasa.gov/HumanExplore/Exploration/EXLibrary/docs/ApolloCat/Part1/ALSEP.htm* (accessed 8 November 2005).
80 Although the Lunar Roving Vehicle had a total range capability of 91 kilometers (57 miles), the astronauts would not have the time nor safety margin

to fully use this capacity. On Apollo 15, the LRV was driven a total of 27.7 kilometers (17.2 miles) with a maximum range from the Lunar Module being 5 kilometers (3.1 miles). On Apollo 16, the total was 26.7 kilometers (16.6 miles) and reached a distance of 4.5 kilometers (2.8 miles) from the LM. The Apollo 17 LRV accumulated 35.9 kilometers (22.3 miles) with a maximum distance of 7.5 kilometers (4.7 miles) from the LM.

81 Mogan and Mintz, p. 55.
82 Robert Barnes interview from *Vital Links*.
83 See Note 77.
84 Helen T. Wells, Susan H. Whiteley, Carrie Keregeannes, *Origins of NASA Names* (National Aeronautics and Space Administration Special Publication-4402, 1976), p. 109. The name 'Skylab,' a contraction connoting 'laboratory in the sky,' was suggested in 1968 by Lt. Colonel Donald L. Steelman of the United States Air Force while assigned to NASA. He later received a token award for his suggestion.
85 W. David Compton, Charles D. Benson, *Living and Working in Space: A History of Skylab* (National Aeronautics and Space Administration Special Publication-4208, 1983), Chapter 14.
86 Mogan and Mintz, p. 61.
87 Russell P. Patera, William H. Ailor, "The Realities of Reentry Disposal," The Aerospace Corporation, p. 6, 2001.
88 Carolynne White, GSFC News Release 89-2, 13 January 1989, Folder Number 8818, NASA Historical Reference Collection, NASA History Division, NASA Headquarters, Washington, DC.
89 Ibid.
90 "Chronology: Beginnings of ASTP," NASA History Division, *http://history.nasa.gov/astp/chrono.html* (accessed 09 November 2005).
91 Edward Clinton Ezell and Linda Newell Ezell, *The Partnership* (National Aeronautics and Space Administration Special Publication-4209, 1978).
92 Mogan and Mintz, pp. 63–64.
93 Bill Wood email.
94 Donald H. Martin, "Communication Satellites, Fourth Edition," (The Aerospace Corporation Press, 2000), Chapter 1.
95 On Apollo-Soyuz, a transportable station with a 40-foot automatic tracking antenna was set up near Madrid to supplement the 100-foot antenna located at Buitrago, serving as the primary ground relay point between the ATS-6 satellite and the rest of the network.
96 See Note 94.
97 "U.S. Tracking Soyuz 16," JSC News Release No. 74-272, 03 December 1974, Folder Number 8781, NASA Historical Reference Collection, NASA History Division, NASA Headquarters, Washington, DC.
98 Watson interview.

99 Rosenthal, pp. 189–190.
100 Ibid.
101 Kenneth Gatland, *Manned Spacecraft* (Macmillan Publishing Co., Inc., New York, 1976), pp. 249–252.
102 In what could have turned the first international space mission from an unparalleled success into tragedy, a series of events during reentry led to the three American astronauts accidentally inhaling noxious fumes from the Command Module's Reaction Control System. This led to an unanticipated two week hospital stay for the crew in Honolulu. For Deke Slayton, it also meant the discovery of a small lesion on his left lung. An exploratory operation indicated it was a nonmalignant tumor. After a short convalescence, Slayton joined the four other ASTP crew members for two celebratory tours, one in the Soviet Union and the other in the United States.
103 See Note 78.
104 Report by the Subcommittee on Aeronautics and Space Technology of the Committee on Space Sciences and Astronautics, United States House of Representatives Ninety-Third Congress Second Session, September 1974.
105 Mogan and Mintz, pp. 56.

Chapter 6

1 Jane Van Nimmen, Leonard C. Bruno and Robert L. Rosholt, *NASA Historical Data Book Volume I: NASA Resources 1958–1968* (National Aeronautics and Space Administration Special Publication-4012, 1988), pp. 129.
2 Ihor Gawdiak with Helen Fedor, *NASA Historical Data Book, Volume IV: NASA Resources 1969–1978* (National Aeronautics and Space Administration Special Publication-4012, 1994). Contrary to the popular notion that NASA's budget peaked in 1969 and started to decrease only after the first lunar landing was achieved in July of that year, the Agency's funding actually peaked four years earlier when the FY 1965 NASA appropriation was $5,249,700,000. This had dropped to $3,994,993,000 by FY 1969. FY 1974 funding was $3,039,700,000 after which the NASA budget again began to increase so that by FY 1978, it reached $4,063,701,000.
3 Watson interview.
4 Ibid.
5 Gawdiak and Fedor, p. 68 and Judy A. Rumerman, *NASA Historical Data Book Volume VI 1979–1988* (National Aeronautics and Space Administration Special Publication-4012, 2000), p. 468.
6 Tracking and Data Acquisition Program, Subcommittee Report on Aeronautics and Space Technology, Ninety-Third Congress, Second Session, September 1974, Folder Number 8781, NASA Historical

Reference Collection, NASA History Division, NASA Headquarters, Washington, DC.
7   Bill Wood interview from *Vital Links*.
8   Kathleen M. Mogan and Frank P. Mintz, *Keeping Track* (NASA Goddard Space Flight Center, 1992), p. 68.
9   Ibid., pp. 56–58.
10 *Washington Evening Star*, Tuesday, 23 June 1970.
11 "Antigua Tracking Station Closing," 22 June 1970, NASA Press Release Number 70-101, Folder Number 8806, NASA Historical Reference Collection, NASA History Division, NASA Headquarters, Washington, DC.
12 Gerald M. Truszynski to Assistant Administrator for International Affairs, "Closing of Antigua Tracking Station," 20 February 1970, Folder Number 8806, NASA Historical Reference Collection, NASA History Division, NASA Headquarters, Washington, DC.
13 Gerald M. Truszynski to George M. Low, "Summary Status of Activities at Guaymas Station," Memorandum for File, 30 September 1970, Folder Number 8817, NASA Historical Reference Collection, NASA History Division, NASA Headquarters, Washington, DC.
14 "Guaymas Tracking Station," NASA News Release 70-198, 13 November 1970, Folder Number 8817, NASA Historical Reference Collection, NASA History Division, NASA Headquarters, Washington, DC.
15 Charles A. Taylor to Carl Mautner," Extension of Intergovernmental Agreement with Span for the Operation of the Canary Island Station," 6 July 1973, Folder Number 8809, NASA Historical Reference Collection, NASA History Division, NASA Headquarters, Washington, DC.
16 Gerald M. Truszynski to Colonel Daniel Oliver Osana, Technical Note, 22 January 1975, Folder Number 8809, NASA Historical Reference Collection, NASA History Division, NASA Headquarters, Washington, DC.
17 T. O. Paine, Letter to the Honorable Robert P. Mayo, Director, Bureau of the Budget, 20 February 1970, Folder Number 8813, NASA Historical Reference Collection, NASA History Division, NASA Headquarters, Washington, DC.
18 Ibid.
19 Ibid.
20 Douglas J. Mudgway, *Big Dish* (University Press of Florida, 2005), pp. 29–31.
21 South Africa first became involved in space research in 1957 during the International Geophysical Year. When the United States announced its intention to launch an artificial Earth satellite, South Africa, who was actively pursuing radio research, readily agreed to cooperate with the Smithsonian Astrophysical Observatory to establish and operate an optical tracking station (this at a time when no one was as yet fully convinced that artificial satellites could even get off the ground, not to mention be

seen or photographed from the ground once it was in orbit.) Thus when the State Department officially approached South Africa to operate a Minitrack station on behalf of the Naval Research Laboratory, it was received with great enthusiasm. South Africa knew that such a station on its soil would be the first to observe a satellite launched from Cape Canaveral after it had crossed the Atlantic.

22 *South African Panorama*, April 1986: 30-31, Folder Number 8849, NASA Historical Reference Collection, NASA History Division, NASA Headquarters, Washington, DC.
23 R. D. Trishman, "South African Situation," Memorandum for Assistant Director, Space Flight Operations, 04 April 1960, Folder Number 8849, NASA Historical Reference Collection, NASA History Division, NASA Headquarters, Washington, DC.
24 Edmond C. Buckley to the Office of International Programs, "Republic of South Africa," 9 November 1962, Folder Number 8849, NASA Historical Reference Collection, NASA History Division, NASA Headquarters, Washington, DC.
25 *New York Times*, 28 June 1965, Folder Number 8849, NASA Historical Reference Collection, NASA History Division, NASA Headquarters, Washington, DC.
26 Ibid.
27 *South African Sunday Times*, 22 April 1962, p. 1, Folder Number 8849, NASA Historical Reference Collection, NASA History Division, NASA Headquarters, Washington, DC.
28 *Defense/Space Daily*, 22 March 1974.
29 "South Africa: NASA Inches Out of a Segregated Tracking Station," *Science*, 27 July 1973: 331-332, 380, Folder Number 8849, NASA Historical Reference Collection, NASA History Division, NASA Headquarters, Washington, DC
30 "NASA to Phase Out African Stations," *Defense Space Daily*, 11 July 1973: 47, Folder Number 8849, NASA Historical Reference Collection, NASA History Division, NASA Headquarters, Washington, DC.
31 See Note 29.
32 Letter from James C. Fletcher to Dr. C. v.d.m. Brink, President of the Council for Scientific and Industrial Research, 21 October 1975, Folder Number 8849, NASA Historical Reference Collection, NASA History Division, NASA Headquarters, Washington, DC.
33 Letter from James C. Bavely, Chief of Network Operations Branch, Office of Tracking and Data Acquisition to the Director of Supply Division, Office of the NASA Comptroller, 19 September 1974, Folder Number 8849, NASA Historical Reference Collection, NASA History Division, NASA Headquarters, Washington, DC.

34 Letter from R. H. Curtin, Director of Facilities to GSFC Director of Administration and Management, 12 December 1975, Folder Number 8849, NASA Historical Reference Collection, NASA History Division, NASA Headquarters, Washington, DC.
35 See Note 28.
36 Gerald M. Truszynski, "Summary of Discussions with CSIR Regarding South African Station,"Memorandum for the Record, 20 September 1973, Folder Number 8849, NASA Historical Reference Collection, NASA History Division, NASA Headquarters, Washington, DC.
37 Ibid.
38 "Pay Up or Lose Station Madagascar Tells NASA," *Washington Star*, 12 July 1975, Folder Number 8815, NASA Historical Reference Collection, NASA History Division, NASA Headquarters, Washington, DC.
39 NASA Daily Activities Report, 8 April 1980, Folder Number 8815, NASA Historical Reference Collection, NASA History Division, NASA Headquarters, Washington, DC.
40 George Low to James C. Fletcher, Memorandum for Record, 23 July 1973, Folder Number 8815, NASA Historical Reference Collection, NASA History Division, NASA Headquarters, Washington, DC.
41 Mogan and Mintz, pp. 56–57, 68.
42 Force interview.
43 Ibid.
44 Jane Van Nimmen, Leonard C. Bruno and Robert L. Rosholt, *NASA Historical Data Book Volume I: NASA Resources 1958–1968* (National Aeronautics and Space Administration Special Publication-4012, 1988), pp. 57–59.
45 Force interview.
46 Ibid.
47 Linda Newman Ezell, *NASA Historical Data Book, Volume III: Programs and Projects, 1969-1978* (National Aeronautics and Space Administration Special Publication-4012, 1988) pp. 405–406.
48 Mogan and Mintz, pp. 56–57, 68.
49 "NASA Selects Tracking Contractor," NASA News Release 75-315, 17 December 1975, Folder Number 8781, NASA Historical Reference Collection, NASA History Division, NASA Headquarters, Washington, DC.
50 The second satellite, LAGEOS 2, was deployed from Shuttle flight STS-52 on 22 October 1992. It was a joint program between NASA and the Italian space agency ASI, which built the satellite using LAGEOS 1 drawings and specifications as provided by NASA. LAGEOS 2's orbit was selected to provide more coverage of seismically active areas, such as the Mediterranean Basin and California, designed to help scientists understand irregularities noted in the motion of LAGEOS 1. Tracking stations were located in many countries, including the U.S., Mexico, France,

Germany, Poland, Australia, Egypt, China, Peru, Italy, and Japan. Data from these stations was made available worldwide to investigators studying crustal dynamics. LAGEOS 1 also contained a special message plaque addressed to humans in the far distant future, showing maps of the surface of Earth from three different eras: postulated appearance from 268 million years ago, present day and what the appearance might be at 8 million years in the future—the satellite's estimated decay date. (See Note 51.)

51 "LAGEOS 1, 2 Quicklook," Jet Propulsion Laboratory Mission and Spacecraft Library, *http://msl.jpl.nasa.gov/QuickLooks/lageosQL.html* (accessed 03 January 2006).
52 Mogan and Mintz, pp. 59.
53 Rosenthal, p. 198. Goddard Space Flight Center had 1,400 people working the network. The tracking stations had 1,110 people of whom some 500 were foreign nationals.
54 Rosenthal, p. 196.
55 Wes Bodin interview from *Keeping Track*.
56 Harry McKeehan interview from *Vital Links*.
57 The popular American TV program *60 Minutes* aired a program about 1989 called "Isle of Spies" about the Seychelles, in which CBS alleged NASA had a station in the Seychelles. The Seychelles *was* an important location since it was in the right place to view orbit adjustment maneuvers for polar launches from Vandenberg Air Force Base. Office of Space Communications Associate Administrator Bob Aller, through NASA's Public Affairs Office, subsequently advised *60 Minutes* that although NASA used the Air Force station on the Seychelles as appropriate, NASA did not and had never actually had a station there. The Agency never felt apologetic about using Air Force resources. In planning to rerun the story the following summer, *60 Minutes* contacted the new Associate Administrator Charles Force to verify Aller's information. For whatever reason, when the rerun aired, *60 Minutes* seemed to go out of their way in the lead-in to again call it a NASA station! NASA never felt defensive, nor did anyone seem to care. (Charles Force email, 24 September 2005.)
58 Judy A. Rumerman, *NASA Historical Data Book Volume VI 1979-1988* (National Aeronautics and Space Administration Special Publication-4012, 2000), pp. 342–349.
59 *www.wstf.nasa.gov/WSSH/* (accessed 05 January 2006). In addition to the dry lakebeds of Edwards Air Force Base and the 15,000-foot runway at the Kennedy Space Center, a secondary backup landing site is located at the White Sands Space Harbor in southern New Mexico. Two 35,000-foot hard-packed, gypsum landing strips provide what the Space Shuttle needs in an emergency: long, forgiving runways. The Tula Peak Station served the communication needs in the event of a White Sands landing. TULA was closed-out following STS-2 but reopened briefly to

provide just such a capability when STS-3 landed at White Sands on 30 March 1982 (the only Shuttle landing there to date).

60 Mogan and Mintz, p. 68.
61 Joseph P. Allen, Assistant Administrator for Legislative Affairs to The Honorable Roy A. Taylor, 24 May 1976, Folder Number 8820, NASA Historical Reference Collection, NASA History Division, NASA Headquarters, Washington, DC.
62 NASA Daily Activities Report, 23 February 1981, Folder 8820, NASA Historical Reference Collection, NASA History Division, NASA Headquarters, Washington, DC.
63 Rumerman, pp. 342–349.
64 James M. Beggs, Letter to the Subcommittee on Space Science and Applications, Committee on Science and Technology, 04 January 1985, Folder Number 8805, NASA Historical Reference Collection, NASA History Division, NASA Headquarters, Washington, DC.
65 Ibid.
66 "Quito Tracking Station Shuts Down," Goddard News, 15 December 1981: 4, Folder Number 8801, NASA Historical Reference Collection, NASA History Division, NASA Headquarters, Washington, DC.
67 "Quito NASA Satellite Tracking And Data Acquisition Facility: A Story of Excellence," Ecuadorian Services Company, 2006.
68 Benson interview and Force email (24 September 2006).
69 See Note 67.
70 See Note 66.
71 R. E. Smylie to the Administrator, "Closure of the Quito STDN Station," 1 September 1981, Folder Number 8801, NASA Historical Reference Collection, NASA History Division, NASA Headquarters, Washington, DC.
72 The landing of STS-3 on 30 March 1982 is, to date, the only Shuttle landing ever at the White Sands Space Harbor.
73 Rumerman, pp. 342-349. [Pioneer (DSS 11), the original Goldstone station, was deactivated in 1981 and has been designated a National Historic Landmark by the United States Department of the Interior.]
74 "The Fresnedillas MSFN Station," *www.honeysucklecreek.net/other_stations/fresnedillas/main.html* (accessed 10 January 2006) and "Madrid Space Station," NASA Publication P72-223JPL, 31 August 1972.
75 Kerrie Dougherty email to Glen J. Nagle, 7 March 2006.
76 Saxon interview.
77 Hamish Lindsay, "Tracking Apollo to the Moon" (Springer, 2001).
78 Letter from Joseph F. Ada to James Fletcher, 21 December 1988, Folder Number 8813, NASA Historical Reference Collection, NASA History Division, NASA Headquarters, Washington, DC.
79 Henry Iuliano, "NASA Phases Down Guam Tracking Station," Goddard News, September 1989, Folder Number 8813, NASA Historical Reference

Collection, NASA History Division, NASA Headquarters, Washington, DC.
80 See Note 55.
81 Carolynne White, GSFC News Release 89-2, 13 January 1989, Folder Number 8818, NASA Historical Reference Collection, NASA History Division, NASA Headquarters, Washington, DC.
82 Ascension and Guam were both quite remote. A large power converter once failed at a station and analysis revealed a mechanical design flaw. A modification was quickly designed and teams were dispatched to all stations around the world to install the modification—to all stations, that is, except Ascension and Guam. These two stations received in the mail a small bag of parts with some instructions! (Charles Force email, 24 September 2006)
83 John F. Murphy, Letter to Senator Pete V. Domenici, 07 July 1982, Folder Number 8807, NASA Historical Reference Collection, NASA History Division, NASA Headquarters, Washington, DC.
84 Jeffrey M. Lenorovitz, "ESA Seeks U.S. Tracking Agreement, New Downrange Station for Ariane," Aviation Week & Space Technology, 7 January 1985: 117.
85 Ibid.
86 Thomas L. Matlick, "Ascension Closure," TN-88-310, 29 December 1988, Folder Number 8807, NASA Historical Reference Collection, NASA History Division, NASA Headquarters, Washington, DC.
87 David Harris email.
88 Thomas L. Matlick, "Closure of the Ascension Island Tracking Station," TN-89-023, 30 January 1989, Folder Number 8807, NASA Historical Reference Collection, NASA History Division, NASA Headquarters, Washington, DC.
89 Thomas L. Matlick, "Ascension Island Tracking Station," TN-89-055, 21 February 1989, Folder Number 8807, NASA Historical Reference Collection, NASA History Division, NASA Headquarters, Washington, DC.
90 See Note 55.
91 "Questions and Answers on the STDN," NASA News Release Number 89-6, 1 February 1989, Folder Number 8781, NASA Historical Reference Collection, NASA Headquarters, Washington, DC. (The personnel breakdown of contractors by location in 1989 were as follows: Ascension 43, Bermuda 77, Dakar 16, Guam 102, GSFC 227, Hawaii 67, MILA/PDL 129, Santiago 75, Wallops 4, White Sands 375; for a total of 1,115.)
92 Judy A. Rumerman, *NASA Historical Data Book Volume VI 1979-1988* (National Aeronautics and Space Administration Special Publication-4012, 2000), pp. 342–349.

Chapter 7

1. Arthur C. Clarke, "Extra Terrestrial Relays: Can Rocket Stations Give Worldwide Radio Coverage?" *Wireless World*, October 1945: 305–308.
2. David J. Whalen, "Communications Satellites: Making the Global Village Possible," unpublished article. The geosynchronous or geostationary orbit is known in some circles as the Clarke Orbit in honor of Arthur C. Clarke. However, it is not exactly clear that his article was actually the inspiration for modern telecommunications satellites. Pierce has stated that the idea was "in the air" at the time and certain to be developed regardless of Clarke's publication.
3. Ibid.
4. Glen E. Cameron, "Ground Station Design and Operation," pp. 7–13, 14, Applied Technology Institute, Clarksville, Maryland, 1998.
5. The first major three-axis stabilized geosynchronous satellite project was the DOD's classified Advent communications satellite. It was large and heavy. At over 500 pounds, at the time it could only be launched by the Air Force's Atlas-Centaur launch vehicle. The Advent never flew, primarily because the Centaur upper stage was not fully reliable until 1968. When the program was canceled, it was seen by many as the death knell for the three-axis, stable-platform, geosynchronous satellite. This was a premature prognosis, as many characteristics of the Advent would end up becoming commonplace in satellites less than 10 years later.
6. See Note 2.
7. There has often been confusion with respect to the acronym "TDRS," which refers to the Tracking and Data Relay Satellite itself, versus "TDRSS," which refers to the Tracking and Data Relay Satellite System, including the ground terminals. Both are pronounced the same ("tid-dres").
8. Ken Atchison, "NASA to Change Tracking and Data Acquisition Operations," News Release Number 89-172, 7 December 1989, NASA Headquarters.
9. Kathleen M. Mogan and Frank P. Mintz, *Keeping Track* (NASA Goddard Space Flight Center, 1992), p. 74.
10. A ground terminal situated outside of United States territory could have offset the coverage loss, but NASA at the time deliberately sacrificed technical performance for the benefit of security, locating the crucial ground terminal well within continental United States soil.
11. Ken Atchison, "NASA to Change Tracking and Data Acquisition Operations," News Release Number 79-172, 7 December 1979, NASA Headquarters.

12   Carolynne White, GSFC News Release 89-2, 13 January 1989, Folder Number 8818, NASA Historical Reference Collection, NASA History Division, NASA Headquarters, Washington, DC.
13   William C. Schneider, Statement Before the Subcommittee on Science, Technology and Space of the Committee on Commerce, Science and Transportation of the U.S. Senate, 28 February 1979, NASA Historical Reference Collection, NASA History Division, NASA Headquarters, Washington, DC.
14   Linda Neuman Ezell, NASA Historical Data Book, Volume III: Programs and Projects, 1969–1978, p. 424 (NASA SP-4012, Washington, DC, 1988), p. 424.
15   Edmond C. Buckley to George E. Mueller et al., "Data Relay Satellite System Requirements and Interface Panel," 1 September 1967; and Gerald M. Truszynski to John F. Clark, "Data Relay Satellite System Studies," 3 December 1968.
16   Richard L. Stock to record, "Senate Space Committee Inquiry Re: TDRSS," 07 September 1973.
17   James C. Fletcher to Roy L. Ash, 28 September 1973; and Thomas V. Lucas to Robert Lottmann, "TDRSS Economic Benefit/Cost Analysis," 5 April 1974.
18   Neuman Ezell, p. 425.
19   Gerald M. Truszynski, Statement Before the Subcommittee on Science, Technology and Space of the Committee on Commerce, Science and Transportation of the U.S. Senate, 16 September 1977, NASA Historical Reference Collection, NASA History Division, NASA Headquarters, Washington, DC.
20   Truszynski to record, "TDRSS Possible Bidders," 27 February 1975; and Fletcher to record, "Selection of Contractor for Tracking and Data Relay Satellite System Services (TDRSS)," February 1977.
21   "Harris Antennas and Hardware for White Sands," Harris Fact Sheet #1, Harris Corporation, Melbourne, Florida, Folder 8781, NASA Historical Reference Collection, NASA History Division, NASA Headquarters, Washington, DC.
22   Judy A. Rumerman, *NASA Historical Data Book Volume VI 1979-1988* (National Aeronautics and Space Administration Special Publication-4012, 2000), p. 314.
23   Robert Spearing interview.
24   Donald H. Martin, "Communications Satellites 1958–1992," The Aerospace Corporation, El Segundo, California, 1991, pp. 186–189. In July 1983, Fairchild Industries, Incorporated of Germantown, Maryland and the Atlanta, Georgia based Continental Telecom, Incorporated, announced that they had become owners of Space Communications by completing their purchase of Western Union's half-interest in Spacecom.

Western Union's share for liquidating its 50 percent interest amounted to $29 million. Prior to the transaction, Fairchild and Continental Telecom each owned 25 percent of Spacecom. The agreement followed a change in the contract giving NASA total use of TDRSS. Originally, a portion of the satellite network capacity was to have been reserved, available for commercial use by Western Union. The three members of the former partnership received $35 million from NASA under the terms of the buyout.

25   Aller, Robert O., "TDRSS Lessons Learned. Presentation to NASA Management, Office of Management and Budget, and Congress," Annapolis, August 18019, 1989.

26   See Note 23.

27   Although he thought there must have been a better word, Charles Force, then Deputy Project Manager for TDRSS, first used the term "segment" to describe both the space and ground components of TDRSS, realizing neither the spacecraft nor the ground station people would ever accept the term "sub-system" to describe those parts of the system. As used here to describe space and ground components of satellite systems, "segment" is a term that has since come into common use.

28   See Note 9.

29   See Note 23.

30   Harold G. Kimball to Marvin Skeeth, Electromagnetic Compatibility Analysis Center, Technical Note 2077, 15 July 1978, Folder 8781, NASA Historical Reference Collection, NASA Headquarters, Washington, DC. Electromagnetic compatibility (or more precisely, interference) was a serious issue. In 1991, for instance, TDRS-3 interfered with two major U.S. commercial, communication satellites, the result of which was disruption of cable television services across the country. On October 22, as Goddard was repositioning TDRS-3, it came near the Hughes Galaxy 1 communications satellite disturbing half of its C-band transponders. The next day, users of GE Americom's Satcom 1R experienced similar trouble. The result was hours of intermittent noise, snow and ghost images for those watching CNN, ESPN, TNT and several movie channels. Hughes official said NASA had not told industry that it was moving the TDRS spacecraft and the company had to call the North American Aerospace Defense Command (NORAD) to find out what caused the snafu. Associate Administrator Charles T. Force said at the time that Goddard engineers suspected a hardware problem on the TDRS which resulted in errant transmissions. (Aviation Week and Space Technology, 4 November 1991, p. 19.)

31   Joe McRoberts, Western Union TDRSS Contract Modified, NASA News Release 79-19, NASA Historical Reference Collection, NASA History Division, NASA Headquarters, Washington, DC.

32 This explains why there are no photographs showing any of the actual satellites fully deployed, only drawings or computer generated graphics. Photographs that do show a TDRS fully deployed on Earth are actually showing a scaled mockup.
33 NASA's Tracking and Data Relay Satellite System, NASA Press Release 82-186, December 1982, NASA Historical Reference Collection, NASA History Division, NASA Headquarters, Washington, DC.
34 "Tracking and Data Relay Satellite Description: Electrical Power System," *http://msp.gsfc.nasa.gov/tdrss/eps.html* (accessed 25 January 2006).
35 Ibid.
36 "Tracking and Data Relay Satellite Description: Attitude Control System," *http://msp.gsfc.nasa.gov/tdrss/attitude.html* (accessed 25 January 2006).
37 "Tracking and Data Relay Satellite Description: Propulsion System," *http://msp.gsfc.nasa.gov/tdrss/prop.html* (accessed 25 January 2006).
38 "Tracking and Data Relay Satellite Description: Thermal Control System," *http://msp.gsfc.nasa.gov/tdrss/tcntrl.html* (accessed 25 January 2006).
39 Space Network User's Guide (SNUG), 450-SNUG, Revision 8, NASA Goddard Space Flight Center, Greenbelt, Maryland, June 2002.
40 See Note 33.
41 Ibid.
42 Ibid.
43 "Tracking and Data Relay Satellite Description: Tracking, Telemetry & Command System," *http://msp.gsfc.nasa.gov/tdrss/ttc.html* (accessed 25 January 2006). As somewhat of a footnote, there is actually one other communication system. TDRS 1 through 6 actually hosted a Ku-band commercial payload which was never activated, and a commercial C-band antenna and payload package which was then operated by Western Union dedicated to their Westar satellite program.
44 Dennis R. Jenkins, "Space Shuttle: The History of the National Space Transportation System" (Ian Allan Publishing, Ltd., 2001), p. 270.
45 A Ku-band system provides a much higher gain, stronger RF signal with a smaller antenna than is possible with a S-band system.
46 *National Space Transportation System Reference Manual* (NASA Kennedy Space Center, 1988).
47 Alfred Rosenthal, *Vital Links* (NASA Goddard Space Flight Center, 1983), p. 210.
48 See Note 46.
49 "Pioneer NASA Spacecraft Celebrates 20 Years of Service," NASA New Release 03-130, 03 April 2003, *http://www.nasa.gov/home/hqnews/2003/apr/HP_news_03130.html* (accessed 10 March 2006).
50 Ibid.
51 Transcript from the Operational Recorder recovered from onboard the vehicle showed that at least some of the crew was aware something

was wrong immediately before the vehicle broke up. Pilot Mike Smith uttered an ominous "Uh oh," the last thing recorded just before loss of all data. Wreckage eventually retrieved from the bottom of the Atlantic also showed evidence that the crew (at least some) may have survived the initial breakup of *Challenger* as some switches activating each crew member's emergency personal oxygen kit had been manually toggled. Today, the remains of the Shuttle *Challenger* are permanently entombed inside a sealed Minuteman missile silo at Complex 31 on Cape Canaveral Air Force Station.

52 William P. Rogers, Chairman, "Presidential Commission on the Space Shuttle Challenger Accident, Final Report," 6 June 1986.
53 Jenkins, pp. 291–296.
54 After the loss of *Challenger*, the issue of crew escape systems was reexamined. A telescoping slide-pole concept was selected. The method, useful only below velocities of 370 kilometers (230 miles) per hour and altitudes below 9,150 meters (30,000 feet) during a controlled glide, had a 3 meter (9.8 foot) long curved aluminum pole which could be extended from the Orbiter's side hatch. During an evacuation, crew members would attach themselves with special parachute harnesses and slide down the pole, directing them underneath the Shuttle's left wing. A crew of seven could theoretically evacuate the Orbiter in 90 seconds. This would be the last resort prior to ditching the vehicle in a catastrophic emergency.
55 SNUG, p. 2–6.
56 Force interview.
57 Ibid. There was also some concern for earthquake damage and storage mishaps if the spacecraft were stored on the ground as opposed to in orbit.
58 "Space Network Online Information Center," *http://scp.gsfc.nasa.gov/tdrss/* (accessed 10 March 2006).
59 Joseph P. Allen to the Honorable Harold Runnels, Letter C:lgh:N237410f, 25 August 1975, Folder 8781, NASA Historical Reference Collection, NASA History Division, NASA Headquarters, Washington, DC.
60 Rumerman, p. 315.
61 See Note 22.
62 Ground Network Users Guide (GNUG).
63 NASA Daily Activities Report, 6 December 1977, Folder 8818, NASA Historical Reference Collection, NASA History Division, NASA Headquarters, Washington, DC.
64 Robert E. Smylie, Speech at the Whites Sands TDRSS Ground Terminal Acceptance Ceremony, 17 August 1981, Folder 8818, NASA Historical Reference Collection, NASA History Division, NASA Headquarters, Washington, DC.
65 Ibid.

66 Michael Braukus, "New NASA Ground Station Keeps Pace with Spacecraft Technology," Goddard News, Volume 33:9, September 1987, NASA Goddard Space Flight Center, Folder 8781, NASA Historical Reference Collection, NASA Headquarters, Washington, DC. and Carolynne White, GSFC News Release 89-2, 13 January 1989, Folder Number 8818, NASA Historical Reference Collection, NASA History Division, NASA Headquarters, Washington, DC.

67 Wes Bodin interview from *Keeping Track*.

68 "NASA Dedicates Communications Terminal in White Sands, New Mexico," Goddard New, February 1990, p. 6, Goddard Space Flight Center, Greenbelt, Maryland, Folder 8818, NASA Historical Reference Collection, NASA History Division, NASA Headquarters, Washington, DC.

69 Ibid.

70 Jim Elliot, "Dignitaries Participate in Event," Goddard News, Volume 33:9, September 1987, NASA Goddard Space Flight Center, Folder 8781, NASA Historical Reference Collection, NASA History Division, NASA Headquarters, Washington, DC.

71 David Batcho, "Excavation for Second Ground Terminal Opens Window Into Past," Goddard News, Volume 33:9, September 1987, NASA Goddard Space Flight Center, Folder 8781, NASA Historical Reference Collection, NASA History Division, NASA Headquarters, Washington, DC.

72 Ibid.

73 See Note 68.

74 Dwayne C. Brown, "NASA Ground Terminals Receive Native American Names," NASA News Release 93-83, NASA Headquarters, Washington, DC, 13 May 1993, Folder 8818, NASA Historical Reference Collection, NASA History Division, NASA Headquarters, Washington, DC.

75 Ibid.

76 Ibid.

77 Launched on 25 August 2003, the Spitzer observatory was the only one of the four not launched by the Space Shuttle. It was originally intended to, but after the *Challenger* accident, the Centaur liquid-oxygen/liquid-hydrogen upper stage that would have been used to push it into its intended high-Earth orbit was banned from Shuttle use.

78 "The CGRO Mission (1991–2000)," *http://cossc.gsfc.nasa.gov/docs/cgro/index.html* (accessed 06 February 2006).

79 Earth is surrounded by a close-to-spherical magnetic field, the magnetosphere. According to what we know today, it is being generated by actions deep in Earth's interior where conducting liquid metals are kept in motion by the forces of convection, Coriolis (centrifugal), and gravitation. Just as the charged windings in the coil of a generator puts out a magnetic field, these masses create Earth's magnetic field. This field protects the planet from space radiation by deflecting high energy particles from deep space

or by capturing them in the Van Allen Belts. Of these belts, discovered by the first U.S. satellite, Explorer 1 in 1958, there are two, one closer and the other farther away. Both surround Earth like a doughnut. However, at a certain location over the South Atlantic Ocean between Brazil and Africa, the shielding effect of the magnetosphere is not quite spherical but shows a "pothole" or a dip, which scientists explain as a result of the offset of the center of the magnetic field from the geographical center of Earth (by some 450 kilometers or 280 miles), as well as the displacement between the magnetic and geographic poles of Earth. For low-Earth orbiting satellites inclined between 35 and 60° with respect to the Equator, this oddity, called the South Atlantic Anomaly, becomes important since spacecraft in those orbit periodically pass through that zone of reduced natural shielding and thus spend a few minutes during each orbit exposed to much higher cosmic particle flux. Thus, vehicles in such orbits require higher shielding for the crew. It is also of concern in the design of space-hardened electronics which are degraded faster by higher particle fluxes. The design of the International Space Station, for instance, takes this effect into account.

80 Frank Stocklin, "*GRTS: An Experience of a Lifetime*," NASA Goddard Space Flight Center, Greenbelt, Maryland.

81 The actual amount of coverage provided by TDRSS varied depending on the attitude of the Compton observatory at any one time. Its attitude was science dependent on the mission. This sometimes led to the blockage of the high-gain antenna and line-of-sight to TDRSS.

82 See Note 80.

83 Ibid.

84 Ibid, and *NASA Opens Ground Station for Compton Gamma-Ray Observatory*, undated Goddard Space Flight Center New Release, Greenbelt, Maryland.

85 Ibid.

86 *New NASA Facility Will Complete Worldwide Communications Coverage*, Goddard Space Flight Center Press Release 98-122, 13 July 1998, Folder Number 8818, NASA Historical Reference Collection, NASA History Division, NASA Headquarters, Washington, DC.

87 "*Summary of Resources Requirements*," Mission Support: Fiscal Year 1998 Estimates Budget Summary, Office of Spaceflight, Space Communications Services (*http://www.hq.nasa.gov/office/codeb/budget/PDF/spacecom.pdf*), accessed 10 February 2006.

88 "Guam Remote Ground Terminal," *http://msp.gsfc.nasa.gov/tdrss/guam.html* (accessed 10 February 2006).

89 A. B. Comberiate et al., "Global, High Data Rate Ka-Band Satellite Communications: NASA's Tracking and Data Relay Satellite System," NASA Goddard Space Flight Center, TDRSS Project Office.

90 Ibid.

91 "Tracking and Data Relay Satellite H, I, J: The Next Generation," *http://msp.gsfc.nasa.gov/tdrss/tdrshij.html* (accessed 15 February 2006).
92 Ibid.
93 Ibid.
94 Ibid.
95 Ibid.
96 Ibid.
97 Tracking and Data Relay Satellite System: Chronology of Events, NASA Goddard Space Flight Center brochure.
98 "NASA to Take Control of TDRS-H Satellite," Boeing News Release, 10 August 2001, El Segundo, California.
99 See Note 39.
100 See Note 89.
101 Ibid.
102 Ibid.
103 Kathleen M. Mogan and Frank P. Mintz, *Keeping Track* (NASA Goddard Space Flight Center, 1992), p. 66.
104 Watson interview.

Chapter 8

1 "The Hawaii Story," Bendix Field Engineering Corporation, Folder Number 8814, NASA Historical Reference Collection, NASA History Division, NASA Headquarters, Washington, DC.
2 "Kokee Park Geophysical Observatory," Goddard Press Release, undated, Folder Number 8807, NASA Historical Reference Collection, NASA History Division, NASA Headquarters, Washington, DC.
3 Ibid.
4 Watson interview.
5 Dennis R. Jenkins, "Space Shuttle: The History of the National Space Transportation System" (Ian Allan Publishing, Ltd., 2001), p. 315.
6 "Bad Weather Could Delay Shuttle Launch," Cable News Network, Inc., 18 November 1997.
7 Ibid.
8 A deep space capability was also added at Merritt Island when the Jet Propulsion Laboratory Deep Space Network Compatibility Station (DSS-71) at Cape Canaveral was closed in 1974 and its systems relocated to MILA. This facility was renamed MIL-71 and provided support to planetary and deep space missions conducted by JPL. The first mission supported by MIL-71 was the Helios 1 Sun probe launched on 10 December 1974.

9   "The MILA Story," *science.ksc.nasa.gov/facilities/mila/milstor.html* (accessed 11 January 2006), and Bill Watson interview.
10 Ibid.
11 "MILA Station Tracks Space Shuttle," Spaceport News, p. 13, 27 October 2000, NASA Kennedy Space Center.
12 Ibid.
13 "Wallops Flight Facility," *http://www.wff.nasa.gov/about/* (accessed 17 February 2006). See "Wallops Station and the Creation of the American Space Program" (NASA SP-4311, 1997) by Harold D. Wallace, Jr. for a short history of the Wallops Flight Facility.
14 Rumerman, pp. 402–403.
15 Ibid. and "Scout," *http://www.fas.org/spp/military/program/launch/scout.htm* (accessed 17 February 2006).
16 "Wallops Orbital Tracking Station Becomes Operational," NASA News Release Number 86-12, 03 April 1986, NASA Goddard Space Flight Center, Folder 8823, NASA Historical Reference Collection, NASA History Division, NASA Headquarters, Washington, DC.
17 "Wallops Mobile Range Control System Provides Remote Support," Inside Wallops: XIX-99:16, NASA Wallops Flight Facility, Wallops Island, Virginia, 26 April 1999, Folder 8823, NASA Historical Reference Collection, NASA History Division, NASA Headquarters, Washington, DC.
18 Ibid.
19 FAA/AST is the governing body for all U.S. commercial launch activities. It promotes, sets guidelines, regulates and grants licenses for all non-government launches by U.S. companies.
20 Watson interview.
21 "Antarctic Connection: Robert F. Scott (1868–1912)," *http://www.antarcticconnection.com/antarctic/history/scott.shtml* (accessed 21 February 2006).
22 "Center for Astrophysical Research in Antarctica," *http://astro.uchicago.edu/cara/vtour/mcmurdo/* (accessed 20 February 2006).
23 All U.S. science activities in Antarctica are sponsored by the National Science Foundation.
24 "The McMurdo Ground Station (MGS)," *http://amrc.ssec.wisc.edu/MGS/history.html* (accessed 20 February 2006), and Watson interview.
25 Watson interview.
26 "Poker Flat Research Range General Information," *http://www.pfrr.alaska.edu/pfrr/index.html* (accessed 21 February 2006).
27 "DataLynx," *http://www.honeywell-tsi.com/datalynx/aboutus.shtml* (accessed 27 February 2006).
28 Watson interview.
29 "Welcome to Svalbard and SvalSat," Konnesberg Satellite Services information brochure, 2003.
30 Watson interview.

31 Joseph H. Rothenberg, Testimony before the House Subcommittee on Space and Aeronautics, Committee on Science, 11 March 1999. Many companies, including DataLynx, have agreements for support from Svalbard, operated by the Konnesburg Satellite Corp, or KSAT. NASA contracts with HTSI under the NENS contract for service from Svalbard.
32 Force interview.
33 Other centers represented on the SOMO were the Ames Research Center, Dryden Flight Research Center, Kennedy Space Center and Lewis Research Center.
34 See Note 31.
35 Spearing interview.
36 See Note 31.
37 Ibid.
38 Spearing interview.
39 Ibid.
40 Costrell interview.
41 The five contracts that made up the Space Mission Communications and Data Services were: 1) GSFC: Mission Operation and Mission Support—mission operations support for Goddard; 2) GSFC: Near Earth Networks Services—Goddard's tracking and data acquisition for near-Earth missions; 3) KSC: Kennedy Integrated Communications Services—communication services to support the Space Shuttle and other space operations; 4) MSFC: Unified NASA Information Technology Services—development, implementation and management of information technology services; and 5) MSFC: Huntsville Operations Support Center—voice, video and data services in support of simulations, near real-time and real-time flight mission support. There were other pieces of CSOC that were combined into a contract at JSC and separately competed through the JPL that were not included in the Space Mission Communications and Data Services procurement action. These assumed a substantial amount of the value of the old contract.
42 Michael Hardy, "NASA Alters IT Outsourcing Strategy," *Federal Computer Week*, 21 April 2003.
43 Ibid.
44 Force interview.

Chapter 9

1. Kathleen M. Mogan and Frank P. Mintz, *Keeping Track* (NASA Goddard Space Flight Center, 1992), p. 77.
2. Bill Wood interview from *Vital Links*.
3. Henry Clements interview from *Vital Links*.
4. Vern Stelter interview from *Vital Links*.
5. Ozzie Covington interview from *Vital Links*.
6. Cliff Benson interview.
7. Murray Weingarten interview from *Vital Links*.
8. Gerald Truszynski interview *from Vital Links*.
9. Jon Z. Walker, "Space Shuttle Return-to-Flight and International Space Station Status," NASA Goddard Space Flight Center Code 450, 10 June 2004.
10. Ibid.
11. Automated Transfer Vehicle (ATV) - European Space Agency (ESA), *http://hsf.honeywell-tsi.com/atv.html* (accessed 4 March 2006).
12. See Note 9.
13. David Harris email, 11 May 2006.
14. James Costrell, "IOAG Charter," 7 December 2004.
15. David Harris email.
16. Bill Watson interview.
17. Ibid.
18. Phil Liebrecht and Roger Clason, "GSFC Vision for Future Space Communications," NASA Goddard Space Flight Center, September 2005.
19. There is of course no physical boundary between what is called near-Earth and Deep Space. There is, however, a logical breakpoint based on radio frequency allocations. According to the Federal Communications Commission, the Deep Space Network allocation for Ka-band is 32 gigahertz, whereas communications below two million kilometers, the FCC thinks of it as near-Earth operations at 25 to 27 gigahertz. (Watson interview.)
20. "Integrated Near-Earth Network," NASA Headquarters Briefing, 2005. The term "L2" refers to an unstable equilibrium point on the other side of either body from a line segment joining two bodies (Earth-Moon, Earth-Sun). The term "L1" refers to the unstable equilibrium point on the line segment between two bodies. The L2 point for the Sun-Earth pair is 1.5 million kilometers from Earth, on the side opposite from the Sun. The L1 point for Sun-Earth is 1.5 million kilometers from Earth, in the direction of the Sun. The L1, L2 concept is hard to grasp, but for the Earth-Moon system, it is associated with high ocean tides—both on the side of Earth facing the Moon and on the opposite side of Earth. Thus, a high tide occurs every 12 hours.
21. See Note 11.

22 See Note 16.
23 Bob Spearing interview.
24 See Note 16.
25 See Note 23.

BIBLIOGRAPHY

Letters and Correspondences from NASA Historical Reference Collection, Washington, DC

Ada, Joseph F. to James Fletcher, 21 December 1988, Folder 8813
Administrator's Briefing Memorandum, 22 January 1964, Folder 8824
Allen, Joseph P. to the Honorable Harold Runnels, Letter C:lgh:N237410f, 25 August 1975, Folder 8781
Allen, Joseph P. Assistant Administrator for Legislative Affairs to The Honorable Roy A. Taylor, 24 May 1976, Folder 8820
Aller, Robert C. to Noel Hinners, TN-85-165, 9 August 1985, Folder 8801
Apollo Instrumentation Ships Technical and Management Problems, 12 June 1967, Folder 8792
Bavely, James C. "Staffing at Malagasy for Satellite and Gemini Programs," 13 October 1964, Folder 8815
Bavely, James C. to H. R. Brockett "Ascension Island Operations," 26 April 1967, Folder 8807
Bavely, James C. Chief of Network Operations Branch, Office of Tracking and Data Acquisition to the Director of Supply Division, Office of the NASA Comptroller, 19 September 1974, Folder 8849
Beggs, James M., Letter to the Subcommittee on Space Science and Applications, Committee on Science and Technology, 04 January 1985, Folder 8805

Brockett, N. R. to Mr. Buckley "Status of Zanzibar," 13 April 1964, Folder 8824

Brockett, N. R. to Carl Jones "Antigua Apollo Station," 28 July 1966, Folder 8806

Brockett, N. R. to Robert C. Seamans, Jr., "A/Ria Aircraft," 29 May 1967, Folder 8798

Buckley, Edmond C. "Requirement for Establishment and Operation of 40-foot Parabolic Antenna at the Existing Minitrack Station in Santiago, Chile," Memorandum for the Office of International Programs, Folder 8799

Buckley, Edmond C. to Director, Office of Space Flight Programs "Review of Some of the Inter-governmental Agreements for the Mexican Station," 23 April 1961, Folder 8817

Buckley, Edmond C. to the Office of International Programs "Republic of South Africa," 09 November 1962, Folder 8849

Buckley, Edmond C. to Dr. Hugh L. Dryden "Visit of Governor Manuel F. L. Guerrero Territory of Guam," 10 June 1964, Folder 8813

Buckley Edmond C. to Dr. Harry J. Goett, Technical Note, 22 July 1964, Folder 8810

Buckley, Edmond C. "NASA/DOD ARIA Management Agreement," 26 January 1965, Folder 8798

Buckley, E. C. "Ascension Island Power Plant," Technical Note, Director of the Office of Tracking and Data Acquisition to AFETR, Patrick Air Force Base, 18 June 1965, Folder 8807

Buckley, Edmond C. to Morton Berndt "Guaymas Station Importance," 6 August 1965, Folder 8817

Buckley, Edmond C. to Harry J. Goett "Requirements for Instrumentation Aircraft for Support of Apollo," 8 January 1966, Folder 8798

Buckley, Edmond C. to George E. Mueller et al., "Data Relay Satellite System Requirements and Interface Panel," 1 September 1967, Folder 8798

Curtin, R. H. Director of Facilities to GSFC Director of Administration and Management, 12 December 1975, Folder 8849

Director, Office of Tracking and Data Acquisition, "Unified S-Band Facility Bermuda; Project 9831 Revised," Administrator's Briefing Memorandum, 10 March 1965, Folder 8808

Dryden, Hugh L. to Keith Glennan "Mercury Tracking Station in Mexico," 17 September 1959, Folder 8817

Fletcher, James C. to Dr. C. v.d.m. Brink, President of the Council for Scientific and Industrial Research, 21 October 1975, Folder 8849

Fletcher to record "Selection of Contractor for Tracking and Data Relay Satellite System Services (TDRSS)," February 1977, Folder 8809

Folsom, W. L. "Clarification of NASA Requirements for ARIA," 02 May 1966, Folder 8798

Goett, Harry J. to Ira H. A. Abbott "Interim Report on Operation of Research Steering Committee on Manned Space Flight," 17 July 1959, Folder 8819

Griffin, L. F. to Edmond C. Buckley, "Bendix Field Engineering Corporation," 11 March 1964, Folder 8824

Hooker, Ray W., Assistant Chief, Engineering Service Division "Tracking and Ground Instrumentation Systems for Project Mercury, Special Report on African Sites," Memorandum for File, 20 October 1959, Folder 8819

"Interagency Meeting to Discuss U.S. Requirements for Canton Island," Office of Tracking and Data Acquisition, Memorandum for File, 05 August 1964, Folder 8810

Kerrigan, E. J., undated Memorandum of Record, Folder 8810

Kimball, Harold G. to Marvin Skeeth, "Electromagnetic Compatibility Analysis Center," Technical Note 2077, 15 July 1978, Folder 8781

Letter from the Director, Office of Tracking and Data Acquisition to the Office of International Programs "Establishment of a Station in South Africa to Provide Supplementary Coverage to a Station Located in Zanzibar," 3 March 1964, Folder 8815

Letter from NASA Administrator to the Secretary of State, 19 March 1964, Folder 8824

Letter from NASA Headquarters to Dr. H. J. Goett, "Zanzibar Station," 23 May 1963, Folder 8824

Low, George to James C. Fletcher, Memorandum for Record, 23 July 1973, Folder 8815

Lucas, Thomas V. to Robert Lottmann, "TDRSS Economic Benefit/Cost Analysis," 5 April 1974, Folder 8809

Matlick, Thomas L. "Ascension Closure," TN-88-310, 29 December 1988, Folder 8807

Matlick, Thomas L. "Closure of the Ascension Island Tracking Station," TN-89-023, 30 January 1989, Folder 8807

Matlick, Thomas L. "Ascension Island Tracking Station," TN-89-055, 21 February 1989, Folder 8807

Minutes from the Panel for Manned Space Flight, 24, 30 September and 1 October 1958, Folder 8819

Minutes from Meeting of Research Steering Committee on Manned Space Flight, 25-26 May 1959, Folder 8819

Mueller, George E. to E. C. Buckley, "Instrumentation Aircraft Support," 11 May 1964, Folder 8798

Murphy, John F. to Senator Pete V. Domenici, 7 July 1982, Folder 8807

Owen R. to OWU "Evolution of Telemetry, Command and Voice on the MSFN," 29 January 1991, Folder 8809

Paine, T. O. to Senator Clinton P. Anderson, 28 October 1969, NASA Office of the Administrator, Folder 8792

Paine, T. O. Paine to the Honorable Robert P. Mayo, Director, Bureau of the Budget, 20 February 1970, Folder 8813

Pozinsky, N. Chief, Facilities and Station Implementation to Director, Office of Tracking and Data Acquisition "Guaymas Station Land Requirements for Apollo USB," 10 May 1965, Folder 8817

Pozinsky, Norman "Memorandum of Communication: Agreement to Establish the Apollo Station on Antigua, BWI," 20 July 1966, Folder 8806

Robinson, Lorne M. to T. V. Lucas "Chronology of Requirements for Unified S-band Systems at Grand Bahama Island (GBI) and Grand Turk Island (GTI)," 9 February 1966, Folder 8811

Seamans, Robert C., Jr., to Harold Brown "Memorandum of Agreement on Operation and Support of NASA Instrumentation Facilities on Antigua and Ascension," 22 May 1965, Folder 8807

Smylie, R. E. to the Administrator "Closure of the Quito STDN Station," 1 September 1981, Folder 8801

Taylor, Charles A. to Carl Mautner "Extension of Intergovernmental Agreement with Span for the Operation of the Canary Island Station," 6 July 1973, Folder 8809

Trishman, R. D. "South African Situation," Memorandum for Assistant Director, Space Flight Operations, 4 April 1960, Folder 8849

Truszynski, Gerald M. to Donald Crabill "The Need for Apollo Ships," 5 March 1964 Folder 8792

Truszynski, Gerald M. "Construction of Apollo Unified S-Band Facilities on Antigua B.W.I. and Grand Bahama Island," 16 June 1965, Folder 8806

Truszynski, Gerald M. to John F. Clark, "Data Relay Satellite System Studies," 3 December 1968, Folder 8805

Truszynski, Gerald M. Truszynski "Fatal Accident at NASA's Alaska Tracking Station," Memorandum to the Administrator, 7 May 1969, Folder 8805

Truszynski, Gerald M. to Assistant Administrator for International Affairs "Closing of Antigua Tracking Station," 20 February 1970, Folder 8806

Truszynski, Gerald M. to George M. Low "Summary Status of Activities at Guaymas Station," Memorandum for File, 30 September 1970, Folder 8817

Truszynski, Gerald M. "Summary of Discussions with CSIR Regarding South African Station," Memorandum for the Record, 20 September 1973, Folder 8849

Truszynski, Gerald M. to Colonel Daniel Oliver Osana, Technical Note, 22 January 1975, Folder 8809

Truszynski to record "TDRSS Possible Bidders," 27 February 1975, Folder 8809

Webb, James to E. S. Terlaje, 07 April 1967, Folder 8813

Wood, Clotaire "Morning Report to the Administrator," 2 January 1959, Folder 8807

Press Releases from NASA Historical Reference Collection, Washington, D.C.

"Antigua Tracking Station Closing," 22 June 1970, NASA News Release 70-101, Folder 8806

"Dignitaries Participate in Event," Goddard News, September 1987, Folder 8781

"Excavation for Second Ground Terminal Opens Window Into Past," Goddard News, September 1987, Folder 8781

GSFC News Release 89-2, 13 January 1989, Folder 8818

"Guaymas Tracking Station," NASA News Release 70-198, 13 November 1970, Folder 8817

"Kokee Park Geophysical Observatory," undated Goddard News Release, Folder 8807

"Mexico, U.S. Extend Agreement for Operation of Guaymas Tracking Station," NASA News Release 65-76, 4 March 1965, Folder 8817

NASA Daily Activities Report, 8 April 1980, Folder 8815

NASA Daily Activities Report, 23 February 1981, Folder 8820

"NASA Dedicates Communications Terminal in White Sands, New Mexico," Goddard News, February 1990, Folder 8818

"NASA Ground Terminals Receive Native American Names," NASA News Release 93-83, 13 May 1993, Folder 8818

NASA News Release 63-240, 24 October 1963, Folder 8820

NASA News Release 63-279, Folder 8815

"NASA Opens Ground Station for Compton Gamma-Ray Observatory," undated GSFC News Release, Folder 8818

"NASA Phases Down Guam Tracking Station," Goddard News, September 1989, Folder 8813

"NASA Selects Tracking Contractor," NASA News Release 75-315, 17 December 1975, Folder 8781

"NASA's Tracking and Data Relay Satellite System," NASA News Release 82-186, December 1982, Folder 8815

"NASA to Change Tracking and Data Acquisition Operations," NASA News Release 79-172, 7 December 1979, Folder 8815

"New Bermuda Radar Selected for Apollo Moon Mission Support," GSFC News Release No. G-9-65, 8 April 1965, Folder 8808

"New NASA Facility Will Complete Worldwide Communications Coverage," GSFC News Release 98-122, 13 July 1998, Folder 8818

"New NASA Ground Station Keeps Pace with Spacecraft Technology," Goddard News, September 1987, Folder 8781

"Questions and Answers on the STDN," NASA News Release 89-6, 1 February 1989, Folder 8781

"Quito Tracking Station Shuts Down," Goddard News, 15 December 1981, Folder 8801

"U.S. Tracking Soyuz 16," JSC News Release No. 74-272, 3 December 1974, Folder 8781

"Western Union TDRSS Contract Modified," NASA News Release 79-19, Folder 8813

"Wallops Orbital Tracking Station Becomes Operational," NASA News Release Number 86-12, 3 April 1986, Folder 8823

"Wallops Mobile Range Control System Provides Remote Support," Inside Wallops: XIX-99:16, 26 April 1999, Folder 8823

Other Letters and Correspondences

Davey, R. C. to The Secretary of the Department of Science "Australian Contribution to NASA Activities in Australia," 70/703, 17 October 1974, Historical Records Box 18, NASA Canberra Office, Yarralumla, Australian Capital Territory

Hunter, Wilson H. to Robert A. Leslie "GSFC Reorganization," 28 January 1971, Folder 10/1/1 part 1, NASA Canberra Office, Yarralumla, Australian Capital Territory

Lee, William A. to Mgr. ASPO, "The case for television transmission during LEM descent and ascent," 27 April 1964

Leslie, Robert A. to Wilson H. Hunter "Funding MSFN Wing M&O at Tidbinbilla," 680-5-23, 17 April 1969, Folder 680/5/23, NASA Canberra Office, Yarralumla, Australian Capital Territory

Leslie, Robert A. to The Minister of Science "An Annual Contribution by Australia to the Cost of the U.S. NASA Tracking Stations in Australia," 74/2967, 4 February 1976, Historical Records Box 18, NASA Canberra Office, Yarralumla, Australian Capital Territory

Phillips to Administrator, "Apollo 5 Mission (SA-204/LM-1) Post Launch Report #1," 12 February 1968

Phillips to Manager ASPO "Apollo On-board TV," 10 April 1968

"Organization for Civil Space Programs," Memorandum for the President, Executive Office of the President, President's Advisory Committee on Government Organization, 5 March 1958, Eisenhower Library

Stock, Richard L. to record, "Senate Space Committee Inquiry Re: TDRSS," 7 September 1973

Yschek, Henry P. MSC to North American, "Contract Change Authorization 95," 24 September 1963

News Articles

"Bad Weather Could Delay Shuttle Launch," *Cable News Network, Inc.*, 18 November 1997

Clarke, Arthur C. "Extra Terrestrial Relays: Can Rocket Stations Give Worldwide Radio Coverage?" *Wireless World*, October 1945: 305-308

Defense/Space Daily, 22 March 1974

Gray, George "The UNIVAC 418 Computer," *Unisys History Newsletter*, 4:2, August 2000

Hardy, Michael "NASA Alters IT Outsourcing Strategy," *Federal Computer Week*, 21 April 2003

Henize, Karl G. "The Baker-Nunn Satellite Tracking Camera," *Sky and Telescope*, 16, 1957: 108–111

Howe, Russell *The Washington Post*, 3 December 1961

Lenorovitz, Jeffrey M. "ESA Seeks U.S. Tracking Agreement, New Down-range Station for Ariane," *Aviation Week & Space Technology*, 7 January 1985: 117

Mengel, John T. and Herget, Paul "Tracking Satellites by Radio," *Scientific American*, January 1958, 198, No. 1

"MILA Station Tracks Space Shuttle," *Spaceport News*, 27 October 2000

"NASA to Phase Out African Stations," *Defense Space Daily*, 11 July 1973: 47

"NASA to Take Control of TDRS-H Satellite," *Boeing News Release*, 10 August 2001, El Segundo, California

New York Times, 28 June 1965

"Pay Up or Lose Station Madagascar Tells NASA," *Washington Star*, 12 July 1975

Russell, Christine "Watch Ends for Sputnik Spotters," *Washington Star*, A3, 26 June 1975

Simons, Howard "LBJ Orders Space Station Out of Zanzibar," *The Washington Post*, 08 April 1964

"South Africa: NASA Inches Out of a Segregated Tracking Station," *Science*, 27 July 1973: 331–332, 380

South African Panorama, April 1986

South African Sunday Times, 22 April 1962

Tanner, Henry "Nigeria Tracking Station Set Up in Desolate Area Near Aged City," *The New York Times*, 21 February 1962

The Royal Gazette, p.1, 11 April 1961

Washington Evening Star, p.1, 23 June 1970

Briefings, Pamphlets, Papers, Reports and Testimonies

Aller, Robert O., "TDRSS Lessons Learned," Presentation to NASA Management, Office of Management and Budget, and Congress, Annapolis, 18–19 August 1989

"Apollo 8 Mission Commentary," NASA Tapes 242-3, 277-1, 24 December 1968

Apollo and the Dish Down Under (Commonwealth Scientific and Industrial Research Organization Media Release 2000/266, 12 October 2000)

"A Study of Requirements of NASA Space Flight Programs for Instrumentation Ship Support for Telemetry & Tracking,"Office of Tracking and Data Acquisition, NASA Headquarters, Washington, DC, March 1962 (Folder 8792, NASA Historical Reference Collection, NASA History Division, NASA Headquarters, Washington, DC)

Cameron, Glen E. *Ground Station Design and Operation* (Applied Technology Institute, Clarksville, Maryland, 1998)

Canberra Space Center Fact Sheet (Commonwealth Scientific and Industrial Research Organization)

"Carnarvon Tracking Station," Technical Secretariat Group, Weapons Research Establishment, Department of Supply, 1963

Comberiate, A. B. et al., "Global, High Data Rate Ka-Band Satellite Communications: NASA's Tracking and Data Relay Satellite System," NASA Goddard Space Flight Center, TDRSS Project Office

Corliss, William R. *History of the Goddard Networks* (NASA Goddard Space Flight Center, 1969)

____, *Histories of the STADAN, MSFN and NASCOM* (NASA Goddard Space Flight Center, 1974)

Costrell, James A., "Interagency Operations Advisory Group (IOAG) Liaison Statement to CCSDS Management Council Generated at the IOAG Meeting During May 20–22, 2003," Office of Space Communications, NASA Headquarters, May 2003

____, "Return to Flight and International Space Station," Office of Space Communications, NASA Headquarters, 9 June 2004

____, "United States Vision for Space Exploration," Office of Space Communications, NASA Headquarters, 9 June 2004

____, "IOAG Charter," Office of Space Communications, Office of Space Communications, NASA Headquarters, 7 December 2004.

Covington, Ozro M.,"Apollo Communications and Tracking," 27 August 1964

"Data Acquisition Facility, Rosman, NC," NASA Goddard Space Flight Center Information Brochure

Discovery: Canberra Deep Space Communications Complex (Commonwealth Scientific and Industrial Research Organization)

Dougherty, Kerrie and Sarkissian, John "Dishing Up the Data: The Role of Australian Space Tracking and Radio Astronomy Facilities in the Exploration of the Solar System," AIAA/IAC-02-IAA.2.3.01, 2002

"Fact Sheet: Advanced Range Instrumentation Aircraft," Office of Information, United States Air Force Systems Command, November 1976

"Final Project Report to NASA: Project Mercury," NAS1-430, Western Electric, Inc., June 1961

Ground Network User's Guide (GNUG), 453-GNUG Revision 1, (NASA Goddard Space Flight Center, February 2005)

"Harris Antennas and Hardware for White Sands," Harris Fact Sheet #1, Harris Corporation, Melbourne, FL

Hocking, William M. "The Evolution of the STDN," *International Telemetering Conference,* 16, 80-01-03

"Integrated Near-Earth Network," Office of Space Communications, NASA Headquarters, 2005

Liebrecht, Phillip and Clason, Roger, "GSFC Vision for Future Space Communications," NASA Goddard Space Flight Center, September 2005

"Madrid Space Station, INTA/NASA Information Brochure," JPL P72-223, 31 August 1972

Mogan, Kathleen M. and Mintz, Frank P. *Keeping Track: A History of the GSFC Tracking and Data Acquisition Networks 1957 to 1991* (NASA Goddard Space Flight Center, 1992)

National Space Transportation System Reference Manual (NASA Kennedy Space Center, 1988)

Network Operations Orientation Course 853 (NASA Goddard Space Flight Center, 1974)

Orbiting Solar Observatory Satellite OSO-1 (NASA, SP-57, 1965)

Pan American World Airways, Inc., Aerospace Services Division, Public Relations MU517

Patera, Russell P. and Ailor, William H. "The Realities of Reentry Disposal," The Aerospace Corporation, 2001

Proceedings of the Apollo Unified S-Band Technical Conference (NASA, SP-87, 1965)

Project ATS-6 Support Instrumentation Requirements Document, Revision 1 (NASA Goddard Space Flight Center, 1975)

Report by the Subcommittee on Aeronautics and Space Technology of the Committee on Space Sciences and Astronautics, United States House of Representatives Ninety-Third Congress Second Session, September 1974

Rogers, William P. *Presidential Commission on the Space Shuttle Challenger Accident, Final Report,* 06 June 1986

Rosenthal, Alfred *Vital Links: The First 25 Years of NASA's Space Tracking, Communications and Data Acquisition 1958–1983* (NASA Goddard Space Flight Center, 1983)

Rothenberg, Joseph H. "Statement Before the Subcommittee on Space and Aeronautics, Committee on Science, House of Representatives," 11 March 1999

"Satellite Tracking and Data Acquisition Facilities Fiscal Year 1962 Construction and Facilities Estimates, Rosman Data Acquisition Facility, Project No. 3379," 14 Feb 1962 (Folder 8820, NASA Historical Reference Collection, NASA History Division, NASA Headquarters, Washington, DC)

Schneider, William C. "Statement Before the Subcommittee on Science, Technology and Space of the Committee on Commerce, Science and Transportation of the U.S. Senate," 28 February 1979

Smylie, Robert E. "Speech at the Whites Sands TDRSS Ground Terminal Acceptance Ceremony," 17 August 1981 (Folder 8818, NASA Historical Reference Collection, Washington, DC)

"Space and Ground Networks Overview Presented to CMC-IOAG Joint Session," NASA Goddard Space Flight Center Code 450, 7 December 2004

Space Network User's Guide (SNUG), 450-SNUG Revision 8, (NASA Goddard Space Flight Center, June 2002)

"Space Transportation Costs: Trends in Price per Pound to Orbit 1990-2000," Futron Corporation, Bethesda, MD, 6 September 2002

"STADAN Facility, Orroral Valley, ACT," Information Brochure, The Department of Supply, Melbourne, Victoria

Stocklin, Frank, "GRTS: An Experience of a Lifetime," NASA Goddard Space Flight Center

"The Bermuda Story," Bendix Field Engineering Corporation, 1971

"The Corpus Christi Story," Bendix Field Engineering Corporation, 1971

"The Fairbanks, Alaska Story," Bendix Field Engineering Corporation, 1971

"The Hawaii Story," Bendix Field Engineering Corporation, 1971

"The Manned Space Flight Tracking Network (MSFTN)," description brochure, NASA Goddard Space Flight Center, 1965.

"The Worldwide Deep Space Network," description brochure JPL 400-326, NASA Jet Propulsion Laboratory, May 1989

"Tracking and Data Acquisition Facility Ascension Island," description brochure, NASA Goddard Space Flight Center, 25 June 1965

Tracking and Data Acquisition Program, Subcommittee Report on Aeronautics and Space Technology, Ninety-Third Congress, Second Session, September 1974

"Tracking and Data Relay Satellite System Chronology of Events," NASA Goddard Space Flight Center, 2003.

Truszynski, Gerald M. "Statement Before the Subcommittee on Science, Technology and Space of the Committee on Commerce, Science and Transportation of the U.S. Senate," 16 September 1977

Walker, Jon Z. "Space Shuttle Return-to-Flight and International Space Station Status," NASA Goddard Space Flight Center Code 450, 10 June 2004

_____, "Cross-support Resources," NASA Goddard Space Flight Center Code 450, 10 June 2004
_____, "Space Network Functional Overview," NASA Goddard Space Flight Center Code 450, 10 June 2004
Watson, Bill, "Bill's Trip to Antarctica," January 1999
_____, and Devito, Dan, "Svalbard Fiber Dedication," January 2004
"Welcome to Svalbard and SvalSat," information brochure, Konnesberg Satellite Services, 2003
Whalen, David J., "Communications Satellites: Making the Global Village Possible," unpublished paper
Zale, "Fast Summary: Minitrack System," unpublished paper, 1958

Websites

"1947 History" White Sands Missile Range Fact Sheet
http://www.wsmr.army.mil/pao/FactSheets/1947his.htm
(accessed 16 November 2005)

"American Experience: Race to the Moon,"
http://www.pbs.org/wgbh/amex/moon/peopleevents/e_telecasts.html
(accessed 7 November 2005)

"Antarctic Connection: Robert F. Scott (1868–1912),"
http://www.antarcticconnection.com/antarctic/history/scott.shtml
(accessed 21 February 2006)

"Apollo 15 Camera Equipment,"
http://www.lpi.usra.edu/expmoon/Apollo15/A15_Photography_cameras. html#COLORTV
(accessed 7 November 2005)

"ARIA 328 Memorial,"
http://www.flyaria.com/memorial/1999/personal.htm
(accessed 27 December 2005)

"Automated Transfer Vehicle (ATV) - European Space Agency (ESA),"
http://hsf.honeywell-tsi.com/atv.html
(accessed 4 March 2006)

"Center for Astrophysical Research in Antarctica,"
http://astro.uchicago.edu/cara/vtour/mcmurdo/
(accessed 20 February 2006)

"Chronology: Beginnings of ASTP," NASA History Division,
http://history.nasa.gov/astp/chrono.html
(accessed 9 November 2005)

"DataLynx,"
http://www.honeywell-tsi.com/datalynx/aboutus.shtml
(accessed 27 February 2006)

"Experiment Operations During Apollo EVAs," ARES Publications,
http://ares.jsc.nasa.gov/HumanExplore/Exploration/EXLibrary/docs/ApolloCat/Part1/ALSEP.htm
(accessed 8 November 2005)

"Fort Stewart History,"
http://www.stewart.army.mil/ima/sites/about/history.asp
(accessed 5 September 2005)

"Guam Remote Ground Terminal,"
http://msp.gsfc.nasa.gov/tdrss/guam.html
(accessed 10 February 2006)

"Honeysuckle Creek Tracking Station,"
http://www.honeysucklecreek.net/
(accessed 9 September 2005)

"Intelsat," Jet Propulsion Laboratory Mission and Spacecraft Library,
http://msl.jpl.nasa.gov/Programs/intelsat.html
(accessed 12 October 2005)

"LAGEOS 1, 2 Quicklook," Jet Propulsion Laboratory Mission and Spacecraft Library,
 http://msl.jpl.nasa.gov/QuickLooks/lageosQL.html
(accessed 3 January 2006)

"Pioneer NASA Spacecraft Celebrates 20 Years of Service," NASA New Release 03-130, 3 April 2003,
http://www.nasa.gov/home/hqnews/2003/apr/HP_news_03130.html
(accessed 10 March 2006)

"Poker Flat Research Range General Information,"
http://www.pfrr.alaska.edu/pfrr/index.html
(accessed 21 February 2006)

"Scout,"
http://www.fas.org/spp/military/program/launch/scout.htm
(accessed 17 February 2006)

"Single Object Tracking Radars,"
http://www.wsmr.army.mil/capabilities/nr/testing/range_inst/radar/sinobj.html
(accessed 30 August 2005)

"Space Network Online Information Center,"
http://scp.gsfc.nasa.gov/tdrss/
(accessed 10 March 2006)

"Summary of Resources Requirements," Mission Support: Fiscal Year 1998 Estimates Budget Summary, Office of Spaceflight, Space Communications Services
http://www.hq.nasa.gov/office/codeb/budget/PDF/spacecom.pdf
(accessed 10 February 2006)

"The CGRO Mission (1991–2000),"
http://cossc.gsfc.nasa.gov/docs/cgro/index.html
(accessed 6 February 2006)

"The Fresnedillas (Madrid, Spain) MSFN Station,"
http://www.honeysucklecreek.net/other_stations/fresnedillas/intro.html
(accessed 2 November 2005)

"The History of Bendix,"
http://www.bfec.us/bfectxt6.htm
(accessed 18 June 2005)

"The McMurdo Ground Station (MGS),"
http://amrc.ssec.wisc.edu/MGS/history.html
(accessed 20 February 2006)

"The MILA Story,"
http://science.ksc.nasa.gov/facilities/mila/milstor.html
(accessed 11 January 2006)

"U.S. Centennial of Flight Commission,"
http://www.centennialofflight.gov/essay/Evolution_of_Technology/NASA/Tech2.htm
(accessed 01 September 2005)

"Tracking and Data Relay Satellite Description: Attitude Control System,"
http://msp.gsfc.nasa.gov/tdrss/attitude.html
(accessed 25 January 2006)

"Tracking and Data Relay Satellite Description: Electrical Power System,"
http://msp.gsfc.nasa.gov/tdrss/eps.html
(accessed 25 January 2006)

"Tracking and Data Relay Satellite Description: Propulsion System,"
http://msp.gsfc.nasa.gov/tdrss/prop.html
(accessed 25 January 2006)

"Tracking and Data Relay Satellite Description: Thermal Control System,"
http://msp.gsfc.nasa.gov/tdrss/tcntrl.html
(accessed 25 January 2006)

"Tracking and Data Relay Satellite Description: Tracking, Telemetry & Command System,"
http://msp.gsfc.nasa.gov/tdrss/ttc.html
(accessed 25 January 2006)

"Tracking and Data Relay Satellite H, I, J: The Next Generation,"
http://msp.gsfc.nasa.gov/tdrss/tdrshij.html
(accessed 15 February 2006)

"Wallops Flight Facility,"
http://www.wff.nasa.gov/about/
(accessed 17 February 2006)

http://www.globalsecurity.org/military/facility/blossom-point.htm
(accessed 2 September 2005)

www.wstf.nasa.gov/WSSH/
(accessed 5 January 2006)

Books

Bate, Roger R., Mueller, Donald D., White, Jerry E., *Fundamentals of Astrodynamics* (Dover Publications, Inc., 1971)
Benson, Charles D., Faherty, William B., *Moonport: A History of Apollo Launch Facilities and Operations* (NASA, SP-4204, 1978)
Bille, Matt and Lishock, Erika, *The First Space Race: Launching the World's First Satellites* (Texas A&M University Press, 2006)
Brooks, Courtney G., Grimwood, James M., Swenson, Loyd S., Jr., *Chariots for Apollo* (NASA, SP-4205, 1979)
Cernan, Eugene, *The Last Man on the Moon* (St. Martin's Press, 1999)
Chaikin, Andrew, *A Man on the Moon* (Penguin Books, 1994)
Chiles, James R., *Inviting Disaster: Lessons From the Edge of Technology* (Harper Collins, 2002)
Clary, David C. *Rocket Man: Robert H. Goddard and the Birth of the Space Age* (Hyperion, 2003)
Compton, W. David, Benson, Charles D., *Living and Working in Space: A History of Skylab* (NASA, SP-4208, 1983)
Cunningham, Walter, *The All-American Boys* (Simon & Shuster, Inc., 2004)
Dodd, Annabel Z., *The Essential Guide to Telecommunications* (Prentice Hall, 2005)
Ezell, Edward Clinton, Ezell, Linda Newell, *The Partnership* (NASA, SP-4209, 1978).
Ezell, Linda Neuman, *NASA Historical Data Book, Volume III: Programs and Projects 1969–1978* (NASA, SP-4012, 1994)
Gatland, Kenneth *Manned Spacecraft* (Macmillan Publishing Company, Inc., 1976)
____, *Missiles and Rockets* (Macmillan Publishing Company, Inc., 1975)
Godwin, Robert, *Friendship 7: The First Flight of John Glenn. The NASA Mission Reports* (Apogee Books, 1999)
____, *Space Shuttle: STS Flights 1-5 The NASA Mission Reports* (Apogee Books, 2001)
____, Whitfield, Steve *Deep Space: The NASA Mission Reports* (Apogee Books, 2005)
Gawdiak, Ihor with Helen Fedor, *NASA Historical Data Book, Volume IV: NASA Resources 1969–1978* (NASA, SP-4012, 1994)
Hacker, Barton C. and Grimwood, James M., *On the Shoulders of Titans* (NASA, SP-4203, 1977)
Hallion, Richard P., Young, James O., *Space Shuttle: Fulfillment of a Dream, Case VIII of The Hypersonic Revolution: Case Studies in the History of Hypersonic Technology, Volume 1, From Max Valier to Project PRIME (1924-1967)* (United States Air Force, 1998)
Hansen, James R., *First Man: The Life of Neil Armstrong* (Simon and Schuster, 2005)
Harrison, Ian, *The Book of Firsts* (Octopus Publishing Group, Ltd., 2003)

Kerrod, Robin, *Hubble: The Mirror on the Universe* (Firefly Books, 2003)

Kyle, Howard C., *Manned Spaceflight Communication Systems: Advances in Communications Systems, Volume 2* (Academic Press, Inc., 1966)

Jenkins, Dennis R., *Space Shuttle: The History of the National Space Transportation System* (Ian Allan Publishing, Ltd., 2001)

Koelle, Heinz H., *Handbook of Astronautical Engineering* (McGraw Hill, 1961)

Labrador, Virgil S. and Galace, Peter I., *Heavens Fill with Commerce: A Brief History of the Communications Satellite Industry* (SATNEWS Publishers, 2006)

Launius, Roger D., *Apollo: A Retrospective Analysis* (NASA, SP-2004-4503, 2004)

Launius, Roger D., and Gillette, Aaron K., *The Space Shuttle: An Annotated Bibliography* (National Aeronautics and Space Administration, Monographs in Aerospace History, Number 1, 1992)

Lindsay, Hamish, *Tracking Apollo to the Moon* (Springer, 2001)

Martin, Donald H., *Communication Satellites, Fourth Edition* (The Aerospace Corporation Press, 2000)

McLaughlin Green, Constance and Lomask, Milton, *Vanguard: A History* (NASA, SP-4202, 1970)

Mudgway, Douglas J., *Big Dish: Building America's Deep Space Connection to the Planets* (University Press of Florida, 2005)

____, *Uplink Downlink: A History of the NASA Deep Space Network, 1957–1997* (NASA, SP-4227, 2001)

Portree, David S. F., *NASA's Origins and the Dawn of the Space Age* (National Aeronautics and Space Administration, Monographs in Aerospace History #10, 1998)

Rumerman, Judy A., *NASA Historical Data Book Volume VI 1979-1988* (NASA, SP-4012, 2000)

Schmitt, Harrison H., *Return to the Moon: Exploration, Enterprise and Energy in the Human Settlement of Space* (Praxis Publishing, Ltd., 2006)

Seamans, Robert C., Jr., *Project Apollo: The Tough Decisions* (NASA, SP-2005-4537, 2005)

Seifert, Howard, *Space Technology* (Wiley, 1959)

Swanson, Glen E., *Before this Decade is Out...* (NASA, SP-4223, 1999)

Swenson, Loyd S., Jr., Grimwood, James M., Alexander, Charles C., *This New Ocean: A History of Project Mercury* (NASA, SP-4201, 1998)

Thompson, Milton O. *At the Edge of Space: The X-15 Flight Program* (Smithsonian, 1992)

Van Nimmen, Jane, Bruno, Leonard C. and Rosholt, Robert L., *NASA Historical Data Book Volume I: NASA Resources 1958-1968* (NASA, SP-4012, 1988)

Wells, Helen T., Whiteley, Susan H., Keregeannes, Carrie, *Origins of NASA Names* (NASA, SP-4402, 1976)

West Reynolds, David, *Apollo: The Epic Journey to the Moon* (Tehabi Books, Inc., 2002)

Zimmerman, Robert, *Genesis: The Story of Apollo 8* (Random House, Inc., 1998)

Interviews

Conducted by Author

Benson, Cliff	18 January 2006	Colorado Springs, CO
Costrell, James A.	26 October 2005	Washington, DC
Dinn, Michael	31 March 2006	Canberra, Australia
Force, Charles T.	14 November 2005	Colorado Springs, CO
Nagle, Glen	30 March 2006	Tidbinbilla, Australia
Newman, Neil R.	29 March 2006	Canberra, Australia
Saxon, John	31 March 2006	Canberra, Australia
Spearing, Robert E.	27 October 2005	Washington, DC
Watson, Bill	27 October 2005	Washington, DC

From Other Sources

By Alfred Rosenthal from *Vital Links*

Barnes, Robert	15 September 1982
Buckley, Edmond C.	Deceased, statements made in 1966
Clements, Henry E. "Pete"	30 November 1982
Covington, Ozro M. "Ozzie"	17 January 1983
Dunseith, Lynwood C.	30 November 1982
Looney, Chesley H., Jr.	7 October 1982
McKeehan, Harry B.	27 October 1982
Kraft, Christopher C.	1 December 1982
Mengel, John T.	20 November 1982
Owen, Robert L.	12 May 1983
Roberts, Tecwyn "Tec"	10 August 1982
Sade, Richard S.	21 July 1982
Schneider, William C.	5 October 1982
Sjobert, Sigurd A.	30 November 1982
Stelter, Lavern R. "Vern"	16 September 1982
Stompf, Steven W.	7 December 1982
Truszynski, Gerald M.	22 September 1982
Weingarten Murray T.	21 September 1982
Wood, H. William	8 April 1983

By Kathleen M. Mogan and Frank P. Mintz from *Keeping Track*

Bodin, Wesley J.	7 November 1990

Hunsicker, G. 7 August 1990
Kronmiller, George 21 April 1991
Morse, Gary A. 9 November 1990
Pfeiffer, William A. 28 November 1990

NASA Oral Histories

Armstrong, Neil A. 19 September 2001 Houston, TX (Stephen E. Ambrose, Douglas Brinkley)

Beggs, James M. 7 March 2002 Bethesda, MD (Kevin M. Rusnak)

Brand, Vance D. 12 April 2002 Houston, TX (Rebecca Wright)

Clements, Henry E. "Pete" 31 August 1998 Melbourne, FL (Carol Butler)

Collins, Michael 8 October 1997 Houston, TX (Michelle Kelly)

Cooper, L. Gordon, Jr. 21 May 1998 Pasadena, CA (Roy Neal)

Fendell, Edward I. 19 October 2000 Houston, TX (Kevin M. Rusnak)

Gerstenmaier, William H. 22 September 1998 Houston, TX (Rebecca Wright, et al.)

Glenn, John H., Jr. 25 August 1997 Houston, TX (Sheree Scarborough)

Haney, Paul P. 20 January 2003 High Rolls, NM (Sandra Johnson)

Heflin, J. Milton 9 March 1998 Houston, TX (Michelle Kelly)

Mott, Michael 23 April 1999 Houston, TX (Rebecca Wright)

Mueller, George E. 27 August 1998 Kirkland, WA (Summer C. Bergen)

O'Neill, John W. 12 July 2001 Houston, TX (Carol Butler)

Seamans, Robert C., Jr.	30 September 1998	Beverly, MA (Michelle Kelly)
Smylie, Robert E.	17 April 1999	Bethesda, MD (Carol Butler)
Thomas, Andrew S. W.	22 July 1998	Houston, TX (Rebecca Wright, et al.)
Thompson, Robert F.	29 August 2000	Houston, TX (Kevin M. Rusnak)
Truly, Richard H.	16 June 2003	Golden, CO (Rebecca Wright)

APPENDIX 1

Acronyms, Abbreviations, and Glossary

AA	Associate Administrator
AAA	Active Acquisition Aid
AAP	Apollo Applications Program
ACS	attitude control system
ACN	Ascension
ACU	antenna control unit
AES	Apollo Extension System
AFSC	Air Force Systems Command
AFSCN	Air Force Satellite Control Network
AGO	Santiago
AIS	Apollo Instrumentation Ship
ALT	Approach and Landing Test
CDR	Critical Design Review
CDSCC	Canberra Deep Space Communication Complex
CSIR	Council for Scientific and Industrial Research
CSIRO	Commonwealth Scientific and Industrial Research Organization
CSM	Command and Service Module
CSOC	Consolidated Space Operations Contract
DAF	Data Acquisition Facility
dB	decibels
DDMS	Department of Defense Manager for Manned Spaceflight

DF	direction finding
DJS	Dzhusaly, Razakhgtan
DLR	Germany's Deutsches Zentrum für Luftund Raumfahrt
DOD	Department of Defense
DOI	Department of the Interior
DOS	Department of Supply
DRSS	Data Relay Satellite System
DSN	Deep Space Network
DSS	Deep Space Station
EGO	Eccentric Geophysical Observatory
EGR	Eglin Gulf Test Range
ELVIS	Enhanced Launch Vehicle Imaging System
ERS	Earth Resource Satellite
ERTS	Earth Resource Technology Satellite
ESA	European Space Agency
ESD	Air Force Electronic Systems Division
ESMC	Air Force Eastern Space and Missile Center
EUMETSAT	European Organization for the Exploitation of Meteorological Satellites
ETR	Eastern Test Range
EVA	extravehicular activity
EUT	Eupatona, Ukraine
FAA	Federal Aviation Administration
FCC	Federal Communications Commission
FDR	Final Design Review
FM	frequency modulation
FY	fiscal year
Gbps	gigabits per second
GDSCC	Goldstone Deep Space Communication Complex
GHz	gigahertz
GLTN	Goddard Laser Tracking Network
GMT	Greenwich Meridian Time
GN	Ground Network
GOES	Geostationary Observational Environmental Satellite
GRARR	Goddard Range And Range Rate
GRGT	Guam Remote Ground Terminal
GRO	Gamma Ray Observatory
GRTS	GRO Remote Terminal System
GSFC	Goddard Space Flight Center
GWM	Guam
HAW	Hawaii
HDTV	high definition television
HEAO	High Energy Astronomy Observer

HST	Hubble Space Telescope
HTV	H-II Transfer Vehicle
IAGS	Inter-American Geodetic Survey
IBM	International Business Machines
IBEW	International Brotherhood of Electrical Workers
ICBM	intercontinental ballistic missile
IDEA	International Space Station Downlink Enhancement Architecture
IGY	International Geophysical Year
IMP	Interplanetary Monitoring Platform
INEN	Integrated Near-Earth Network
INTA	Spanish National Institute of Aerospace Technology
INTELSAT	International Telecommunications Satellite Consortium
IOAG	Interagency Operations Advisory Group
IPF	Image Processing Facility
ISIS	International Satellite for Ionospheric Studies
ISO	Infrared Space Observatory
ISS	International Space Station
IUE	International Ultraviolet Explorer
IUS	Inertial Upper Stage
JAXA	Japan Aerospace Exploration Agency
JDMTA	Jonathan Dickinson Missile Tracking Annex
JPL	Jet Propulsion Laboratory
JSC	Johnson Space Center
JWST	James Webb Space Telescope
KHz	kilohertz
KLP	Kolpashevo, Russia
Kpbs	kilobits per second
KPGO	Kokee Park Geophysical Observatory
KSAT	Kongsberg Satellite Services
KSC	Kennedy Space Center
LAGEOS	Laser Geodynamics Satellite
laser	light amplification by stimulated emission of radiation
LM	Lunar Module
LRC	Langley Research Center
LRO	Lunar Reconnaissance Orbiter
LRV	Lunar Roving Vehicle
M&DO	Mission and Data Operations
MA	Mercury-Atlas
MA	Multiple Access
MAD	Madrid
maser	microwave amplification by stimulated emission of radiation
Mbps	megabits per second
MCC	Mission Control Center

MDSCC	Madrid Deep Space Communication Complex
MESA	Modular Equipment Stowage Assembly
MFS	Manned Flight Support
MGS	McMurdo Ground Station
MHz	megahertz
MILA	Merritt Island Launch Annex
MIT	Massachusetts Institute of Technology
MOBLAS	Mobile Laser Ranging System
MOCR	Mission Operations Control Room
MOTS	Minitrack Optical Tracking System
MRCS	Mobile Range Control System
MSC	Manned Spacecraft Center
MSFN	Manned Space Flight Network
NACA	National Advisory Committee for Aeronautics
NAR	Non-Advocate Review
NAS	National Academy of Science
NASA	National Aeronautics and Space Administration
NASCOM	NASA Communications Network
NASDA	National Space Development Agency of Japan
NC	Network Controller
ND	Networks Directorate
NEN	Near-Earth Network
NFL	Newfoundland
NISN	NASA Integrated Services Network
NOAA	National Oceanic and Atmospheric Administration
NOCC	Network Operations Control Center
NORAD	North American Aerospace Defense Command
NRD	Air Force National Range Division
NRL	Naval Research Laboratory
NSC	Norwegian Space Center
NSF	National Science Foundation
NTSC	National Television System Committee
NTTF	Network Test and Training Facility
OAO	Orbiting Astronomical Observatory
OCC	Operations Control Center
OGO	Orbiting Geophysical Observatory
OMS	Orbital Maneuvering System
ORR	Orroral
OSC	Office of Space Communications
OSO	Office of Space Operations
OSO	Orbiting Solar Observatory
OSTDA	Office of Space Tracking and Data Acquisition
OTDA	Office of Tracking and Data Acquisition

PAL	Phase Alternating Line
PAM	Pulse Amplitude Modulation
PCM	Pulse Code Modulation
PCR	Payload Changeout Room
PCS	Pacific Command Ship
PCU	Power Control Unit
PDL	Ponce de Leon Station
PDR	Preliminary Design Review
PER	Pre-Environmental Review
PLSS	Portable Life Support System
PMR	Pacific Missile Range
PMRF	Pacific Missile Range Facility
POCC	Project Operations Control Center
POGO	Polar Orbiting Geophysical Observatory
PPK	Petropavlovsk-Kamchatskaya, Russia
PSAC	President's Science Advisory Committee
QUI	Quito
RCA	Radio Corporation of America
RCS	Reaction Control System
RF	radio frequency
RFP	Request For Proposal
ROS	Rosman
RTG	Radioisotope Thermoelectric Generator
RTHU	roll to heads-up
RTLS	Return to Launch Site abort
SA	Single Access
SAO	Smithsonian Astrophysical Observatory
SATAN	Satellite Automatic Tracking Antenna
SCAMA	Switching, Conferencing And Monitoring Arrangement
SCAMP	Satellite Command Antenna on Medium Pedestal
SDK	Ussuriysk, Russia
SDO	Solar Dynamics Observatory
SEB	Source Evaluation Board
SECAM	Sequential Color with Memory
SECO	sustainer engine cutoff
SGL	Space to Ground Link
SGLT	Space to Ground Link Terminal
SGS	SvalSat Ground Station
SIRTF	Space Infrared Telescope Facility
SITE	Satellite Instructional Television Experiment
SN	Space Network
SNIP	Space Networks Interoperability Panel
SOMO	Space Operations Management Office

SRB	Solid Rocket Booster
STADAN	Satellite Tracking And Data Acquisition Network
STADIR	Station Director
STDN	Spaceflight Tracking and Data Network
STG	Space Task Group
STGT	Second TDRSS Ground Terminal
STS	Space Transportation System
STTCS	S-band Tracking, Telemetry & Command System
T&DS	Tracking and Data Systems
TAGIU	Tracking And Ground Instrumentation Unit
TAGS	Text And Graphics System
TAL	Transatlantic Abort Landing
TAN	Tananarive
TAT-1	trans-Atlantic telephone cable
TBL	Tbilisi, Georgia
TCS	Thermal Control System
TDRS	Tracking and Data Relay Satellite
TDRSS	Tracking and Data Relay Satellite System
TDSD	Tracking and Data Systems Directorate
TELOPS	Telemetry Online Processing System
TEI	Trans-Earth Injection
TIROS	Television Infrared Observation Satellite
TLI	Trans-Lunar Injection
TLRS	Transportable Laser Ranging System
TM	telemetry
TTS	Test and Training Satellite
TSSG	Tracking System Study Group
TT&C	Tracking, Telemetry and Command
TWT	traveling wave tube
UHF	ultrahigh frequency
ULD	Ulan-Ude, Russia
US	United States
USAF	United States Air Force
USB	Unified S-Band
USN	Universal Space Network
USNS	United States Navy Ship
USSR	Union of the Soviet Socialist Republic
VERLORT	Very Long Range Tracking
VHF	very high frequency
VLBI	Very Long Baseline Interferometry
WART	White Sands Complex Alternative Resource Terminal
WECO	Western Electric Company
WGTC	Working Group on Tracking and Computation

WRE   Weapons Research Establishment
WSGT   White Sands Ground Terminal
WSMR   White Sands Missile Range

Glossary

acquisition: The process of first finding, either visually or electronically, a satellite or spacecraft of interest so it can then be tracked.

antenna feed: The electronic device at the focal point of an antenna through which electromagnetic wave transmissions are received, amplified, and/or transmitted.

apogee: The high point in a trajectory.

array: The electronic combining of antennas pointed at the same object so as to increase the received signal strength.

autotrack: Automatic tracking of a spacecraft by an antenna (or vise versa) where the position is continuously and automatically computed.

bandwidth: The range of frequencies occupied by a radio frequency carrier wave. The more information there is on the carrier (for example, high-definition television), the more bandwidth is required to fully transmit that data. UHF air-to-ground voice, for instance, is a low bandwidth item.

bit errors: The fraction of received digital bits that are errors in a transmission. The lower the bit error, the better the quality of the transmission. Bit errors, or more precisely, bit error rate, is usually expressed in exponential notation such as 1×10^{-6} (one bit error in a million).

bit rate: The rate that digital bits of data are transmitted in a digital stream. The higher the bit rate, the faster the transmission. Bit rates can vary from kilobits per seconds (thousands of bits per second) to gigabits per second (billions of bits per second).

boresight: The focal axis of a directional antenna.

Capcom: By rule, the designated voice contact between Mission Control and the astronauts. By tradition, the Capcom is always another astronaut. Others may also, on occasion, speak with the astronauts but are never referred to as "Capcom."

carrier frequency: The selected frequency used to transport radio signals.

collimation tower: A tower, usually located a few kilometers from the main ground station antenna, equipped with a radio frequency emitter used as an aim-point to checkout and calibrate the automatic tracking capability of the antenna.

data rate: The rate of downlink or uplink between a spacecraft and its ground station. Usually measured in bits per second.

demodulate: The removal of the modulation on a carrier frequency using a series of electronic filters so as to isolate the data from its carrier.

digital: A transmitted radio frequency signal or scheme comprised only of ON and OFF pulses (0's and 1's).

eccentricity: The amount of oblateness in a spacecraft's orbit. A perfectly circular orbit has zero eccentricity while a high apogee orbit that swings around, for example, to the back side of the Moon is of high eccentricity.

electromagnetic waves: Electromagnetic (EM) waves or radiation is a self propagating wave in space with electric and magnetic components. Such waves carry energy and momentum. All energetic media such as heat, light and radio transmissions are part of the electromagnetic spectrum.

elevation: The angle above the horizon. An antenna pointed perfectly at the horizon has zero elevation. Ninety-degrees is directly overhead.

equatorial orbit: An orbit which has no inclination or tilt with respect to the Equator. Thus a spacecraft in true equatorial orbit will always revolve over the Equator.

field of view: The amount of sky that can be viewed at any one time. For an optical device, it is what can be seen at any moment with a given aperture. For an electronic device like an antenna, it is also the region where a radio frequency link can be reasonably detected.

Figure of Merit: Commonly referred to as G/T, or "G-over-T", a ground station's Figure of Merit is a fundamental quantitative measure of its overall capability to enhance the received signal with respect to noise. The higher the G/T, the more sensitive the ground station is.

g: A measure of the equivalent gravitational acceleration experienced by an object or a person. A person standing at mean sea level experiences 1 g. Apollo astronauts returning from the Moon experienced around 8 g during reentry.

gain: An increase in electromagnetic signal strength due to any of several sources, resulting in the output signal being measurably stronger than the input signal. Types of gain include amplifier gain due to active electronic components such as High Power Amplifiers or Low Noise Amplifiers, and antenna gain due to antenna features such as large dish aperture and parabolic shape.

geode: The mathematical, gravitational model of Earth characterizing its local variations in shape, size, and mass concentrations, used by computers to calculate the orbit and trajectory of a spacecraft.

geosynchronous orbit: A zero-inclination, circular orbit at an altitude of 35,900 kilometers (22,300 miles) above the Equator. In such an orbit, a spacecraft's rate of revolution round Earth is the same as the rotation rate for a point on the Equator. The craft would thus appear to be hanging stationary over a given location. The term is often used synonymously with 'geostationary orbit.'

Go/No-Go: A decision point during a mission when Mission Control has to determine whether to proceed or abort.

Ground Network: A network of NASA ground stations organized under the Science Mission Directorate of Headquarters. These consist of stations in Alaska, Antarctica, Florida, Norway, and Virginia. The Ground Network (GN) also includes support from the Network Integration Center located at the Goddard Space Flight Center and the GN scheduling and VHF systems at White Sands, New Mexico.

ground station: A location on the ground comprising of electrical, mechanical, and/or optical subsystems used for communicating with and retrieving data from space assets.

ground track: The path that a spacecraft traces on the ground.

GSTDN: The original ground elements of the remaining Spaceflight Tracking and Data Network (STDN) as the Tracking and Data Relay Satellite System (TDRSS) was being brought operational in the 1980s.

high inclination orbit: An orbit whose plane is highly inclined with respect to the Equator. Such orbits have ground tracks that enable them to pass over or observe a greater amount of Earth's surface than low inclination orbits.

high power amplifier: An electronic device usually located at the base or the back of an antenna which greatly amplifies its transmitting signal strength for establishing command uplink with a spacecraft.

housekeeping telemetry: Data from a spacecraft used only for assessing the performance, health, and status of the spacecraft itself. Typical data include voltages, temperatures, propellant tank pressures, etc.

Lagrange Points: The Lagrangian points are five positions in the Earth-Moon-Sun interplanetary system where a small object affected only by gravity can theoretically appear stationary relative to two larger objects (such as a satellite with respect to Earth and the Moon). Such an object would appear to be in a "fixed" position (or on a relatively small "Figure 8" orbit) in space rather than in a traditional orbit in which its position changes continuously.

launch azimuth: The direction that a rocket is launched in, usually measured in degrees relative to due North. For example, a rocket launched due East from Cape Canaveral has a launch azimuth of 90°.

"Lights-out" operations: A trend in ground station operations in which a station operates autonomously in an unattended fashion or with minimal staffing. Lights-out can take many forms such as nine to five workday operations with automated service at all other times. It can also be a centralized, fully staffed operations center continuously monitoring a suite of multiple, unattended remote ground stations.

line of sight: The straight line between a transmitting antenna and a receiving antenna. The two can only communicate when a line of sight has been established.

multipath: The propagation phenomenon that results when radio signals reach the receiving antenna by two or more paths. Causes of multipath include ionospheric reflection and refraction, and reflection from mountains, buildings, ocean, and the ground. The effects of multipath include constructive and destructive interference, errors, and phase shifting of the signal. A common multipath phenomenon is ghosting of television images, for example.

"Picket Line": The seven original primary Minitrack stations located approximately along the 75th West meridian, spanning North and South America. The picket line had a better than 90 percent chance of capturing every pass of a satellite in low inclination, low-Earth orbit.

polar orbit: An orbit whose plane is inclined 90° to the Equator. Thus, as its name implies, a satellite in polar orbit travels over, or near, both the North and the South pole. A great advantage of a satellite in such an orbit is its ability to observe the entire surface of the globe over time.

max-q: The occurrence of maximum aerodynamic pressure (q) during a rocket's ascent or a spacecraft's descent through the atmosphere. Knowing when max-q occurs is important as it factors into the structural stress experienced by the vehicle.

parabolic: Surface shape of an object, like an antenna reflector or the objective mirror in a telescope, based on the mathematical curve $Y=X^2$ which focuses incoming rays to a single point.

radio frequency: The number of oscillation cycles per unit time that an electromagnetic wave propagates through space at. Frequency is measured in hertz (Hz).

radio interferometry: The use of separate antennas to receive a radio signal at slightly different times so as to determine phase differences in the radio waves. These differences can then be used to calculate position solutions for an object. Radio interferometry had the advantage of yielding highly accurate tracking angles and could be used under virtually any atmospheric condition.

Schmidt camera: A telescopic, astronomical camera designed to provide wide fields of view, typically used as a survey instrument in which a large amount of sky must be covered.

solar wind: A stream of charged particles (plasma) which are ejected from the upper atmosphere of a star. It consists mostly of high-energy electrons and protons that are able to escape the star's gravity. Many phenomena can be explained by the solar wind, including: geomagnetic storms that knock out power grids on Earth; auroras, and why the tail of a comet always points away from the Sun.

sounding rocket: A small, usually unguided rocket launched into the upper atmosphere for conducting experiments and scientific research.

Space Network: NASA's constellation of geosynchronous Tracking and Data Relay Satellites and their associated ground segment. The ground segment consists of the White Sands Complex, the Bilateral Ranging Transponder System, the Merritt Island Launch Area TDRSS Relay, and the Data Services Management Center. The Space Network (SN) is run by the Goddard Space Flight Center under management of the Space Operations Mission Directorate at Headquarters.

state vectors: The set of position and velocity measurements of a traveling object as a function of time, particularly those of a spacecraft in orbit or a rocket on a ballistic trajectory.

telemetry: Electronic data measurements downlinked or transmitted from a spacecraft to the ground.

teletype: A now largely obsolete electro-mechanical typewriter which was used to communicate typed messages from point to point through a simple electrical communications channel, often just a pair of wires. Later versions used a screen instead of a printer.

tracking: Collection of spacecraft position and velocity measurements so that its orbit or trajectory can be determined.

wavelength: The distance traveled by an electromagnetic wave during one cycle of oscillation. Radio frequency wavelengths ranges from millimeters to meters.

Wing Station: A support or backup ground station located near the designated primary station on a given mission. Deep Space Network sites were often used to support nearby Manned Space Flight Network sites in this manner.

yagi antenna: An antenna consisting of an array of linear elements, such as a common rooftop television antenna. The antenna achieves a distinct response bandwidth determined by the length, diameter, and spacing of all the individual elements. Its overall gain is proportional to its length, rather than simply the number of elements. Yagis can range in size from small (like TV antennas) to very large (dozens of feet long with multiple elements).

APPENDIX 2

Maps

Each location has been plotted by the author on blank Robinson Projections.

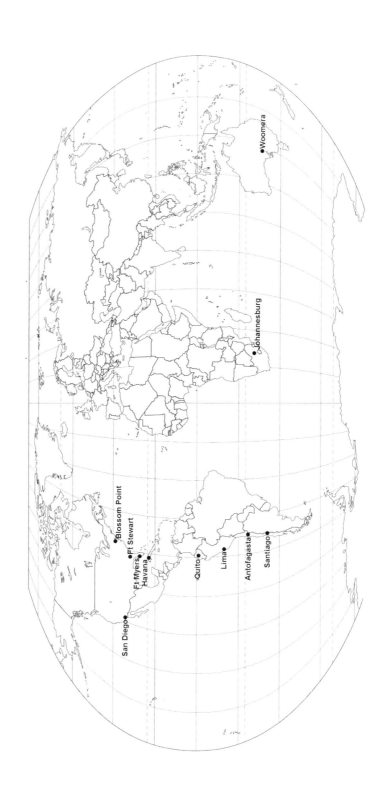

Appendix 2 \ Maps 417

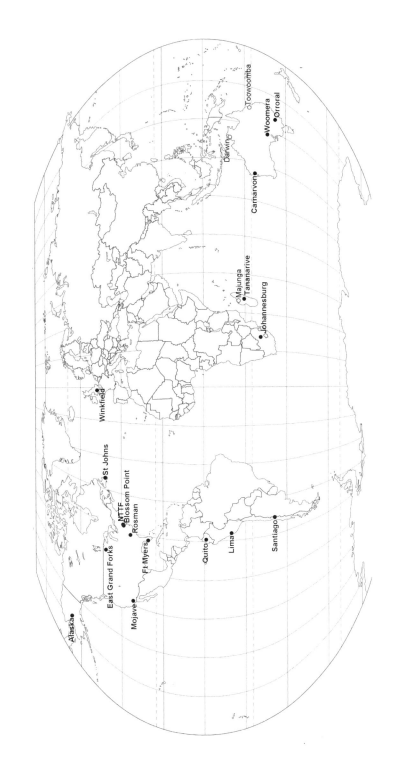

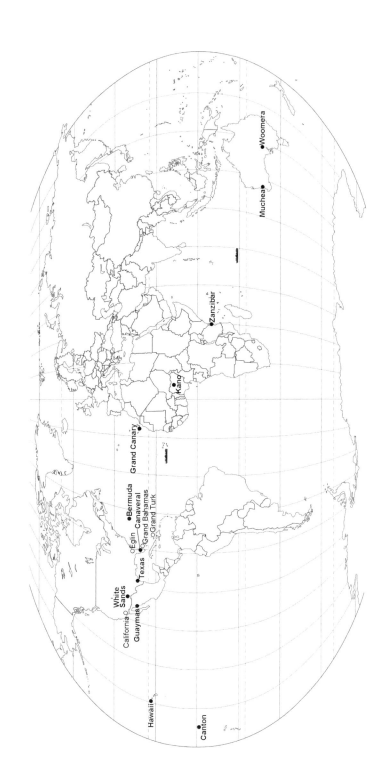

Appendix 2 \ Maps 419

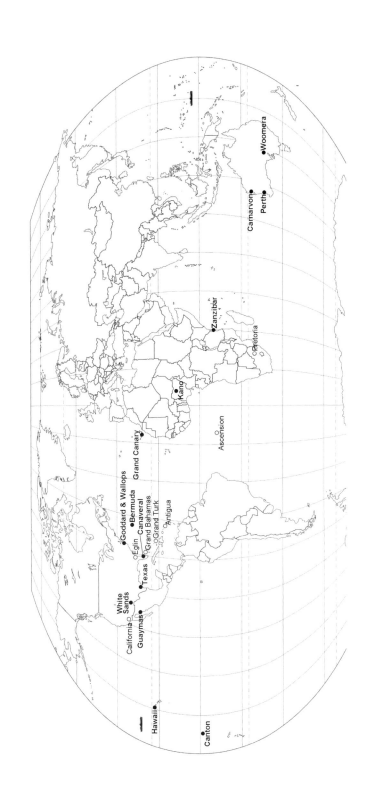

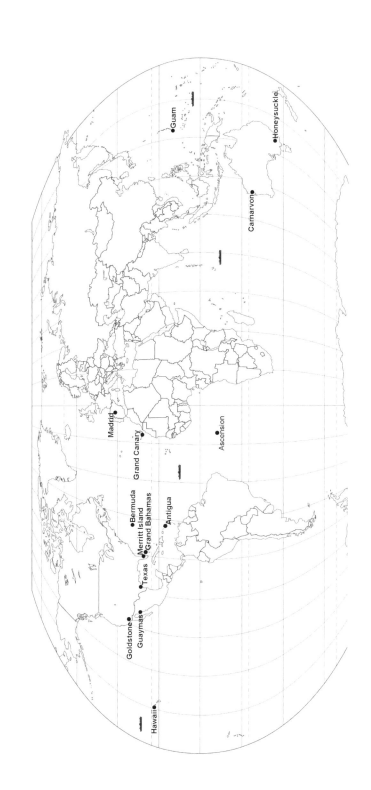

Appendix 2 \ Maps 421

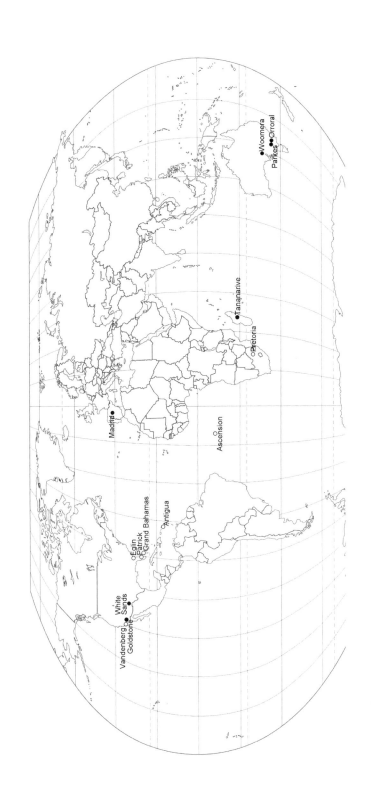

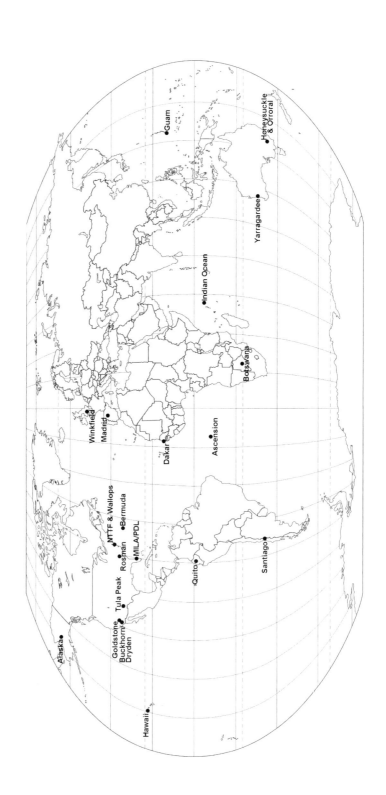

The Spaceflight Tracking and Data Network (mid 1970s–early 1980s)

Appendix 2 \ Maps 423

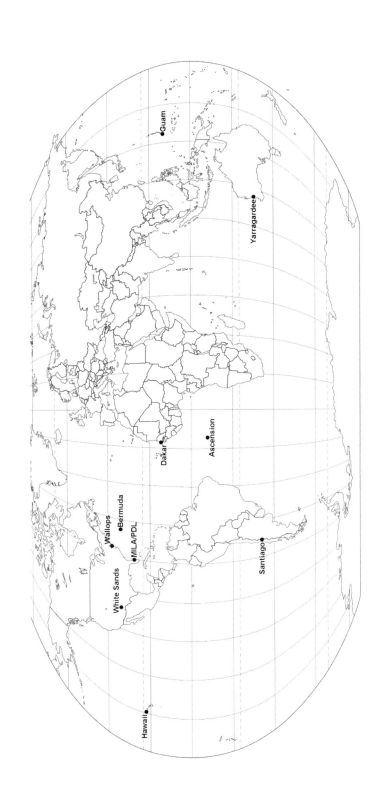

The Spaceflight Tracking and Data Network (mid 1980s–early 1990s)

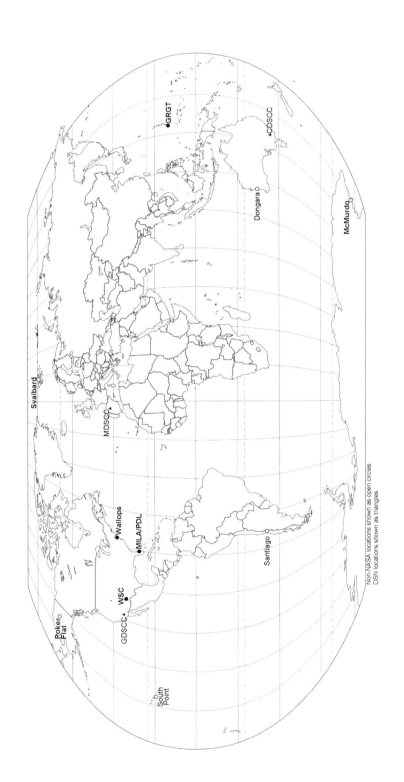

APPENDIX 3

Radio Frequency Chart

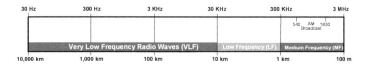

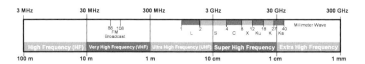

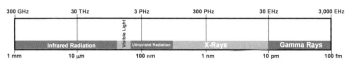

Source: Federal Communications Commission (FCC)

APPENDIX 4

Honeysuckle Station Log for Apollo 11

The actual mission log as recorded at the Honeysuckle Creek Tracking Station (HSK) is reproduced here courtesy of John Saxon. When he donated these scans to the Honeysuckle tribute Web site (*www.honeysucklecreek.net*) in 2003, John wrote:

> It may seem strange, but Apollo support sites were not required to keep a formal log. Usually, particular times were reported to Houston as they happened or when the action was completed. Some times were reported in post-pass teletype messages.
> At Honeysuckle we did try to maintain an overall log of mission events as we observed them, and the operating position on the right hand side of the main station Operations console got lumbered with the job. This position was also required to monitor anything up to six or seven voice loops simultaneously and respond as required. Also, there were two 25 key CAMs (Computer Address Matrix) used to control some aspects of the Command and Telemetry Computers (send commands to the spacecraft if data communications were lost between the site and Houston, etc.), operated by this position. So keeping a written log as well was sometimes difficult.

Most sections of the station (USB, TLM, Computer, Comms, etc.) also made note of significant times and events, and so this helped the main log keeper because they could be reported later in 'easy time.' Times were sometimes guessed when there was a chance to jot them down, so normally they were noted to the nearest minute unless they were deemed to be very significant!

The log pages presented here were maintained for the entire Apollo 11 mission 24 hours per day by my opposite number, Ken Lee (otherwise known as "the silver fox") and myself as we worked alternating 12-hour shifts. It should be noted that they were "borrowed" by me at the end of the Apollo program. All the rest of the mission files, which contained many other documents on each mission, were consigned to the local landfill site (Tharwa Tip) when we were not looking.

Appendix 4 \ Honeysuckle Station Log for Apollo 11

0357
836

_____ DSIF LOG SHEET HSK SITE

DATE 15/16 JUL 69 (GMT)   LOG KEEPER SAXON

MISSION NCG 725 TERMINAL COUNT & LAUNCH.

GMT	EVENT
2200	LM BER's complete & Motor cals
2310	Start Ø1
231730	Ø1 complete
0114	298 & 249 low at wing during line checks
0237	CMP MAG TAPE handler suspect, running diag on CMP
0300	" " " now green
0320	Slow aborting some wing CSM BER to run Ø3
0335	Patching troubles in TLM during Ø3 slowing things up
0344	Start FE 600 BER's
0350	Ø3 complete & FE 600's
0351	Configured for cadf.ss
0401	TLCd on of TLM 1/F & CAD CVE
0402	USB B/RA HELO CVE
0406	Go on TLM Cadf.ss
0409	" " " 1/F / LBR
0410	" " Prime B/S
0414	Cue APP Prime go on B/S
0422	Go on prime B/S A
0425	Cue on Wing B/S
0432	Started CMD 1/F and FM/FM Go on wing B
0434	FM/FM 1/F Go
0440	Go on CMD 1/F & A/G
0450	Prelaunch loading
0455	→ " complete INV Sent
0600	Tape rolling for HBR Cadf.ss/Decom 1/F
0603	Cued HBR TLM cadf.ss
0604	Go on Cadf.ss/decom
0710	K.S.L on
0746	Prime Msite on line followed by wing –
0830	Holding at T-3.30 (scheduled)
0836	Commencing confidence check

JPL 0307 DEC 60

DSIF LOG SHEET ____HSK____ SITE

DATE 16 Jul (GMT)   LOG KEEPER LEE

MISSION NC6-725 Terminal Count.

GMT	EVENT		
0936	Status to NC + confirmed by TWX		
1030	Doing 2 hr confidence checks - found faulty Mux Modem for Syst 4 Verification - replaced & all check- OK.		
1148	Status check.		
1149	PA.2 control circuitry faulty - further investigation. PA2. to go manual on beam volts 16 kV then vary power by varying drive.		
1202	PA.2. Beam volts set now Green.		
1223	level checks OK on CSM/LM -8 + noise -48.		
1230	Starting Confidence checks.	14.32.28	
1253	Final Status to NC for launch.	14.36.28	
1332	L/O.		
1344	101.4/103.6. Go for orbit		
REV.1. 1357	H-35 start		
1400	U/L checks + 2/4v. OK	Fm Remote	14 29 40
	Cmptrs did not have ROS-1	TX ON. CSM	14 31 01
	entered.	" " 10	14 31 31
1418	Decoms uninhibited.	TX OFF (both)	14 39 31
1421	H/S on cam 888 to MCC.		
1426	USB in operate		
AOS	143232. Pr. 143250. 6FC		
	A/G very good.		
LOS	143705. 143728.		
1440.	A/G + USB Stat + H/S data off.		
1440	CSM I history sent.		
1449	CSM II H/S E EOF.		
1624	Successful TV burn.		
1816	Queried GOPS for 4th time re release. - msg forthcoming 1 min		
1825	Released - no P/B req'd.		
1828	Release MSG. rec'd.		
1830	Started SRT		

Appendix 4 \ Honeysuckle Station Log for Apollo 11

1250 GET
RTN O

OPN GCTR/NOM
INFO HNET/NC
IT OPS.

CF0035R
cabin flow
1%

51.14.1

DSIF LOG SHEET __HSK__ SITE

DATE __16/17 Jul__ (GMT) LOG KEEPER __LEE__

MISSION __NCG-725 TLC#1 DAY 01__

(3)

10
24 00
23 23

37

GMT	EVENT
1839	Ø1 A/G Completed
1915	OPS 2 H/O JS ON
2000Z	CMD mag tape handler logic problem
2009	Running CSM cost both computers. Changing cards in CMP Mag tape logic
2043	CMP 1540 logic green card swapping cured it
2052	AMQ MinPrx problem? Rcvr 7 AGC voltage ranging. MSC B/P filter out took ½
2055	M25 Green – red approx 1900 Z blowing fuses
2223	Pen ON Rcd. S SB1 red -117
2231	Ø3 complete configed for I/F
	TLC #1
2232	SCM RCD Wing 3 way on IV !
2303	Starting FE 600 checks
2304	FE 600's comp
2306	Start CMD & F M/FM Net 5
2315	Go on all I/F Biomed line poor 5 sec cyclic drop outs
2326	Wing IV 29 pt Rcd after cue to G/OPS. EST AOS @SMP 24 00 00
2346	M25 cannot support again Wing released from " " /W 0070 SC
	IV support – put on CSM
2353	'DD' prime L/S on line
2356	Biomed on slight AOS's — Too soon only 2-way & B/U sites remote
2357	AOS Rcvr 1 S/C config for 10 mins of TV. no one queried.
17/0001	4CA int decom lock. 30 370
0005	Autotrack -113 FM TV Mod. Solid HBR
0008	Told TIC about HBR
0017	Receiving poor quality TV.
0019	JS'P L/s Header to line
0021	FM LOS & TV
0027	R# 1 lock lite u/s on console RE 1 solid lock, no light
0040	Wing solid lock
0049	-121 dbm omni for PTC

JPL 0307 DEC 60

```
1250 GET
RTN = 0                                           (7)
                    _____ DSIF LOG SHEET __HSK__ SITE

                    DATE _17 JUL 69_ (GMT)   LOG KEEPER __SAXON__
130 start
131 int.    MISSION __NCG-725  TLC #1   DAY 01__
```

GMT	EVENT
0052	New SCM rcd. 2 way & monitor LM D/W at 110000 GET
0054	RH EKG noisy (crew eating)
0123	New SOM on H/S Neg TRK
0128	Rcdr 5 at wing no timing for 7 mins now green
0132	H/O GDS → HSK A/G remote. Stat sent. very smooth H/O
0135	Valid Range. All sigs Sol'd
0141	3 omni B on cm D's. 1st report due TLM dropouts. 2nd HSD err 05 prob due 28 mwp & ½ sec delay MCC → HSK, 3rd Div'd ON
0226	M/O mode 1 for RTN to 0 from N/C — wilco'ed. M25 green
0309	LOS due low signal reacquired with CMD mod ON RTC request.
0330 THR 0402	Range Correlation test
0358	Data had 430 Hz set on FM/FM I/D incorrectly TLC said don't change
	21 Gps SPS 3 secs 2640 GET for MCC #2.
0502	Wing Rcdr #5 still int on 5 and also on Rcdr #2
0500	Approx PK's acquired
	-115 HSK Ω -100 at PKS Total PWR
0522	SB1 110.5 db SB2 TOTAL PWRs
0525	SB1 - SB2 PKS -101 TOTAL PWRs
	Coh AGC 146 Coh AGC 115
0550	Voice check goss comp with comm tech
0619	New SOM Prime H/S
0711	H/S data backing up at CP. Tracking & histories
0720	KSL ON. St. 500 Nm @ 1830 local
0749	CSM 2 H/S & EOF.
0755	Track reports Veh 10 from wing showed 10 (Ch 3.)
0803	TLC asks for check on Biomed - her 5 - passing ACSW.
0821	CSM 2 H/S & EOF — Correlation test @ 0802 to 0807
0843	" " "
0907	WB #5 Sig Con Red cannot - no sync

Appendix 4 \ Honeysuckle Station Log for Apollo 11

```
I/F 2250
00 18 31
35 30 GET
```

DSIF LOG SHEET HSK **SITE**

DATE 17 July NCG-725 (GMT)   **LOG KEEPER** LEE

MISSION NCG-725 TLC #1 DAY-01

GMT	EVENT		
0904	H/S Hist. & EOF		
0915	Re acq Rng - suspect invalid number	57,000 NB1	5030/ps
0919	" " 2300 g, +30 X -10 Y		
0937	Scm rec'd.	0930	my mod
0940	H/S Hist + EOF		
10203	LOS	1010 Z	90 seq / 5785
10 21 40	AOS GFC.		
1032	Omni H/O to MAD. deremote H/S + Biomed.		
1037	USB Safe.		
1044	H/S Hist/EOF		
1045	LOS.		
1047	H/S TLM OFF by TLC		
1137	RLS MSG rec'd.		
1720	Started SRT Ø1		
1732	Ø1 comp.		
1916	OPs 2 H/O JS on.		
1921	Problem with ant prime X servo error resolved.		
2019	Possible range doppler fault SB1.		
2148	A/G I/F run with MCC, Start Ø3		
2157	Ø3 complete.		
2217	Set up for cadfiss. Start sent.		
2222	Sent main stat SRT stat, & manning msg		
2231	Start FE 600's nets 4 & 6.		
2233	End FE 600's all nets normal.		
2250	Start CMD I/F		
2252	Start Biomed.		
2257	Go on CMD & Biomed.		
23010	Cadfiss deleted - problems no. c end.		
2321	NC called re HSK & PKS recording capability TV - set him right		

JPL 0307 DEC 60

noon acq msg
+ as many more
ass poss to Pks.
HA/

DSIF LOG SHEET ___HSK___ SITE
DATE 18 JUL 69 (GMT)   LOG KEEPER SAXON
MISSION NCG-725   TLC #2   DAY 02

+ .15 °g/sec

1920

GMT	EVENT
2334	Update to SCM B/site later at 351000 GET AOS 003600Z GET
0015	SB1 1st AOS Prelimit
0030	Still in prelimit HGA HBR no demod lock yet
0032	S/C on omni GDS Mars -126 dbm
0037	Good Autotrack & Solid HBR announced AOS & Solid HBR
0751	Tic'd on
0042	'JJ' on prime t/s Wing AOS. S/C now on HGA
0043	Biomed on Net 5
0100	D/V +12 → 0 on pas normal test tone + ⊘
0107	VCO drift on N/B sig cond 1, card replaced no data lost ran on decom 3 prime for 5 mins
01263O	VDB to operate
0132	HGA H/O GDS to HSK A/G remote stat sent — was to be, carrier held at GDS
0135	" panic " " " GDS LOS 1st approx 80 secs 1 way A/G remote stat sent
0137	Requested Track for 2 more 29 pts for parkes
0152	A/G excellent at -125 dbm
0156	23.307 144 Rx VCO for PKS
0159	NST VSB called re wing time correlation
0246	WB #3 cannot, staggering tapes stat sent ETO 0400Z MCC #8 1fps MCC#4 <2.5 No intent at this time to move EVA up from FD debrief always possibility. Celeste sys problem lunar stay or crew rested
0647	Goss voice check MCC Comtech we re
0652	occasional dropouts -125 prime -128 → 129 at wing
0659	SB1 Rec'd 2 pen 8 Red not used to be fixed post pass
0708	New SOM prime & Wing L/s - JK
0715	KSV un. PA↓ NO ETO
0716	WB Red's # 3 Queens Pd3 ½
0719	L/s Init + EOT
0845	PA↓ No ETO, PA3 ETO 30 min Rvr L 10.47 ✓
0922	P/s failure

JPL 0307 DEC 60

Appendix 4 \ Honeysuckle Station Log for Apollo 11

DSIF LOG SHEET — HSK SITE

DATE 18 JUL (GMT)   LOG KEEPER LEE

MISSION NGG-725 DAY 02 TLC #2

IF 2255
   2224
      31

1018

(7)

GMT	EVENT
0833	hng LOS 11721 S/v P/L 1027 updated to TRK
	MAD P/L 1058 Z D/L 1032
0916	SCM acid
0917	PA 3 green PAH ETO 1 week.
0924	EOF not written in time
1032	H/O omni → MAD — no problem 123 V/C
1032	A/S History + EOF
1027	HSK pre-limits.
1110	Informed NST/Status of no-change SYD Keen B.1.
111725	Wing LOS
	48:00 { E 160760/3529 ; M 64115/
1513	Get an updated Stat for P/A 3 hours hrug at 0800 K
1700	Started SRT
1930	OPS 2 H/O JS ON 725 DAY 03 TLC #3
2051	Sys 2 deemed reversed problem - USB Cost. No post on 2nd run KM vcp voltage marginally lower than 2 pulser runs O.K.
2140	Dump voice sig/noise lvls from wing peculiar (recorded too low)
2205	03 cmp confirmed for cadfiss stabs sent.
2243	FR 600's complete
2254	Start CMD I/F
2257	Go on " " loss Sys 4
2259	Start FM I/F Problem with data from ACSW to GSFC took 2 mins
2303	Go on FM I/F.
2305	Start Cadfiss TLM
2308	" prime & wing L/s B/s wing dual to be selected
2313	Go on TLM no go on SB 2 B/s to be rerun'd
2315	Go on prime B/s rerun'd for wing
2319	Failed wing B/k again then switched Sys 4 to LM freq.
2326	A/G I/F complete
2331	Go on wing B/s with expected errors, go on prime APP

JPL 0307 DEC 60

```
                    DSIF LOG SHEET    HSK    SITE
21/0032z
                          18/19
                    DATE  19/20 JUL  (GMT)   LOG KEEPER  SAXON     (8)

            MISSION    NCG-725   DAY 03   TLC # 3
```

GMT	EVENT
	Ext Wing AOS = 0046z
0005	→ " Prime AOS = 0020z
0006	TDP prime L/s to line 'DD'
0015	Connected SCM red putting us back up at 5930 GET but moving H/s data later disregarding
0020	USB 1 1st AOS's
0032	" " H/s TDP 'JJ' header.
0034	TIC'd ON
0043	SB2 RF contact SB 1 still well down
0046	SB1 solid lock still in manual pos".
0048	Auto track in prelimit SB1
0052	Demod lock TM e Voice ut HBR decom lock on HBR
0102	B/U site FM/FM on
0115	-112 Prime Mars -112
0136	s/c to OB D/L e LBR. No voice subcarrier.
0147	Back to PM O2 Had to CMD reset. S/c e select opposite omni
0200	PK's AOS
0300	processing PK's TLM Thr-Hold looks approx 2½dB better on prime and 2 dB worse than 'wing'.
0327	USB operate
0332	Omni H/O GDS to HSK A/G remote stat sent.
	Prime
	PC -122 247 ampl
0434	Started wing Rx S/decom sync validity cal -1295 +0.5 / -1.0 220 / 217
0448	Checking CSM O/P seems very quiet comparing 212
	with LM e B/U Vegas 0450 -129
0545	Start parkes BER cal 0501 -130
	23:30 7158 Rx VCO for PK's µwave P/B9 lui
	had problems happen ACSU
0609	AOS) omni switch late
0616	AOS
0649	PK's -111 db total is average figure
0657	Through with parkes test Going to prime.

Appendix 4 \ Honeysuckle Station Log for Apollo 11   437

Early EVA. 105.30

2255
2157
58

79 4847
80 34

DSIF LOG SHEET __HSK__ SITE

DATE __19 Jul__ (GMT) LOG KEEPER __LEE__

MISSION __NC6 - 725   TLC #3   DAY 003.__

(9)

Prime path problem M/W at Williamslake.
both jitter was a pwr supply.

GMT	EVENT		
0720	KSL on ! Mode 1.	-127.5 /128.	(0729)
0801	TLM test completed 27 950 /3818	-127	(0739)
0820	Revr 5 AGC back thru AMQ.	-127	(0749)
0908	Momentary LOS. Prior to Ant. reset	-127	(0500 JD)
	P/L 1052Z 2150K 5.7 S.L Data		
	MADX 1076Z LOS 1184 Wrng		
1022	Scm H/O 69.15.00 (1047Z) → MADX		
1034	5 S/C rejects on Ant. RST (-140).		
1052	S/C called & tried to MCC but not in Voice Up mode. informed		
	MCC Comm tech - went to MO 6 - but H/O to Mad		
1102.	LOS P - expected LOS here as 1124		153 J/L
1126	LOS. W. H/S TLM OFF		
1155	Verbal release.		
1200	LLS TWX rec'd - doing LM D/L test.		
1729	Start ∅1 SRT		Decom 3 2FM1
1737	∅1 completed		
1751	LOI-1 was successful. 61.6/169.5		
1920	Ops 2 H/o J.S. on. PCM cost & LM mode 9 BER's in progress.		
1949	LM mode 9 prime figures being repeated. Not much better.		
2117	Start ∅3 CSM & LM		
2128	End ∅3		
2140	PKs finally manned.		
2157	Started setting PKs subcarriers cadfiss start sent.		
2223	Set up for Cadfiss still checking PKs thresholds Cadfiss will be LM powered up		
2232	FE 600's Nets 7 & 6 completed. comptrs on line.		
2255	Start CMD 1/F		
2255	Start Biomed 1/F.		
2259	Go on CMD & Biomed.		
2305	Start prime & wing L/S B/S & TLM cadfiss.		
2313	Go on TLM		
2316	Go on Prime & Wing 4/S B/S starting prime APP		
2323	Go on prime APP		

JPL 0307 DEC 60

DSIF LOG SHEET ___ 14 SK ___ SITE (10)

DATE 19/20 JUL 69 (GMT)   LOG KEEPER SAXON

MISSION MCG-725 L/O REVS 4, 5, 6, 7, 8, 9 & 10. DAY 04

GMT	L/O #4 EVENT
2327	lost net 1 transmit between ACSN & AAW switch.
2332	Go on A/G
	112:25 → Cancelled CROW. CANCELLED } M MCC TV.
	107 GET for TV I/F Aus To U$ 21/0032 } Wing rise
	earliest possible 105:30 GET. 20/2300 Z
2351	Verbal update to SCM B/U now 83:30. Wing AOS 00:$854
	Prime Prelimit 0107 (11 5)
20/0034	'JT' to line prime H/s
0048	SB 1 AOS CSM
0051	CSM Voice demod lock
0053	SB 2 AOS LM
005330	T.C'd on
0055	LM LBR int Lock & int CSM Hi lock. LM comm checks in progress
	GDS & HAW w/R LM
0101	SB 1 Autotrack
0109	Problem w/R faulty 29 pt msg not reported until 30 mins had gone by.
011540	CSM LOS
	L/O #5
0157	Keying Check Net
0159	JT H/s to line prime.
020203	CSM AOS P & W LBR Then 4BR
0206	FM carriers & Subcarr + 250 KHz carr dev.
020740	32:1 Dump ON + 300 KHz " "
020930	32:1 " OFF
021020	FM carrier OFF
0217 6,34	LOS } went to omni from HGA
021726	AOS GDS glitch on U/L ? 70 KHz Subcarriers turnaround lost he/a carrier
021930	LOS
022040	AOS
0224	HGA again
0225	IN LM AOS At wing?

JPL 0307 DEC 60

Appendix 4 \ Honeysuckle Station Log for Apollo 11

```
SRT at 1 AM
PKs I/F at
  04 D018
  33 47 18
  ─────
     13 00

  05 12 25
  64 46 25
  ─────
     26
```

DSIF LOG SHEET __HSK__ SITE (11)

DATE __20 JUL 69__ (GMT)   LOG KEEPER __SAXON__

MISSION __NCG-725 L/O Revs 5 6 7 8 9 & 10 Day O7__

GMT	EVENT
	L/O #5 [cont]
0240	PK's AOS
0245	GDS O/L off for HGA auto reacquisition. Wing LOS. Ren AOS
0245	" " ON
024926	" " off " " " "
025030	" " ON short loss.
0253	PKs data on Decom 2 (inhibited)
031801	LOS all links.
	L/O #6
0340	ODB to operate.
0345	Prime HS TTY on & PA 1 ON A/G not remoted GDS P/B on Net 1
0350	A/G remote.
0357	" keying check. USB 1 did not have t/c on line
040245	AOS Go for CMD — Late auto reacq. not too good.
	& P. 00.07
040715	FM carriers and Subs — 98 dbm prime — 88 pks
0905	Start Dump. #1 : 1
041358	" " 32 : 1
041591	End Dump.
041626	FM off
0421	-103 SB1 -102 SB2 -88.5 db Parkes
0442	SB1 -103 dbm SB 2 -102 dbm PKS - 89 dbm Total
0502	" -103 " " -102.5 " " -88.5 "
051221	LOS all systems 87 38 39 DSE recording start before occultation
0515	FM/FM FMT 16 CSM PM
	84:40 00 10 mins.
	40.14 D (2)
0531	P/B on line after cals.
0540	" complete.

JPL 0307 DEC 60

DSIF LOG SHEET __HSK__ SITE (12)

DATE 20 JUL 69 (GMT)   LOG KEEPER SAXON

MISSION NCG-725 L/O's 7 8 9 & 10

GMT	L/O #7 EVENT
0543	Carriers ON. Prime H/S Data on line
0548	leading edge of moon tape looks good.
0556	Wing H/S to line.
055842	AOS all stat sent.
0602	FM ON. -97
060465	Start Dump 32:1
06058	End dump
060612	FM OFF
0610	Data had 1/3 irig to 400Hz after FM remoting P/B reset
0622	SB1 103 SB2 102 Pks -88
0645	after " 103 " 102 " -88.5
	Final LOS & N/c will be coming up out of conf to HSK GWM & HAW **
0707	SB1 103 SB2 102 PKS -88
071046	LOS All sys Stat sent
0713	LSTTY Hist sent Bad stat sent. H/s Hist taken.
	L/O 8
0720	KSV on. Param list + HSP Prev + FMTS OK
0742	Prime H/S to line.
0752	VG6HA in line (GWM B/U)
075716	AOS Stat sent -103 -88.5 (PKS)
07585	FM on.
080043	FM carrier (-98R) (-92 W)
08 0138	Dump Start. Dump Stop. 080330. 32:1
08 0337	FM off
0823	P-103 W-102 PK -89.
0838	Small clutter on PKS data (reported by DATA)
0854	Signal levels unchanged.
090608	LOS all systems.

JPL 0307 DEC 60

Appendix 4 \ Honeysuckle Station Log for Apollo 11   441

(107.45 108.15)

1/F 2320
    2120

2320
2135
1.95

(13)

_____ DSIF LOG SHEET ___HSK___ SITE

DATE __20 JUL__ (GMT)   LOG KEEPER ___LEE___

5060 076/136

MISSION __NE6-725   40°S 9 & 10__

GMT	EVENT
0946	4/09 H/S TLM ON. Cam 888
0952	Voice / Key checks on Good.
	40KHZ sweep
100130	-102 AOS 5.5 mins late
1004	FM ON.
100602	FM carr/ Sub. c.
100800	Dump Start 100968. Stop.
110730	LOS.
	MG Voice #3. Let 5. P/B FMT 16 CSM
1116	H/S history & EOF 92:20 → End of don
1122	Wrong 10 Seg. in FMT 16
1127	Data on line for FM P/B. (49 RTC U/L)
1135	RLS msg released.
1142	Data had wrong GET chart time on P/B 92.14
	92.23 30 S/C AGE CT 0620.
	92.2 0640.
1154	P/B completed.
1517	Start Ø1 SRT
1755	Noisy PLSS tape + Astro Sig. Gen U/S

1024 49
    29 11
    12

1905	OPS 2 H/S GS ON SRT in progress loading ops programs.
1932	Ran LBREMOD due Mode 08 conf turnaround w/R wing. Comms patched
	in a scope during transmission & killed it.
	PLSS Touch
	Armstrong heartrate 110 156
20 1741	T⊙UCHDOWN ! !
2088	Doing PCM cost & ISA cost.
2049	ISA & PCM costs completed
2055	Start Ø3.
2015	End Ø3 CSM wing TV only remaining
2216	EARLY EVA AT 01157 APPROX.

JPL 0307 DEC 60

107:30. (12g)   108:15 FM on Net3
EVA 0115Z   _____ DSIF LOG SHEET HSK SITE   (14)
approx.
                DATE 20/21 JUL (GMT)   LOG KEEPER SAXON
014251
108 11   MISSION NCG 725 LUNAR SURFACE CSM REVS
10748.   17 Thru 22 Day 5.
                                                      1 06:15
                                                      Start prep

GMT	EVENT
2247	Prime Rise B/S 108:11 Wing Moon rise 107:48. 1 Passed to track 106:11 AOS CSM TV Lunar surface 107:30.
2259	Still progressing Rates 1/F EM 107:30 107:30 1/F. 108 Befor Wing
2319	Start FE600 checks. 108 HRS Hatch open.
2320	Start CMD I/F & Biomed.
2326	Biomed is go.
2333	CMD GO
2334	A/G Go Start B/S Prime H/S & L/S & TLM Cadfiss fmts 5 & 11
2342	Start Wing B/S H & L
	06 11 GET crew starting EVA prep.
	CH 35 Bit 3.
0003	Running APP prime after B/S rerun prober
0016	cleared on cadfiss wing have bit 130 RRDQ Ex 3 problem Relay changed No further problems.
0049	Moon rise wing 0115. CSM 17 AOS wing 0140 08
0055	107:54 H/O to wing CSM Moon rise wing 0115Z Constant key grinder bypass from wing. LM AOS Prime 0142 ← 15 mins 03 D/L CSM. 0147 B/U on LM
0106	LM 09 or 10 from track.
0108	PAM's
LM 0112	Prime AOS LM Int Signal. PM 014 108
0118	TIC'd on. HBR LM solid LM EKG only. 3308
0124	Confirm LM relay to CSM after AOS from N/C EVA preps in progress.
0130	FM reempted TIC request CSM L/O 17
0135	LM to 09 or 10.
0137	LM TLM & TV sync.
0138	LM EKG off
0139	EVA 1 & 2 EKG
0141	CSM AOS Wing.

JPL 0307 DEC 60

Appendix 4 \ Honeysuckle Station Log for Apollo 11

```
10 9 :21
110 0735
05 53 12
02 88 12                DSIF LOG SHEET __14 SK__ SITE        23255289
  75     725
                DATE  21 JUL  (GMT)   LOG KEEPER  SAXON
9 93            MISSION  Lunar Surface & CSM revs 17 thru 22
033924
030824
  36    GMT                      EVENT
```

	GMT	EVENT
LM		CSM L/O 17 (cont)
Surface	0143	Nascom relay verified.
	0146	CSM FM dump - 96 from 014313 THRU. 15° Elev prime 014905.
	0150	FM off
	0152	TV off test signals ON
	0154	Go for cabin depress.
	0201	CSM LOS wing
	0202	" AOS " } To LBR & B/U voice. worked pseudo nascom
	021127	HS LD 00222 0904 2105
	021906	Inv Sum MCC
PLSS	0221	0.15 Deg offset in A
1	0223	-140 CSM dropping
9 8	622330	CSM LOS wing
80 50		-0.25 X -0.5 Y
12	0228	CSM AOS HGA
	0244	-100 FM on LM
		EVA EKG2 worse on LBR.
	0251	CDR on LM porch
	025400	TV ON.
	025626	CDR on Moon ! ! ! Processing HSK & GDS TV alternately
	0306	Starting up using their TV & PAM -90 dbm
	0325	Video recorder VR1100 red cannot ETO V.K
		L/O 18.
	0340	AOS stat
	034252	CSM FM ON CMD.
	034440	FM carrier.
	034525	Dump mod 32:1
	0346	2106 LM H/S & 0025
	03475	LOS & FM &
	0349	Nixon uplinking

JPL 0307 DEC 60

```
111.51
111.24
  .27
          053743
0451 31                                                    (16)
0405 31      ____ DSIF LOG SHEET ___ HSK ___ SITE
 46 00                                              0502 E0
             DATE 21 JUL (GMT)   LOG KEEPER SAXON

             MISSION 725 EVA L/os 18 Thru 22


     GMT                    EVENT
              L/O 18 (CONT)
     0354    8 HS Lds Cleaned
     0355    LOS CSM wing
     035630  AOS CSM go for CMD
     0357    LOS CSM wing
     0403¹⁴⁷ CSM B/U voice              S/C AGC 8.7 V/S 2-75 LM FM
     0425802 uplink at wing to CSM
     0426    CSM to HGA
     0427    06 uplink
     0427    Normal voice down
     045130  CSM LOS A/G local

     0457    EVA 2 Ingress
             L/O 19
     0510    112:10 H/O Vice 1/1:51 LM GDS sops   CSM AOS 053743 2
     0528    HS LD 0024                Fm A/G problems M/o LM 0542
     0524    LOS both PAM's
             21.893750 U/L LM freq based for doppler
             CSM & LM to Net 1
     053756  CSM AOS A/G remote                                      GDS
     0536    FD 2107 Rec'd           TVH 118 10                      3772
     0539    Data Lad CSM decom inhibited
A16→ 0542    H/O LM GDS → HSK
     0544    U/L RD 0024 CSM
     0544    1205 U/L seq
     0552    U/L verify LM on CSM D/L No relay or grinder SDDS Investig 053743
     0602    CSM Vogaa bypassed   CSM D/V -50 dbm LM up bas normal level -7 dbm   1743
     0617    0025 H/S Ld Rcd  & 2108 & 0906                                        20
     0619    CSM AOS
```

Appendix 4 \ Honeysuckle Station Log for Apollo 11

DSIF LOG SHEET __HSK__ SITE

DATE __21 Jul__ (GMT)   LOG KEEPER __LEE__

MISSION __NC6-725   L/O 19__

GMT	L/O 19 cont — EVENT
061950	H/G 4 CSM AOS Go for CMD MD 02 D/L FM ON
0621	All Loads except 0025 0906 1205 & 2108 Cleared
062148	FM carrier & Subcarriers -96
062356	Dump Start 32:1
062723	stop → tH 5 secs of 1:1 at end of
062743	FM off
064850	CSM LOS, 0651, A/G remoted — 109:40 vR 1110 / 109:45
0721	LD 2108 U/L OK Compare
0729	Range Acqn ASAP after H6A re-acq
0731	U/L MD 7 LM @ 111:02:00
L/0 20 07.36.13	HOS CSM Sht cent -103
0739	FM Subcarrier (-96)
	Dump Start 074109. Dump Stop 0743.15
0742	U/L Mode 5 CSM D/L 2
074310	A/G de-remoted
0747	LM crew jettisoning equipt
07.5400	CSM U/L MD 6 - M6 remoted
075750	TV LOS L/M
0759	MD 1 PM D/L LM
080110	CSM U/L MD.5 D/L M0.2
080235	LM U/L MD 6 JJ 080454
0804	L/S from LM, when on line - reacquire range
0805	Sub-carrier - just seen
0807	Repeat Rng Acq every 5 mins until direction from MCC
114:33	C/O LM Sub-carrier BE 114:35:00 lower bandage to centre calibr CK6 114 clone clint heard. m
0819	Cease 5 min range acqs
0821	Quality bit setting on Range Unit - check it
0825	LOS 0026 + 0907 & 2109
0827	LM U/L 7 : D/L 01
0828	Faulty NB 2 cond. - DC restoration problem

DSIF LOG SHEET __HSK__ SITE

DATE __21 JUL__ (GMT)   LOG KEEPER __LEE__

MISSION __NCG-725 L/O 20.__

GMT	EVENT
0830	Loads cleared. 1205, 0025, 2108, 0906.
0830	Terminate L/S from Prime on LM.
0846	Go to MD.7 L/S on DD.
084801	LOS CSM.
0903	MD 5 CSM / 7 for LM. 14 min valid Range
	L/O 21
0934	Remote CSM A/G at AOS
093430	AOS CSM
093853	FM Carr/Scarr. (-96)
098806	Dump Start / 093944 Dump Stop
0945	SCM rec'd for L/O 22.
1018	Clear LDS 0026, 0907, 2109 GOA60
1019	H/S LD 0027, 0908, 2110
104628	CSM LOS
104831	USB Stat sent 1049 CSM L/S sent.
1057	L/S Hist. + EOF Using LOS 1258Z
	L/O 22
113750	AOS CSM 118.30.00 ACN. (2022)
113832	FM Carr/Subcarr, -98.
113934	Dump Start Stop 114112.
1202	H/O to ACN (nom' HGA) - no problem. LM A/G de-rerated
1215	Take down CSM carrier at 119.25.00 (1257Z)
1220	LOS LM P/L 3-way.
124445	LOS CSM
1257	Wing carrier down.
1308	RLS MSG.
1833	Start Ø1 SRT
1753	APS LIFT OFF
1841	SRT Ø1 complete

Annotations: LM HG H/O on FM (MMX)
114 RTC
3 LDS

JPL 0307 DEC 60

Appendix 4 \ Honeysuckle Station Log for Apollo 11   447

```
1/E 2350
   2319
   ──────
    46
```

(19)

DSIF LOG SHEET ___HSK___ SITE

DATE _21 JUL 69_ (GMT)   LOG KEEPER _SAXON_

MISSION __NCG-725 DAY 06 REV 29, 30 TEI & TE 1__

GMT	EVENT
1954	D/L BER's SB1 too good suspected test tx out of spec, carrying on will wing
2123	Track sending acq msgs check
2135	DOCKED !
2156	29 pts U/S can't Y Angles - requested again still U/S
2217	Msgs were good forgot
2317	Ø3 complete
2327	7/9 Net 5 5/6/7/0/0/0/2/3/4 W 01 46 z 132:14
	reading 02 05 132:33
2340	30:09 est on LM sep from N/C 1Hr early? P 02 14 2 132:42
2346	FE 600's complete
2350	N/B sig cond out of spec L Pass amp requires realignment
	Comp alignment under way.
2352	Start Biomed I/F
2353	Go on Biomed I/F
2357	Start CMD I/F LM JETTISON
0006	Go on CMD I/F
0010	Start TLM cadfl/ss Prime having range lock problems
0012	Start Wing BORA L/S
0013	" prime " "
0019	Wing BORA Go
0022 14	No retral go on prime BORRA Cue for Borra
0031	Go on PosA waiting for
0042	Start A/G recording Space-15 Mark - 14
0047	Go on A/G I/F
0119	Prime H/s data on line DD

JPL 0307 DEC 60

(20)

0216
0238

  38
031745
  4745

DSIF LOG SHEET _____ HSK SITE

DATE 21 JUL 69 (GMT)   LOG KEEPER SAXON

MISSION NCG 725 DAY 06 REVS 29 30 TEI & TEI

GMT	LM & CSM Rev 29 EVENT
0136	Wing CSM & LM glitches.
0139	Prime glitches on CSM.
0142	Hi Bit rate LM Tlm solid. CSM LOS 02 3138
0145	Tic'd on still no CSM TLM LM LOS 02 3203
0148	TIG 135 23 41 For TEI 01 56 03 / 36 0
0151	CSM sig level rising — still no tlm or demod lock.
0157	CSM HBR solid from Wing sent to 64285.
0208	Prime 'JJ' to line, other Auto track still in prelimits.
0224	SBI unable to establish B/S either scope out of cal or loss of main gain
023143	LOS CSM
023156	LOS LM
	CSM REV 30 Tic configured to 7 9 2303
0233	Reported to N/C main paramp 10 db low investing.
0248	Main line attenuator had to be decreased Klystron may be on way out. declared it Green to N/C.
0302	Prime TX on & L/S data 'JJ'.
0308	CSM A/G remote.
0312	HS loads 1304 & 0032 Recd.
031245	Fm Fm remote.
0315	INV SUM neg MCC — on line.
0316	L/S Data JJ wing MCC Stat neg BAD — No SCO's.
031745	AOS Go for CMD Stat sent
031747	0033
031934	Fm ON CMD 03 2130 Fm Carrier & Sub
032230	Dump start 32:1
032411	" stop.
0325	FM OFF
0326	Start 0032 U/L OK U/L a single line DSKY correction (comp)
0329	Start 0033 U/L O K comp
0330	" 1304 U/L O.K comp

JPL 0307 DEC 60

Appendix 4 \ Honeysuckle Station Log for Apollo 11

```
       04 32 05
       34 05
    05 06 05
     04 29 47
    03 32 47
       39 00
```

_____ DSIF LOG SHEET ___HSK___ SITE

(21)

DATE 21 Jul 69 (GMT)   LOG KEEPER SAXON

BURN AOS.
135 34 05 MISSION   NCG 725 DAY 06 Rev 30 TEI & T/E #1

NO BURN   LM L/O 32 →
135 44

```
                                                    05 06 05
                                                    04 42 11
                                                       24
```

GMT	EVENT
0339	Reading up TEI pad D/V clear but rather weak
0347	New JJ's prime & wing
0422	1205 H/S LD Rec'd
042948	CSM LOS. 04 2 134:57 08
0435.	Comp 1 - & C2 & real time biomed 30 secs to dump the remote dump biomed.
	T/E #1
050618	AOS Stat sent. Go for cmd. SB2 on S/B
050753	FM ON
050937	FM carriers & subcarriers
051045	Dump pro to line
051052	" Start 32:1
051143	" " 1:1 Voice high background noise
051745	" " 32:1
051832	" End.
051844	FM off
052030	U/L LD 1206 in progress. O.K compare
071442	
064042	0522 TIC'd 007 & 009 107 as BU.
34	0635 OP REC 1 130 amp has 3 volt P.P noise spikes on it invalidating AMQ cals - cals were done on RCVR 2 and msg screwed, had 8 & 98 on it! Trying to think of new cal procedure for
0707	Reacquiring range
0706	LOS 0709 AOS then INT AOS poss main paramp p
	LM L/O 32. Wing 2 Way.
071437	AM AOS. -107.
071957	LGC static AGS line GC 0301 GC 0302 Dec
0726.	LM S/S decreasing
072828.	Los CSM
073038	AOS CSM
073125	LOS LM AOS 073236
073288	AOS CSM.
0736	Configure wing for CSM

JPL 0307 DEC 60

DSIF LOG SHEET ___ HSK ___ SITE

DATE 22 Jul (GMT) LOG KEEPER LEE

MISSION NCG-725 TEC #1 DAY 06

(22)

GMT	EVENT
073926	A/G D/L from Wing now removed
0740	Wing auto – 103 Mode 1 FM/CSM
	Syst 3 08.08.01
0741	No further LM support (TWX)
0742	H/O P → W. actual 074341 A/G voice good
0744	Mode 16 D/L OMNI all the way
0750	Mode 3 U/L CSM Syst 3
0750	CMD bit faulted – up again CBARF
0752	Mode 6 U/L ✓ Syst 3
0757	Jump Key 1 UP on CMP/TM1
0757	Permission to configure warm paramp
0802	Lost CMOA after CBARF – will need OUCH after pass. When A/G D/L is to be inhibited, this should be done at Comms NOT Subs – lose recording otherwise Rx 8 on CSM FM
	Permission to put JUMP KEY 1 UP
0838	B/S Adj + EOF (-124)
085432	AOS P. H/O Sys 3→4 0903252
0930	Configure for EASEP FMT 1 2A 1B
094628	✓ 129 Alsep – data on
1145	CSM on Syst 1/2 105:42 – 1
1152	AOS CSM (local) Easep X +.05
1223	LOS CSM – USB stat – RTC Y –.15
	unable to get USB stat – het 4? W. LOS 13452
122950	AOS CSM
125370	✓ Prelim
1302	H/O → MAD OMNI – cancelled due MAD LOS
130408	" " " – OK
	GET 137.5600 – 138.1800 for OUCH 135 M/L
1342	LOS Wing Wing 100 HA Offset
1344	loading OUCH 20 Dec

JPL 0307 DEC 60

Appendix 4 \ Honeysuckle Station Log for Apollo 11   451

23

2350
2338
12
0010
0128

DSIF LOG SHEET HSK SITE

DATE 22 JUL (GMT)   LOG KEEPER LEE

MISSION NC6-725   TEC# 1   DAY 06

GMT	EVENT
1400	DOCH to line — Completed 1401
1413	Yubne 23/1027 AOS / 22/2350 l/F P - alsep W aP
1501	RLS TWX rec'd
1820	Start SRT ∅1 — Completed 1828.
1919	Rcvr 1 150 amp still faulty TEC# 2 Day 7.
1957	Net 4 alsep RcV & Tx Net 6 from NST TLM
2209	Start apollo ∅3. No apollo format tape mounted
2225	end apollo ∅3 (loading alsep in CMD
2323	end alsep ∅3 after ½ hr fooling with PCM simulator.
2347	We should expect -132 approx from EASEP
2352	Start FM/FM I/F Net 5.
2356	Data not being rec'd at Goddard.
0000	Start wing B/S RA
0006	Start dial line TLM cadliss
0009	Some range rate data errors in Wing H/S G/CMPTRS checking
0010	RRate was operator error.
0012	Go on TLM recvd wing B/S.
0020	155:30 TV pass from N/K report quality to MCC TV.
0026	Go on Alsep CMD
	Crit RTC Group 1 enabled. Alsep AOS approx 0209.
	160 CSM AOS " 0150 156 18
0104	deleted from Alsep by Track configuring
	prime site for Apollo 2 way support req'd at 160:00 GET.
	configuring prime for apollo.
0136	RF contacts CSM wing.
0141	-190 increasing
0145	TIC'd on CSM sol.d
0148	SB 2 Auto track. SB 1 still configuring & calling
0216	157:45 GET KTO to end of reconfig
0235	Paramp & P/A's set up for apollo - going to cadliss

JPL 0307 DEC 60

_____ DSIF LOG SHEET _____ HSK _____ SITE

DATE 22 JUL 69 (GMT)   LOG KEEPER SAXON

MISSION NCG-725 TEC #2 DAY 7 [cont.]

(27)

GMT	EVENT
0250	Cooled paramp oscillating configured to warm advised N/C and Bent stat.
0322	AOS SB1 side band
0329	Good
0383	LOS for 30 secs Low sig level.
04 2050	LOS for 20 secs " " "
0442	FM/FM remoted TIC request
0444	LOS for 20 secs low signal.
0449	A/G keying check
0456	LOS twice. 2nd for 2 mins.
0532	H/O HGA GDS to HSK A/G remote. Strat sent lost off at pr. time R/E 1 head wrong B/W selected
0534	M/S LD 0703 Rcd 2 U/L O.K Ampere time update.
0542	S/C AGC two large drops 1st shortly after H/O when we lost D/L did not loose U/L according to SB2 who held lock. 2nd was poss due to crew switching to Omni on HGA.
	05 32 59 S AGC started to decrease 053279 101
	HGC rapid drop 053259 + Gnd rcvrs 140.5
	switch 05 35 44 101 137
0715	KL ON 35 49
0936	Voice check 5 B Gmo. 35 57
	W LOS 1357 87 11
1309	Y pre limits 135630 1715
1357 01	H/O → MAD OMNI ant. — no glitches! 140 U/L
1401	LOS CSM W o P.
	AOS 24/0114Z I/F PR 24/0001Z 168/38 121450/5100
	H-70 for Apollo then @ 24/0402 Z I/F AlSep 43 dup
	24/1415 Z Apollo
	bring AOS 24/0114Z stay.

JPL 0307 DEC 60

Appendix 4 \ Honeysuckle Station Log for Apollo 11

TIDBINBILLA

DSIF LOG SHEET ___HSK___ SITE

DATE 23 June (GMT)   LOG KEEPER LEC

MISSION NCG-725 TEC #3 DAY 08. & EASEP

GMT	EVENT		
1820	Start SRT ∅1.		
2137	Start ∅3		
2146	End ∅3 ready for CadJiss		
2355	FE600's complete Nets 7&6.	AOS wing 01 59 00	180 27
0002	Start FM/FM I/F	AOS prime 01 55	180 24
0004	Go on FM/FM I/F	LOS wing 16 31-16	194:59:16
0008	Start CMD I/F	LOS prime 16 34 09	194:59:09
0013	End " I/F Start BOR/RA prime & TLM.	0150 → 0220	
0015	Cue on wing BOR/RA	H/O 180.45	
0023	Go on wing & prime BOR/RA		
0024	POS A Cue.		
0031	Go on POS A. Start A/G N-11.8 S-12.8	Neg 8 at MCC	
0035	Go on A/G I/F		
015250	AOS CSM prime		
0154	Tic'd on Int TLM.		
015706	'JJ' prime L/S to line		
021700	H/O GDS to HSK omni A/G remote start sent		
0231	Bad echo on Net 2 since 0210 approx noisy Net 1 Receive at MCC after H/O.		
0233	New SOM prime & wing.		
0238	Inhibited transmit side net 1 to check noise at MCC Then they reconfigured nets. also lost echo on Net 2.		
0345	RTC confirms H/O will be at 182:30 GET to GWM Comtech requests H/O upvoice only, Track says will be on Apollo until 185:00		
0356	A/G Upvoice & CMD HSK → GWM. start sent GWM came up sweeping 20 secs early & then turned off at H/O time back on again 5 secs later we lost 20 secs of data due This. 3rd Aug earliest for ESEP		
0632	Breaking down prime CSM track to configure for Alsep. H/S TLM on Net 5 Stream A Alsep on Net 4 Biomed on Net 8.		
0728	Locked on Alsep & Running		

```
                                                              26
          ____ DSIF LOG SHEET  HSK  ____ SITE
               DATE  24 JUL  (GMT)   LOG KEEPER  LEE
          MISSION    NCG-725   TEC #3    DAY 08
```

GMT	ALSEP #2 EVENT
0742	Enable Gp 1
0800	Bring carrier up MD.9
0925	Go to U/L MD.3
0922	-126.5 dbm
1102	VOGAA disabled
1314	CMD mod on ALSEP
1316	Showed normal
1359	Decom & WB Conv fail - lost data - back on NB & outputting again 1400:11. (-128.)
1411	Mod'n off ALSEP
1416:45	Carrier OFF ALSEP - Stat sent.
1421	ROACH on line - rehaus to GORO
1427	CMD cycling Apollo
1437	A16 D/L inhibited CRO enabled.
1443	Jump key 1 DOWN
1447	AOS CSM - US data on.
1505	CSM LOS AOS 1509
1510	LOS. 151519 AOS P. 2 134 137
1541	LO 0038 in
	194 38.00 -646 1020 W 163116
	194 48.00 -646 gain P 1632
1610	ALSEP PS2N sent
194948	SEP.
1627	LOS morning
	LOS P 1631 44 W 163131
	1651 SPLASH

JPL 0307 DEC 60

Index

A

A-12 satellite: 54
Acquisition aid: 35, 73, 75; on Gemini, 108, 112
Acquisition-of-signal: 47, 52, 75, 84, 128, 129, 169; TDRS deployment, 167-168
Ada, Joseph F.: 236
Adams, John: 96
Adelaide switching center, South Australia: 64
Advanced Range Instrumentation Aircraft (*See* Apollo Range Instrumentation Aircraft)
Advanced Research Projects Agency: 29, 67, 208
Aerobee sounding rocket: 2
Agenzia Spaziale Italiana: 337
Air Force, United States (*See also* Department of Defense and specific test ranges and locations): xxxiii, 2, 5, 21, 67, 71, 74, 78, 82, 85-88, 94, 116, 117, 120, 122, 131, 137, 149, 154, 206, 208, 226, 238, 253, 308; Andrews Air Force Base, 77; ARIA support, 159-162, 203; Bonham Air Force Base (*See* Hawaii Ground Station); Department of Defense Manager for Manned Spaceflight, 94; Eastern Space and Missile Center, 237, 239; Edwards Air Force Base, 59, 162, 223, 232; Electronic Systems Division, 161; Hickam Air Force Base, 160; Holloman Air Force Base (*See* Tula Peak Ground Station); Kindley Air Force Base, 82; Military Airlift Command, 117; National Range Division, 160-161; Patrick Air Force Base, 11, 15, 18, 21, 123, 160, 163; Satellite Control Network, 71, 93; Securing the high ground, 93, 253; United States Air Force Museum, 162; Vandenberg Air Force Base, 2, 85, 87, 163, 202; Wright-Patterson Air Force Base, 161
Alamogordo Bombing Range (*See* White Sands Missile Range)
Alaska Tracking Station: xxxii, 41-43, 50, 53, 55, 64, 103, 219, 221, 222, 227, 228, 281, 337; Poker Flat, 317-319
Albert monkey flights: 2
Aldrin, Edwin E., Jr.: 174, 175, 177, 179, 180, 341
Alice Springs, Australia Northern Territory: 189
Aller, Robert O.: 258, 282, 325, 336
Allied Signal (*See also* Honeywell): xix, 289
Alouette satellite: 43
Ames Research Center: xv, 28
Anchored Interplanetary Platform: 39
Anders, William A.: 171
Anderson, Clinton D.: 25
Anderson, Michael P.: 334, 335
Angular measurement: 11, 13, 14, 35, 37, 38, 52
Antarctica (*See* McMurdo Ground Station)
Antenna Control Unit: 46, 47, 52
Antenna operating modes: 52-53, 112
Antofagasta, Chile: 16, 19, 23, 40
Antigua: 114, 136, 138, 203, 204; Apollo support, 117-121, 146, 149, 151, 162, 163, 169; early-ops support, 18, 21, 40, 98, 112
Apartheid (*See* Johannesburg, South Africa)
Apollo, Project: 1B network, 145-146; V network, 145-146; Antenna pattern pull, 136; Apollo 1, 166, 167; Apollo 4, 131, 147, 156, 158, 166-167; Apollo 5, 147, 167; Apollo 6, 168-169; Apollo 7, 150, 163, 170-172, 331; Apollo 8, 147, 151, 156, 171-173; Apollo 9, 173; Apollo 10, 173-174; Apollo 11, xxxv, 120, 151, 157, 158, 174-180, 181, 186, 196, 203, 205, 206, 221, 235, 342, 427-454; Apollo 12, 158, 182, 183; Apollo 13, 180-182, 203, 234, 250; Apollo 14, 174, 182, 187; Apollo 15, 183; Apollo 16, 168, 169, 234; Apollo 17, 182, 183, 185, 306; Apollo Lunar Surface Experiments Package telemetry, 182-184, 205; AS-201, 118, 145; Bailout decision during powered descent, 174; Command/Service Module, 112, 136, 137, 144, 145, 150, 151, 157, 166, 169, 170, 171, 173, 180, 181, 184, 196, 232, 306, 341; Flight phases, 143-144, 153; Free-return trajectory, 151, 181; Implications of moving the launch window, 157-159; Lunar Module, 112, 136, 144, 145, 147, 151, 152, 166, 167, 168, 173-179, 181-185, 232, 305; Original tracking plans, 145-147; Portable Life Support System telemetry, 175, 180; Reentry and recovery considerations, 150-152, 156-159, 160,

163; S-IVB transponder interference, 181; Saturn 1B launch vehicle, 118, 154, 167, 196; Saturn V launch vehicle, 118, 131, 135-138, 145, 147, 151, 154, 156-159, 166, 186, 187, 224; Trans-Earth Injection tracking, 144; Trans-Lunar Injection tracking, 129, 135, 149, 150, 151, 153, 157-159; Transposition and docking tracking requirements, 151-152, 158

Apollo Range Instrumentation Aircraft: 158-161, 163, 193, 203, 214; Flight 328 crash, 161-162; Joint NASA/DOD responsibility, 159-161; Project Configuration Control Board, 161; Two-ocean versus single-ocean support, 158-159

Apollo-Soyuz Test Project: xl, 135, 190-198, 200, 205, 214, 215, 229, 249, 281, 306, 327; American/Soviet cultural differences, 195-197; Soviet ground stations, 191, 193; Soviet tracking ships, 193; Soyuz 16 network rehearsal, 195; Use of ATS-6, 193-195

Application satellites (*See also* specific satellite names): xxxii, xxxvii, xxxviii, xl, 65, 222, 229, 232, 234, 249, 273, 304, 306

Applications Technology Satellite (*See also* Apollo-Soyuz Test Project): 50, 55-56, 156, 193-195, 205, 206, 214, 215, 219, 227, 248

Applied Physics Laboratory, Johns Hopkins University: 184

Ariane launch vehicle: 238, 239

Argee Corporation: 282

Armstrong, Neil A.: xxxv, 174-178, 180, 341

Army, United States (*See also* Department of Defense): xiii, 6, 8, 9, 19, 21, 23, 24, 40, 42, 60, 74; Ballistic Missile Agency, 29; Corps of Engineers, 7, 15, 31, 236; Map Service, 15; Ordnance Department, 6, 7; Project Orbiter, 15; Signal Corps, 15

Ascension Island: xxxii, 2, 98, 114, 116-119, 123, 124, 149, 160, 162, 163, 189, 192, 220, 237-240

AT&T: 101, 196, 244, 246

Atlantic Missile Range (*See* Eastern Test Range)

Atlas launch vehicle: xxxviii, 2, 69, 81, 99, 106, 139, 145, 260, 295, 297, 299, 306

Attwood, William H.: 127

Australia (*See also* specific locations in Australia): xiv, xv, xxxiv, xxxv, 4, 19, 34, 39, 44, 45, 48-50, 56, 83, 95, 99, 110, 113, 114, 146, 148, 150, 174-176, 188, 189, 193, 203, 208, 212, 220, 223, 226, 227, 232, 233-235, 251, 288-290, 304, 318, 338, 340; Australian Land Information Survey Group (*See* Geoscience Australia); Department of Supply, 21, 44, 45, 115; Department of Territories, 235; Department of Transport and Communications, 235; Geoscience Australia, 235; Weapons Research Establishment, 21, 45, 115

Automated Transfer Vehicle: 335

Autotrack: 39, 40, 51-52, 76, 149, 327

Azusa missile tracking system: 9, 10

B

Badgeless controller: xxxvii, 232
Baghadady, Eli: 37
Baikonur-Tyuratam Cosmodrome: 66, 191, 195
Bailey, Harry: 229
Baker, James G.: 3
Baker-Nunn camera: 3-5, 234, 235
Ball Aerospace Corporation: 66
Bandwidth requirements: xxxviii, 35, 62, 164, 174, 298, 300
Barbados: 120
Barbour, Walworth: 84
Barking Sands, Kauai (*See* Hawaii Ground Station)
Barnes, G. M.: 6
Barnes, Robert: 185
Baumgartner, A. W.: 134
Bavely, James: 124
Beall, J. Glenn: 31
Beijing, China: 318
Beamwidth: 52, 144, 181, 232

Bear Lake, Utah: 223
Bell Telephone: 80, 244, 246, 249
Bender, Ed: 96
Bendix: 21, 57, 80, 86, 156, 161, 222, 227, 228, 229, 331-332; Bendix Aviation Corporation, 21; Bendix Field Engineering Corporation, xix, xxxvii, 29, 88, 96, 118, 121, 122, 124, 126, 127, 129, 204, 214, 237, 240; Pay incentives, 57; Transition to Allied Signal and Honeywell, 289
Benson, Cliff: 229, 331
Berg, Wilfred E.: 16
Bermuda Tracking Station: xxx-xxxi, xxxviii, 64, 73, 80, 81-83, 87, 95, 98-100, 102, 109, 110, 112, 118, 119, 135-138, 144, 146, 149, 151, 162, 164, 168, 192, 220, 305-306
Berndt, Morton: 140
Big Dish (book): xiii
Bit errors and bit error rates: xlii, 118, 217, 313, 341
Blagonravov, Anatoly: 190
Blaha, John E.: 284
Blossom Point, Maryland: 15, 17, 19, 21, 23, 55, 58, 60, 203
Bobko, Karol J.: 270
Bodin, Wesley J.: 31, 40-41, 236,
Boeing: 223, 270, 298, 299; Boeing-Delco, 184
Boller and Chivens Company: 3
Bolton, John: 174, 181
Borman, Frank: 171, 172-173, 286
Botswana Ground Station: 226, 240
British Cable and Wireless Company: 88, 116, 237, 239
Brand, Vance D.: 196
Brinkley, Randy H.: 299
Brockett, Norm: 127
Brown, David M.: 334
Brown Field (*See* San Diego Station)
Brussels, Belgium: 196
Buchli, James F.: 284
Buckhorn Ground Station: 224
Buckley, Edmond C.: xi, xviii, xix, xx, xxxiv, 52, 68, 80, 89, 90, 92-93, 116, 112, 123, 127, 131, 136, 140, 209, 326, 327, 329
Budget discussions: xxxiv, 28, 45, 79, 103, 137, 152, 199, 200, 211, 218, 254, 332; Guam and Hawaii, 206-207
Buitrago, Spain (*See* Madrid)
Bureau of Land Management: 228, 317
Bureau of the Budget: 27, 152, 206
Bush, George W.: 326, 341
Bushuyev, Konstantin D.: 197

C

Cadena, Carlos H.: 229
Calibration aircraft: 14, 76-77; High altitude training, 77
California Tracking Station: 95, 99, 101, 114
Call, Dale: 138
Canary Island Tracking Station: xxxi, 88, 94-97, 100, 114, 118, 119, 138, 146, 162, 163, 205-206
Canberra Deep Space Communication Complex: xiii, xxxv, xxxvi, 34, 46, 50, 175, 212, 222, 231, 235, 288-290
Canton Island Tracking Station: xxxi, 83, 84-85, 95, 114, 121-123
Capcom: xxxi, 95, 100, 164, 171, 271; Communication Technician, 100
Cape Canaveral, Florida: xxxvi, xxxviii, 32, 64, 72, 74, 79, 80, 81, 87, 88, 89, 92, 94, 95, 98, 99, 100, 118, 119, 137, 146, 163, 182, 215, 306, 317
Carlucci, Frank C.: 127

Carnarvon, Western Australia: 39, 45, 50, 56, 87, 114, 128, 136, 138, 146, 149, 150, 162, 189
Carpenter, M. Scott: 101
Carrillo Flores, Antonio: 88
Castro, Fidel Ruz: 41
Cellular telephone: xxi, 225, 307, 327, 330
Centaur upper stage: 260, 295, 299, 306
Central Intelligence Agency: 23, 66
Centre National d'Etudes Spatiales: 238, 337
Certification Program (*See* Training programs)
Chaffee, Roger B.: 166, 167
Chandra X-ray Observatory: 287
Chawla, Kalpana: 334
Chula Vista, California (*See* San Diego, California)
Cincinnati Observatory: 12
Civil service personnel, use of: xviii, xxxvii, 26, 57, 94
Clark, George Q.: 78
Clark, Laurel: 335
Clarke, Arthur C.: 244
Clemence, Gerald M.: 22
Clements, Henry H.: 94, 331
Code Division Multiple Access: xxi
Cold War: xxxvii, 1, 105, 140, 171, 195, 198
College, Alaska (*See* Alaska Ground Station)
Collimation tower: 40, 51, 76, 118, 131, 133, 136
Collins, Michael: 171, 174, 175, 178
Collins Radio Company: 149, 209
Commonwealth Scientific and Industrial Research Organization: xv, 235, 318
Commercialization of stations: xlii, 238, 239, 246, 253, 254, 257, 312, 313, 317, 321–323
Commercial satellite (*See also* COMSAT and INTELSAT): xxxvii, 65, 244-248, 257, 265, 289; Early Bird, xxxvii, 247; Telstar 1, 65, 244, 246
Commercial space transportation: 311
Communication and Navigation Architecture Working Group: 338
Communications blackout behind the Moon: xxxiii, 172, 341
Communications Satellite Act of 1962: 246
Communications Satellite Corporation: xxxvii, 196, 246-247
Compton Gamma Ray Observatory: 272, 285, 286; GRO Remote Terminal System (GRTS), 289-292
Computation and Data Flow Integrated Subsystem: 77
Congress, United States: xxiii, xxxiv, 25, 30, 67, 106, 130, 153, 154, 156, 198, 203, 227, 254-256, 258, 275, 282, 324, 332, 333; Debate over Johannesburg, 208-213; Establishment of NASA, 27-28; House Committee on Foreign Affairs, 210; House Congressional Black Caucus, 210; House Space Committee, 210; House Subcommittee on Aeronautics and Space Technology, 210
Conrad, Charles, Jr.: 110, 187
Consolidated Vultee Aircraft Corporation: 9
Consolidated Space Operations Contract: xxiv, 323, 326
Consultative Committee for Space Data Systems: xxii
Cooby Creek, Australia: 50, 56, 203
Cooper, L. Gordon, Jr.: 98, 102-103
Cooper's Island (*See* Bermuda)
Corliss, William R.: x, xii
Coronal holes: 66
Corpus Christi, Texas (*See* Texas Tracking Station)
Cosmic Background Explorer: 310
COSPAS-SARSAT satellite project: 236-237
Cost reduction: xxii, xxiii, xxv, xlii, 300, 313, 323; Rising cost of station operations, 226-227
Costrell, James A.: 326
Council for Scientific and Industrial Research: 208, 210-213
Courier, air and surface: 22, 36

Covington, Ozro M.: xi, xix, xl, 8, 78-80, 101, 178, 185, 201, 327, 331
Crabill, Donald: 152
Crew Exploration Vehicle: 340
Crippen, Robert L.: 224
Crough, Ed: 96
Crowley, J. W.: 70
Crustal Dynamics Project (*See* Tectonics)
Cubic Corporation: 75
Cunningham, R. Walter: 170
Cushman, Ralph E.: 90, 205

D

Dakar, Senegal: 56, 225, 226, 231, 240
Darwin, Australia: 50, 56, 203
Data Acquisition Facility: 43, 47, 51, 53, 64, 85, 135; Power requirements, 47-49; Staffing and operating cost, 49
Data rate requirements: xxxviii, 36, 201, 265, 298, 300, 301, 335, 338
Deep Space Network (DSN): ix, xii-xvi, xix, xxi, xxxv, xli, 33, 45, 50, 52, 59, 80, 109, 117, 123-124, 128, 144, 146, 147-149, 163, 208, 212, 221, 231, 232, 235, 241, 250-251, 288, 308, 322, 338; As Apollo wing-station, 147-149, 232-233; Ground Communications Facility: xiv; Subnets, xiv; World Net: 208
Delta launch vehicle: 202, 287, 306
Department of Commerce: 121
Department of Defense (*See also* Air Force, Army and Navy, United States): xxxiii, 2, 12, 27, 94, 115, 153; Advanced Research Projects Agency, 67; Committee on Special Capabilities, 9
Department of Interior: 84, 111, 123, 281
Department of State: xxxiii, xxxiv, 20, 42, 88, 90, 94, 97, 123, 125, 127, 152, 205, 209, 214, 230, 236
Deutsches Zentrum für Luftund Raumfahrt: 337
Diggs, Charles C.: 210
Digital telecommunications: xlii, 35, 50, 108, 109, 112, 299, 313, 329, 342
Dinn, Mike: 179-180, 234
Donegan, James J.: xxxviii, 71, 77
Douglas Aircraft Company: 30, 76, 161
Dowling, Jack: 138
Down conversion: 53
Dynamic Explorer Satellite: 228
Dryden Flight Research Center: 207, 223, 224
Dryden, Hugh L.: 26, 90, 106, 131, 205
Duncome, R. L.: 22
Dunseith, Lynwood: xxxviii, 82, 110, 182, 186

E

Early-ops tracking: 19, 40, 112, 139
Earth Observatory Data & Images archive: 119
Earth rotation, effect on trajectory: 2
Earth Resource Technology Satellite: 201, 202, 314
Easter, Bill: 138
Eastern Test Range: 2, 98, 113, 115, 117, 120, 121, 149, 204, 238
East Grand Forks, Minnesota: 42, 55, 203
Easton, Roger L.: 12, 15
Eccentric Geophysical Observatories: 54
Echo satellite: 17, 244, 245, 246, 263
Ecuadorian Services Company: 228-229
Eglin Test Range: 70, 95, 99, 114

Eisenhower, Dwight D.: 9, 24, 25, 27
Eisenhower, Milton S.: 88
Eisele, Donn F.: 170, 171
Electromagnetic waves: 6, 9, 50, 286
Eleuthera, Bahamas: 120
Elliptical orbit: xiv, 35, 43, 54, 111, 166, 169, 200, 232, 270
Engle, Joe H.: 230
Enhanced Launch Vehicle Imaging System: 335
Ervin, Samuel J., Jr.: 52
Esceula Politecnica Nacional: 229
European Broadcast Union: 196
European Earth Resource Satellite: 204, 314
European Space Agency: 237, 239, 324, 333-335, 336, 337
Explorer spacecraft: 24, 29, 33, 65, 203, 244; Explorer 1, xiii, 24; Explorer 6, 65; Explorer 35, 39; Explorer 55, 65
Extravehicular activity: 141, 149, 175, 176, 234, 267

F

Fairbanks, Alaska (*See* Alaska Tracking Station)
Fairchild Corporation: 193, 257
Fariss, George: 138
"Faster, Better, Cheaper": xlii, 321
Federal Aviation Administration: 121, 122, 123, 244, 317; Office of the Associate Administrator for Commercial Space Transportation, 311
Federal Communications Commission: xxii, 244, 260
Fiber optics: 289, 308, 320, 335
Field-of-view: 3, 52, 53, 172, 243, 249, 260, 264
Figure-of-merit: 275, 298
Fish and Wildlife Service, United States: 317
Flaherty, Roger: 252
Fletcher, James C.: 211, 236, 254-255
Flight controller: xxxvi, xxxviii, 61, 68, 99, 102, 109, 110, 164, 172, 176, 182, 184, 271, 278
Ford Aeronutronics: 71
Force, Charles T.: xlii, 131, 134, 138, 217, 218, 229, 230, 236, 275-277, 286, 289, 298, 325, 331
Force, Marilyn: 134
Ford, Gerald R.: 197
Fort Myers: 42, 55, 203, 212, 306
Fort Stewart: 19, 21, 42, 60
Fraleigh, William: 97
Frequency allocation: xxii, 337
Fresnedillas, Spain (*See* Madrid Tracking Station)
Frutkin, Arnold W.: 93
FRW-2 transmitter: 75, 112

G

Gagarin, Yuri A.: 66, 67
Galapagos Islands: 194
Galileo space probes: iv
Garvey, Joe: 138
Gemini, Project: 80, 88, 107, 108-113, 116, 118, 120, 121, 122, 128, 131, 135-136, 140-141, 144, 145, 164, 169, 172, 213; Agena docking target, 108, 112, 140; Gemini 3, 110, 128-129; Gemini 4, 74, 107, 111, 113; Gemini 5, 111, 139; Gemini 6, 141; Gemini 7, 141; Gemini 8, 141; Gemini 11, 141; Gemini 12, 87, 122, 147, 161; Titan launch vehicle, 139, 145
General Dynamics: 154-156

General Electric: 88, 202, 253
Geostationary orbit (*See* Geosynchronous orbit)
Geosynchronous orbit: 37, 52, 162, 194, 243, 244, 246, 248, 250, 253, 261, 265, 270-271, 277-279, 282, 298, 300, 336, 338
Giant Leap tour: 206
Gitlin, Thomas A.: 290
Gilmore Creek (*See* Alaska Tracking Station)
Glenn, John H., Jr.: xxix-xxxi, 85, 100, 101
Glenn Research Center: xxii, 28, 315
Glennan, T. Keith: 31, 36, 84, 85, 88, 91
Global Positioning System: 53
Go/No-Go decision: xxxviii, 73, 82, 100, 115, 129, 135, 149, 151, 164, 305
Goddard Laser Tracking Network: 223, 235
Goddard Range And Range Rate system: 37-39, 42-43, 51, 55-56, 144, 165, 219-220
Goddard, Robert H.: 31-32
Goddard Space Flight Center: iv, x, xiv, xix, xxxvii-xxxviii, xxxix, xl-xli, 30-33, 35-36, 38, 39, 47, 48, 53, 57, 58, 61-63, 64, 65, 68, 74, 77-79, 91, 92, 101, 103, 108, 110, 111, 114, 135, 145, 164-165, 178, 180-182, 185, 187, 192, 194-195, 198, 200, 202, 223, 224, 229, 230, 237, 253-255, 271, 289, 296-297, 311, 316, 322, 323, 324, 326, 331; Advanced Development Division, 35; Communications Division, 59, 330; Competition between Code 500 and Code 800, 217-218; Explorations, Operations, Communications and Navigation Systems Division, 300; Foreign policy work, 215, 226, 323; Image Processing Facility, 217; Manned Flight Operations Division, 80, 138, 201; Manned Flight Support Directorate, 80, 215; Manned Space Flight Support Division, 74; Mission and Data Operations Directorate, 215, 236; Mission Operations and Data Systems Division, 236, 287; Mission Operations Control Center, 215; Multisatellite Operations Control Center, 215; Network Office for International Operations, 215; Network Operations Control Center, xxxix, 201, 215-217, 301; Networks Directorate, 215; Project Operation Control Centers, 201, 215-217, 311; Space Communications Branch, 32; Space Network Project Office, 277; Space Projects Center, 30, 68; Space Sciences Division, 31, 66; STADAN Engineering Division, 40; Suborbital Projects and Operations Directorate, 310, 312; Tracking and Data Systems Directorate, 10, 32, 59, 144, 215
Goetchius, Rod: 80
Goett, Harry J.: 52, 68, 74, 77, 78, 79, 117, 125
Goldin, Daniel S.: 322, 325
Goldstone Deep Space Communication Complex: xiii, xiv, 40, 138, 147, 162-163, 172, 174-176, 178, 181, 184, 203, 208, 212, 219, 221-222, 227, 231, 232, 270
Gomez, Louis: 281
Goodman, Charles J.: 196
Grand Canary Island (*See* Canary Island Tracking Station)
Grand Bahama Tracking Station: 21, 40, 112, 114, 118, 119-120, 136-139, 149, 163, 224, 308
Grand Turk Tracking Station: 18, 21, 40, 95, 100, 112, 114, 136-138
Graves, G. Barry, Jr.: 71, 74, 76, 80, 90
Great Observatories spacecraft: 281, 286-287
Grey, Don: 138
Griffin, L. F.: 126, 127
Grissom, Virgil I.: 166, 167
Ground Network (*See also* individual station locations): iv, xli-xlii, 301, 302, 303, 311, 315, 320-321; DataLynx, 317-318, 321; Universal Space Network, xlii, 318, 321, 323
Ground track, spacecraft: xli, 19, 34, 42, 72, 84, 92, 113, 118, 129, 172
Grumman Aircraft Engineering Corporation: 167
Guam Tracking Station: 64, 129-135, 138, 147, 151, 162, 163, 166, 192, 206-208, 219, 229, 235-236, 240, 321
Guaymas Tracking Station: xxxi, xxxiv, 88-92, 95, 99, 101, 114, 138, 139-141, 146, 147, 162, 165, 204-205
Guerrero, Jose A. Leon: 134
Guerrero, Manuel F. Leon: 130, 131, 237, 292

H

H-II Transfer Vehicle: 335
Habib, Edmund J.: 14, 37
Hagen, John P.: 22
Hainworth, H. C.: 84
Haise, Fred W., Jr.: 181, 182
Haney, Paul P.: 170
Harris Corporation: 256, 257, 280
Harris, David W.: 17, 77, 239, 325
Harris, Matt: 96
Hauck, Frederick H.: 282
Hawaii Ground Station: 84-87, 95, 99, 100, 101, 110, 113, 114, 129, 138, 146, 149, 151, 162, 163, 166, 171, 192, 206-208, 219, 240; Bonham Air Force Base, 85; Honolulu switching center, 64; Kokee Park Geophysical Observatory, 85, 219, 303-304; University of Hawaii, 240, 303
Headquarters, NASA: xviii-xx, xxiii-xxv, xxxiv, xxxix, 17, 20, 59, 70, 74, 77, 80, 82, 84, 90, 97, 111, 122, 125, 134, 140, 155, 164, 165, 170, 189, 194, 205, 209, 210, 213, 215, 227, 236, 239, 255, 298, 323, 324-325, 332, 336, 338, 340; Apollo Spacecraft Program Office, 170; Code O and Code T, xx, 323-325; Office of Manned Space Flight, 123, 128, 158, 203; Office of Procurement, 90; Office of Program Planning and Evaluation, 36; Office of Space Communications, xi, xx, xxi-xxiv, 82, 284, 322, 323-325; Office of Space Flight Programs, 77; Office of Space Operations, xx, 258, 325; Office of Space Tracking and Data Systems, 281, 293, 323; Office of Tracking and Data Acquisition, xix, xx, xxii, xxiii, xxxiv, 52, 80, 111, 120, 121, 122, 124, 128, 129, 131, 151, 152, 153, 156, 188, 199, 253, 254, 325
Healey, Fred: 138
Heller, Niles R.: 74
Herget, Paul: 12, 13, 18, 21, 22
Hewitt, Frank: 213
High apogee orbit: xiv, 35, 37, 43, 54, 111, 145, 166, 200
High eccentricity orbit (*See* High apogee orbit)
High Energy Astronomy Observer: 201
High frequency: 12, 60, 61, 155, 159, 238
High inclination orbit (*See* Polar orbit)
Hinners, Noel W.: 198, 237
Hoff, Hal: 77
Honeysuckle Creek, Australia: xxxv-xxxvi, 129, 138, 148, 162, 172, 175-179, 181, 220, 234, 235
Honeywell: xix, 318
Hooker, Ray W.: 71, 93
Hubble Space Telescope: 216, 286, 290, 293, 298, 340
Hughes Aircraft Company: 52, 245, 248, 254, 255, 294, 298
Hunter, Dan: 138
Hurd, Cuthbert C.: 21
Husband, Rick D.: 334
Hynek, J. Allen: 3

I

Ikonos satellite: 119
Indian Ocean Station: 226
Inertial Upper Stage: 260, 266, 270, 275, 276
Infrared Space Observatory: 39
Instantaneous contact requirement: 70
Institute of Aerospace Technology: xv, 233
Integrated Near-Earth Network: 338–340
Interagency Operations Advisory Group: 337
Inter-American Geodetic Survey: 60

Intercontinental Ballistic Missile: xxxviii, 2, 24, 66, 86, 87
Interference, radio frequency: 81, 102, 165, 181, 184, 188, 260, 261; Rocket plume attenuation, 118, 138, 306-307; Solar activity, 94, 101
International Brotherhood of Electrical Workers (*See* Labor dispute)
International Business Machines: 21-22, 32, 74, 77, 81; IBM 1218 computer, 58; IBM 709 computer, 99
International Council of Scientific Unions: 3
International Geophysical Year: 3, 9, 10, 15, 24, 44, 93, 208
International Organization for Standardization: xxii
International Satellite for Ionospheric Studies: 50
International Space Station: xxxvi, xlii, 184, 241, 251, 252, 272, 278, 282, 304, 310, 331, 333, 334, 340
International Telecommunications Satellite Consortium: 64, 247-248, 289, 315; Apollo support, 162-163, 194; Early Bird, xxxvii, 247
International Telecommunication Union: 34
International Ultraviolet Explorer: 55, 201, 216, 228, 310
Internet: 272, 316, 330, 341; Access to South Pole, xlii, 272, 315
Interplanetary Monitoring Platform: 34, 35, 304, 309
Invisible Network: xl, xlii, 102-103, 330, 331
Ippolito, Tony: 308
Ionosphere: 2, 22, 24, 43, 314
Irwin, James B.: 183
Island Lagoon, South Australia: 45, 47, 234
ISS Downlink Enhancement Architecture: 335
Iuliano, Henry: 189, 250

J

Jackson, Chuck: 138
James Webb Space Telescope: 340
Japanese Institute for Aerospace Technology: 318, 337
Jarvis, Gregory B.: 272
Jet Propulsion Laboratory (*See also* Deep Space Network): xiii-xv, xviii, 24, 28, 29, 33, 117, 165, 209, 228, 232, 235, 308, 322–323, 324, 333
Jochen, Larry: 331
Johannesburg Tracking Station: 19, 21, 41, 42, 56, 114, 124, 128, 149, 152, 208-213, 222, 226, 318, 336, 340
Johnson, Lyndon B.: 28, 106
Johnson Space Center: xi, xix, xxiv, 74, 79, 110, 111, 113, 136, 138, 159, 164, 170, 173, 224, 252, 258, 278, 286, 306, 307, 322, 324, 331, 335, 342; Data Systems and Analysis Directorate, 82, 182; Formation of, 33, 68, 77; Relationship with GSFC, xix, xxxiv, xxxvii, xl, 79-80, 108-110, 306
Jonathan Dickinson Missile Tracking Annex: 308, 335
Jones, Jesse C.: 281
Jordan, B. Everett: 52
Jupiter C launch vehicle: 24

K

Kano, Nigeria: xxxi, 30, 50, 86-88, 93, 95, 97, 114, 124, 125, 127
Karume, Abeid Amani: 127
Kauai, Hawaii (*See* Hawaii Tracking Station)
Kelly, Richard: 96
Kennedy, Edward M.: 211
Kennedy, Gregory P.: 282
Kennedy, John F.: 78, 106, 108, 245
Kennedy Space Center: 64, 114, 123, 135, 136, 137, 143, 151, 166, 194, 196, 204, 224, 225, 232, 239, 267, 272, 273, 275, 306, 308, 309, 326, 340

Kent, Marion: 207
Kentron: 86
Kerrigan, E. J.: 89
Kerwin, Joseph P.: 187, 188
Khrushchev, Nikita S.: 105
Killian, James R., Jr.: 25, 26, 27
Kodak Corporation: 3
Kokee Park, Kauai (*See* Hawaii Tracking Station)
Kosygin, Alexey N.: 190
Kourou, French Guiana: 239
Kraft, Christopher C., Jr.: xix, xxxvi, xxxviii, xl, 79, 174, 179, 182, 197, 205, 224
Kranz, Eugene F.: 110, 180
Kronmiller, George: 37-38
Krugman, Clay: 96
Kubasov, Valeri N.: 196, 197
Kwajalein, Marshall Islands: 2, 87, 98

L

Labor dispute: 227
LaFleur, Walt: 138, 230
LAGEOS satellite: 222
Lagrange point: 340
Lake Victoria, Africa: 194
LANDSAT: 50, 202, 212, 226, 228
Langley Research Center: 32, 63, 67, 68, 70, 71, 74, 78, 85, 89, 90, 91, 93, 218, 245, 310; Langley Aeronautical Laboratory, 28
Las Cruces, New Mexico: 278, 281, 283, 284, 285, 286
Laser tracking: 222-223, 235
Launch azimuth: 72, 119, 225; Apollo missions, 135-137, 151, 153, 158, 203
Lawless, Ed: 189
Leddy, Raymond: 91
Lee, Roger: 96
Lee, William A.: 170-171
Leonov, Aleksei A.: 196, 197
Leslie, Robert A.: 45
Lewis Flight Propulsion Laboratory (*See* Glenn Research Center)
Lights-out operations: 259, 308, 318, 320
Lima, Peru: 16, 19, 23, 40, 56, 149, 163, 203
Lindsay, Hamish: xxxvi, 235
Lindsay, John C.: 66
Line-of-sight limitation: xli, 277, 288, 305
Lissajou orbit (*See* Lagrange point)
Lockheed Missiles and Space Company: 66, 76, 253, 299, 323, 326
London switching center: 61, 64
Looney, Chesley H., Jr.: 12, 16, 23, 35
Lopez Mateos, Adolfo: 88, 89
Loss-of-signal: 72, 84, 124, 129, 172, 307, 309
Lovell, James A., Jr.: 171, 181, 182
Low, George M.: 106-107, 170, 171, 204-205
Lowe, Bryan: 138
Lunar Orbiter space probe: xv
Lunar Reconnaissance Orbiter: 338
Lunar Rover, Apollo: xii, 183-185; Solution for communications time lag, 184; Tracking, 185
Lundy, Wilson T.: 286, 325
Lunney, Glynn S.: 197
Lutz, Russell: 96

M

Madagascar: 30, 39, 41, 43, 53, 54, 56, 59, 114, 124, 128, 129, 146, 149, 152, 163, 213, 214, 215
Madrid Deep Space Communication Complex: xiii, 61, 148, 163, 212, 231, 233, 288, 336, 338
Magnetic tapes, 14-track: 22, 48, 53, 60, 216
Manned Lunar Landing Task Group: 106
Manned Spacecraft Center (*See* Johnson Space Center)
Manned Spaceflight Instrumentation and Communications Panel: 136, 138
Mariner space probe: ix, xv
Mars, exploration of: ix, 212, 336, 338, 339, 340; Mars Reconnaissance Orbiter, xv; Spirit and Opportunity rovers, xvi
Marshall Islands (*See* Kwajalein, Marshall Islands)
Marshall Space Flight Center: xiv, 166, 187, 306, 322, 324, 326, 335, 337
Masers: 124
Massachusetts Institute of Technology: 71, 156
Mathews, Charles W.: 70
Matthes, Chester: 42
Mayer, John: xxxviii
Mayo, Robert P.: 206
McAuliffe, S. Christa: 272
McCool, William C.: 334
McCormack, John W.: 28
McDivitt, James A.: 107, 113
McElroy, John H.: 230
McKeehan, B. Harry: 20, 21, 54, 226
McMurdo Ground Station: 272, 313-316; McMurdo TDRSS Relay System, 314; Working with NSF, 313
McNair, Ronald E.: 272
Mendez Docurro, Eugenio: 90, 205
Mengel, John T.: xxxiii, 10-13, 15, 18, 19, 36, 59, 74, 201, 218, 327; Space Communications Branch, 32
Mercury, Project: xxix-xxxi, xxxiv, xxxviii, 54, 61, 62, 63, 64, 67, 68, 69, 70, 72, 74, 75, 79, 80, 81, 84, 88, 92, 94, 102, 103, 105, 108, 110, 111, 113, 116, 125, 128, 139, 145, 150, 164, 172, 240, 310, 331; MA-4, 85; MA-5, 101; MA-6, 85, 92, 100, 101; MA-7, 101; MA-8, 98, 101; MA-9, 98, 99, 102, 105, 109, 121; Project 7969, 67
Mercury Space Flight Network: 65-103; 12 ground rules, 72-73; Communication gap, 73, 84; Mercury Control, 70, 71, 73, 74, 78, 83, 92, 101, 109; Site selection, 71, 73-75, 80-82; Tracking And Ground Instrumentation Unit, 71-72, 74-75, 79, 84, 90; Tracking System Study Group, 68, 70-71, 103
Mexican National Commission for Outer Space: 204
Microwave relay: xiv, 60, 94, 147, 175, 176, 181, 225, 307, 308, 311, 315, 316
Midway Island: 98, 115
MILA: 136-137, 219, 305, 306, 307-308
Minitrack Network (*See also* Naval Research Laboratory): 9-24; Antenna array, 13-14, 16, 18; Calibration, 14, 19, 22, 76; Data processing, 21-22; Initial cost, 12; Mark II and Project Moonbeam, 15; Minitrack Optical Tracking System, 14, 17; Origin of the name, 11; Picket Line, 19, 40, 42; Role of the Defense Department, 15, 21; Role of the State Department, 20, 59-60; Site selection, 16, 18-19
Mintz, Frank P.: x
Mission Control Center, Houston: xxxiii, 64, 74, 109-113, 135, 151, 155, 159, 164, 165, 169, 175, 176, 178, 182, 184, 185, 186, 188, 215, 224, 225, 234, 271, 305, 306, 307; Mission Operations Control Room, 111, 113, 165; Network Controller, 94, 165
Mobile Laser Ranging System: 223
Mobile stations (*See* Temporary tracking stations)
Mogan, Kathleen M.: x
Mongas Lopez, Ricardo: 90, 91
Montoya, Percy: 96

Morse, Gary A.: 301
Mosquera, Fabian: 229
Moss, Frank E.: 211, 212
MPS-26 radar: 54, 128
Muchea, Western Australia: xxxi, 54, 95, 99, 100, 115, 128
Mudgway, Douglas J.: xii
Multipath interference: 14
Multiplexing/De-multiplexing of signals: 224-225, 265
Musgrave, F. Story: 270

N

Nairobi, Kenya: 127
NASA Communications Network: 37, 53, 59-63, 64, 77, 131, 164, 176, 188, 189, 215, 225, 234, 241, 251; Switching centers, 61, 64, 131, 189
NASA, establishment of: 24-29; Anderson, Clinton D., 25; Atomic Energy Commission, 25, 27; Civilian charter, xxxiv, 24, 93, 244; Department of Science and Technology, 27; National Advisory Committee for Aeronautics, 25-28, 207, 310; National Aeronautics and Space Act of 1958, 28; Select Committee on Astronautics and Space Exploration, 28; Special Committee on Space and Astronautics, 28; President's Science Advisory Committee, 25, 26
NASA Integrated Support Network: xiv, 278, 323
NASA Long-Range Plan document: 36
National Academy of Sciences: 2, 3, 4, 10, 25, 27, 93
National Facilities Study: 337
National Institute of Aerospace Technology (Spain): xv, 233
National Oceanic and Atmospheric Administration: 219, 228, 317, 321, 337
National Science Foundation: 25, 27, 272, 277, 314, 316
National Space Development Agency of Japan: 237, 333, 334
National Telecommunications and Information Administration: xxii
Nationalizing of overseas stations: xxxv, 20, 229, 321
Naval Research Laboratory: 7, 9-10, 12, 15, 18, 19, 21-23, 31, 32, 55, 59, 60, 74, 93; Office of Naval Research, 21, 30
Navy, United States (*See also* Department of Defense and names of specific ranges and ships): xxxiii, 2, 6, 15, 16, 24, 31, 40, 74, 84, 98, 101, 113, 116, 122, 127, 130, 134, 150, 152, 154-157, 240, 292, 303, 304, 313; Blossom Point, 15, 17, 19, 21, 23, 55, 58, 60, 203; Bureau of Yards and Docks, 15, 121; Facilities Engineering Command, 97; Instrumentation Ships Project Office, 154, 156; Naval Academy, 55, 228; Naval Observatory, 22, 280, 304
Network centralization: xxxviii, 109-110, 141, 164, 216, 218, 330
Network Director: 165, 301
Network Operations Control Center (Pasadena): xiv
Network Operations Control Center (Greenbelt): xxxix, 201, 215-216
Network Operations Manager: 165
Network Test and Training Facility: 55, 58, 203, 219, 228, 310, 337
Newberry, Stan C.: 325
Newfoundland Tracking Station: 42, 55, 192, 203
New Mexico Institute of Technology: 281
New Mexico State University: 281, 284
Nimbus meteorological satellite: 36, 39, 43, 54, 194, 228
Nitze, Paul H.: 127
Nixon, Richard M.: 190, 197
North American Aviation: 170; North American Rockwell, 254
North Carolina (*See* Rosman Tracking Station)
North Pole: 272, 318, 319
Northern Mariana Islands: 129, 130, 290
Nunn, Joseph: 3

O

Ocean cables: xiv, xxxii, 60, 64, 88, 98, 99, 110, 113, 129, 225, 238
O'Connor, Gerry: 100
Odenthal, Larry: 138
Office of General Counsel: 90, 239
Office of Manned Space Flight (*See* Headquarters, NASA)
Office of Program Planning and Evaluation (*See* Headquarters, NASA)
Office of Space Communications (*See* Headquarters, NASA)
Office of Space Tracking and Data Systems (*See* Headquarters, NASA)
Office of Tracking and Data Acquisition (*See* Headquarters, NASA)
Onizuka, Ellison S.: 272
Operating cost: xxi, xli, xlii, 20, 152, 160, 161, 203, 211, 212, 214, 217, 226, 228, 239, 249, 260, 290, 301, 306, 311, 312, 320, 321, 322, 323, 335; Ground station, 36, 43, 49, 103, 123, 148, 152, 204, 206, 215, 227, 249; Ship, 128, 150, 152, 156, 214
Operation Moonwatch: 4-7
Operation Paperclip: 2
Optical tracking (*See also* Minitrack Optical Tracking System): 2-6, 8, 10, 326; Acquisition, 4, 6, 10, 13; Optical network: 3-4; Weather and lighting, 6
Orbit determination: 18, 22, 35, 40, 47, 53, 70, 100, 169
Orbit inclination, effect on tracking: 19, 34, 50, 67, 73, 87, 270, 272, 305, 312
Orbiting Astronomical Observatory: 50, 61, 215, 227
Orbiting Geophysical Observatory: 39, 50, 54, 56, 61
Orbiting Geophysical Project: 50
Orbiting Solar Observatory: 50, 66
Orellana, José Rubén: 229
Orroral Valley Tracking Station: 44, 47-49, 53, 56, 192, 220, 221, 222, 233-235
Owen, Robert L.: 181
Owens, Joseph S.: 282

P

Pacific Missile Range: 2, 70, 85, 87, 115, 121, 240, 303
PAGEOS satellite: 219
Paine, Thomas O.: 178, 187, 190, 206, 207
Pakistan: 50, 226
Palermo, Sicily: 146
Parabolic antennas: 4.3-meter, 224, 226, 307, 309; 9-meter, 54, 117, 120, 131, 132, 136, 139, 147, 149, 153, 155, 192, 193, 203, 228, 232, 240, 289, 292, 304, 306, 309, 312; 12-meter, 36, 40, 42, 144, 228, 230, 341; 18-meter, 226, 278, 283, 336, 338; 26-meter, xiv, xxxv, 36, 41, 43, 44, 47, 51, 52, 55-56, 144, 146, 147, 148, 149, 172, 173, 174, 175, 176, 181, 209, 221, 222, 228, 231, 232, 233, 234, 249; 34-meter, xiv, 149; 64-meter, xxi, 149, 174, 176, 177, 181; 70-meter, xiv, 149, 231
Parkes Observatory: xxxv, 174-178, 181
PEACESAT: 304
Performance matrix, MSFN and STADAN: 217
Perkin-Elmer Corporation: 3
Perth, Western Australia: 114-115, 150, 189, 318
Peterson, Donald H.: 270
Phillips, Samuel C.: xx, 137, 139, 167, 171
Phoenix Islands: 276, 289, 300
Picard, Fredrick: 127
Pickett, Eugene: 134
Pierce, John R.: 244
Pioneer space probe: iv, 147, 158, 172; Pioneer 3, xiii, xv; Pioneer 4, xiii, xv; Pioneer 6-9, xv; Pioneer 10, xv; Pioneer 11, xv
Polar orbit: 34, 41, 50, 228, 315, 319, 320

Polar Orbiting Geophysical Observatory: 54
Ponce de Leon Tracking Station: 273, 306, 307-309, 335
Portugal: 209
Pretoria, South Africa (*See* Johannesburg Tracking Station)
Puerto Rico: 98
Pulse Amplitude Modulation: 75
Pulse Code Modulation: 108, 109, 159

Q

Quann, John J.: 282
Quito, Ecuador: 16, 19, 42, 56, 192, 220, 222, 228-231; CLIRSEN agency, 231; Ecuadorian Services Company, 229; Mount Cotopaxi, 228, 230

R

Rabasa, Oscar: 91
RADARSAT: 314
Radar tracking: 2, 6, 8, 9, 15, 70, 71, 73-75, 87, 95, 100, 101-102, 108, 135, 144, 153, 162-163, 219-220, 261; Beacon track, 74, 94; FPQ-6, 58, 111-112, 128, 135; FPS-16, 11, 74, 98, 112; MPS-26, 54, 128; Skin track, 74, 111; Very Long Range Tracking, 74-75, 112
Radio Corporation of America: 8, 43, 57, 71, 72, 74, 80, 135, 173, 244, 253, 255, 278
Radio interferometry: 9, 11, 33, 39; Operating theory, 13; Azusa missile tracking, 9, 10; Tracking Viking guided missile, 10-11
Ramon, Ilan: 335
Rangel, Charles B.: 210, 211
Ranger space probes: xv
Ratsiraka, Didier: 213, 214
Raytheon: 289
Reagan administration: 238, 274
Reeves Instrument Corporation: 74, 80
Relay communication satellite: 244
Remote sensing: 201, 202, 217, 281, 300
Rendezvous and docking, tracking requirements: 108, 112, 129, 141, 144, 178, 190, 195
Resnik, Judith A.: 272
Roberts, Tecwyn: xl, 138
Robledo, Spain (*See also* Madrid Deep Space Communication Complex): xiii, 148, 221, 233
Rodd Naval Auxiliary Air Station: 115
Rogers Commission (*See* Space Shuttle Program, *Challenger*)
Rosen, Milton W.: 10
Rosenthal, Alfred: x, xx, 33
Rosman Tracking Station: 39, 42, 43, 50, 51-53, 55, 64, 192, 194, 219, 221, 227-228, 340
Rouillier, Charles: 96, 138
Roy, Melba: 38
Rusk, D. Dean: 127

S

S-band (*See also* Unified S-band): 36-39, 51, 52, 64, 74, 95, 123, 145, 165, 185, 201, 262, 264, 265, 267, 269, 271, 291, 292, 293, 298, 304, 306, 312, 315, 318, 340; S-band ranging, 35; Saturn V plume attenuation, 118
Sade, Richard S.: 230
Samet, Arthur: 207
San Diego Tracking Station: 18, 19, 21, 23, 40, 55, 60, 203
Sanford, J. Terry: 52
San Nicholas Island: 98, 102

Santiago, Chile: 16, 19, 23, 40, 42, 56, 192, 220, 221, 222, 236, 304-305, 318, 321
Sardinia, Italy: 209
Satellite Automatic Tracking Antenna: 39, 42, 43, 44, 51, 53, 55-56, 131, 219-220, 311
Satellite Instructional Television Experiment: 194
Saxon, John: 234, 427
Switching, Conferencing And Monitoring Arrangement: 113
Satellite Command Antenna on Medium Pedestal: 131, 311
Scheer, Julian W.: 170
Schmidt camera (*See* Baker-Nunn camera)
Schneider, William C.: xxxii-xxxiii, 112, 325
Schirra, Walter M.: 69, 102, 170, 171
Schultz, Hank: 138
Schulz, Gary: 331
Scobee, Francis R.: 272
Scott, David R.: 183
Seamans, Robert C., Jr.: 106, 117, 125
Seasat program: 281
Seaton, Fredrick A.: 85
Seychelles (*See* Indian Ocean Station)
Shepard, Alan B., Jr.: 33, 82
Ships, range and tracking: xxxii, xxxiii, xxxvi, 40, 64, 98, 101, 113-116, 123, 128, 144-147, 150-158, 159, 162-163, 173, 193; Apollo Instrumentation Ships, 150-151, 156, 203; *American Mariner*, 98; *Coastal Sentry Quebec*, 95, 98, 113, 114, 128, 146, 150; Cost versus land station, 152-153, 214-215; *Huntsville*, 98, 151, 153, 156, 157, 162, 163, 166, 173, 178; Insertion tracking, 151, 152, 154; *Mercury*, 151, 158, 162; *Range Tracker*, 98, 113, 114; *Redstone*, 151, 157, 158; Reentry tracking, 123, 151, 153, 157, 173; Retrofit, 153-155; *Rose Knot Victor*, 95, 98, 113, 114, 150; *Twin Falls Victory*, 98; *Vanguard*, 151, 154, 155, 158, 163, 188; *Watertown*, 98, 150, 153, 156, 157, 162
Short arc solution: xxxviii
Sicily, Italy: 146, 209
Side-tone ranging: 37
Siepert, Albert F.: 117
Silverstein, Abe: 70, 77, 80
Simpson, George L., Jr.: 52
Simulations: xiv, 19, 59, 73, 76-77, 100, 111, 179, 180, 234; Integrated sim, 77, 278
Siry, Joseph W.: 22
Site survey: 7, 16, 50, 74, 81, 85, 120, 129, 130, 131, 209, 288
Skylab, Project: xl, 135, 161, 186-189, 306; Reentry tracking, 188-189
Slayton, Donald K.: 110, 196
Smith, Albert E.: 87
Smith, Glenn: 96, 331
Smith, Michael J.: 272
Smithsonian Astrophysical Observatory: 2-6, 208, 234
Smylie, Robert E.: 230, 281, 325
Solar Dynamics Observatory: 338
Solar Maximum Mission: 236
Solrad 1: 65
Soule, Hartley: 68
Sounding rocket: 1, 2, 32, 66, 140, 204, 310, 312, 317, 332
South Africa (*See* Johannesburg)
South Atlantic Anomaly: 287
Southern Rhodesia (*See* Zimbabwe)
South Pole (*See* McMurdo Ground Station)
Soviet Union: xxi, xxxvii, 1, 2, 22, 23, 24, 36, 66-67, 105, 141, 215, 236; ASTP, 190-198
Space Electronics Company: 71
Space Network (*See* Tracking and Data Relay Satellite System)
Space Networks Interoperability Panel: 336-337
Space Operations Management Office: xxiv, 322-326; 1995 Zero Base Review, 322; Affect on Code

O and GSFC, 323; Consolidation of management structure, 323, 325; Consolidated Space Operations Contract, xxiv, 322, 325; Shift in program responsibilities from Headquarters to JSC, xxiv, 322

Space Race: xxxvii, 28, 171, 198, 321, 340

Space Shuttle: xii, xx, 64, 198, 200, 225, 226, 231, 232, 240, 250, 254-255, 257, 266, 271, 278, 286, 287, 298, 299, 304-305, 331; Approach and Landing Test support, 223-224; *Challenger* accident, 237, 271-273, 276, 282, 299, 292, 303; *Columbia* accident, 332; De-orbit and reentry coverage, 226; OMS-1 burn tracking, 225; OMS-2 burn, 226; Plume attenuation, 138, 306; SRB, 306-309; Return-to-Launch-Site abort tracking, 307; RTHU maneuver affect on data coverage, 305; Schedule conflict with Ariane 9, 238; Spacehab, 281; Spacelab, 281; STS-1, 224, 226, 308; STS-2, 228, 235; STS-3, 232; STS-4, 275; STS-6, 267; STS-8, 224, 282; STS-26, 275; STS-29, 276, 277, 284; STS-31, 286; STS-37, 285, 287; STS-43, 276, 277; STS-54, 266, 276, 277; STS-61, 286; STS-70, 276, 277; STS-87, 307; STS-93, 287; STS-114, 335; TDRS launch mode, 260-261, 299; Text And Graphics System, 271

Space Station *Freedom*: 282

Space Task Group: 32-33, 68, 70, 72, 77, 310

Space Telescope Science Institute: 216

Space Transportation System (*See* Space Shuttle)

Spearing, Robert E.: 82, 236, 257, 259, 260, 324, 325, 342

Speed, transmission: xxxviii, 60, 61, 64, 224, 330

Spencer, Tom: 125

Spin stabilization: 66, 203, 247

Spintman, Daniel A.: 230

Spitzer Space Telescope: 287

Sputnik: xxxvii, 23-24, 25, 28, 66, 244

Stafford, Thomas P.: 196

Stapp, John P.: 282

Station Director: 45, 57, 86, 123, 129, 138, 140, 236

Station equipment, disposition of: 40, 42, 47, 54, 55-56, 123, 224, 235, 236, 240, 289, 306; Johannesburg, 212; Tananarive, 127, 158, 204, 214

Station performance, measure of: 217

Station workload: 87, 222

St. Incia: 120

St. John's, Canada (*See* Newfoundland Tracking Station)

Stelter, Lavern R.: 59, 60, 61, 331

Stevens, Mike: 230

Stocklin, Frank J.: 290

Stockwell, E. J.: 124

Stromberg-Carlson: 80

Suarez Dias, Jorge: 90

Submarine cables (*See* Ocean cables)

Subpanel on Launch Area Instrumentation (*See* USB Implementation Subpanel)

Surveyor lunar lander: xv, xxxviii

Svalbard Ground Station: 318-321; EUMETSAT, 320; Konnesberg Satellite Services, 318; Norwegian Space Center, 319, 320

Swigert, John L., Jr.: 181

Syncom satellite: 52, 245, 247

T

Tananarive Tracking Station (*See* Madagascar Tracking Station)

Taylor, Roy A.: 52, 227

Teague, Olin E.: 210, 333

Technology Readiness Level: 254

Tectonics: 222-223, 240, 304

Telemedicine: 272, 327

Telemetry Online Processing System: 216
Telephone, network usage: 60, 100, 113
Teletype: xxxvi, xxxviii, 22, 47, 48, 53, 58, 60, 64, 83, 100, 113, 159, 271
Television: 35, 62, 64, 109, 164, 165, 170-173, 176-178, 184, 191, 194, 196, 205, 341; Color, 174; Digital video, 112, 298; First live broadcast to the Rocky Mountains and the Appalachians, 195; High Definition, 217, 335; Scanner Converter Reversing Switch, 178; Signal protocol, 196; Slow scan, 173, 176, 177;
Television Infrared Observatory Satellite: 39, 65
Telstar: 65, 244, 246
TELTRAC: 112, 131
Temporary tracking stations: 41, 50, 55-56, 98, 203
Ten-minute "dead time" rule: 72, 109
Tension between the networks: xxxiii, 123, 124, 217-218
Tereshkova, Valentina V.: 105, 141
Terlaje, E. S.: 166
Test and Training Satellite: 147, 172
Texas Tracking Station: 95,101, 114, 162, 204
Thiele, Otto: 138
Thompson, Floyd L.: 78
Thompson, Henry: 78
Tidbinbilla, Australia (*See* Canberra and Orroral Valley)
Timing requirements: xxxviii, 12, 14, 53, 173, 279; WWV, 53
Tindall, Bill: xxxviii
Tinian: 129
Torres, Julio: 229
Tousey, Richard: 10
Tourist attraction, ground station as: 50, 87, 150
Town Hill (*See* Bermuda Tracking Station)
Townsend, John: 32, 284
Tracking and Data Relay Satellite, First Generation: xxxix, 251, 261-270; Attitude Control System, 261, 263; General specifications, 262; Launch and deployment of, 268-270; Multiple Access antenna system, 254, 262, 265, 298, 301; Omni Antenna, 262, 266-267, 269, 270; Power system, 261, 264; Reaction Control System (See Attitude Control System); Single Access antenna system, 254, 262, 264-265, 266, 270, 271, 298, 300; Space-to-Ground Link Antenna, 262, 265-266, 267, 269-270, 291; TDRS-1, 228, 261, 267, 270-272, 274, 276, 277, 282, 287, 288, 289, 290, 296, 316; TDRS-2, 272-273, 276, 296; TDRS-3, 274-276, 297; TDRS-4, 276, 277, 284, 297, 304; TDRS-5, 276, 277, 297; TDRS-6, 276, 277, 297; TDRS-7, 276, 277, 290, 293, 297; TDRS-East, 275, 277, 281, 305; TDRS-West, 275, 277, 281; Thermal Control System, 263-264
Tracking and Data Relay Satellite, Second Generation: xxxix, 292-301; Critical Design Review, 299; Data capacity, 298, 300; Definition studies, 293, 296-297; General specifications, 291; Impact of *Challenger* accident on launch mode, 274, 295, 299; Initial MA low performance problem, 299; Ka-band efficiencies, 252, 293-294, 298, 300, 337; Launch using Atlas II-A/Centaur, 295, 299; Phase A Preliminary Analysis, 293, 296-297; Phase B Definition Study, 293, 296-297; Power supply, 294-295, 298; Pre Phase A Advanced Study, 293, 296; Propulsion system, 295; Storage locations on-orbit, 275-276, 300; TDRS-H, 297, 299; TDRS-I, 297, 299; TDRS-J, 297, 300
Tracking and Data Relay Satellite System: xxi, xli, 193, 194, 237, 241, 249-301, 305, 322, 323, 334, 340-342; 100% viewing, 250, 272, 275, 292; Availability of the system, 275-277; "Bent-pipe" repeater, 250-252; C-band service, 292, 293; Columbia Communications Corporation, 292; Contel, 257, 271; DRSS Requirements and Interface Panel, 253; Electromagnetic Compatibility Analysis, 260; Fairchild, 193, 257; Forward link, 271, 278; Guam Remote Ground Terminal, 290, 292, 296-297; Ka-band, 252, 293, 294, 298, 300, 338; Ku-band, 257, 262, 264, 265, 278, 280, 292, 293, 335; Launch mode, 261, 272, 295, 299; Leased service approach, 254, 258-259, 292; Lessons learned, 258-259; Loading analysis, 253-254; McMurdo TDRSS Relay System, 315-316; Multiple Access service, 265, 280, 300 301; Naming contest (Cacique and Danzante), 284-286; Network Control Center, 278-279; Operations Control Center, 280; Origins of, 243-249; Public Law 95-76, 256; Remote sensing support, 281, 300; Request for Proposals, 255, 298; Return link, 278, 292; S-band, 262, 264, 265, 267, 269, 291, 292, 293; S-band Tracking, Telemetry

and Command System, 280; Second TDRSS Ground Terminal, 282-286, 289, 290, 296-297; Single Access service, 265, 280; Shared system, 256-257, 259, 293; Spacecom (*See* Western Union Space Communications); Space Network, xli, 250, 252, 272-278, 289, 300-301, 308, 312, 322, 335; Space-to-Ground Link Terminal, 280; Tightly coupled system, 259; Timing accuracy, 280; U.S. territory-based solution, 290; Western Union Space Communications, 255-259, 278, 292, 293, 299; White Sands Complex (*See* White Sands Ground Terminal); White Sands Complex Alternative Resource Terminal, 277; White Sands Ground Terminal, 252, 259, 277-281, 282, 287, 296-297; Zone of Exclusion, 272, 288, 292

Training programs: 19, 20, 58-59, 210, 228, 229, 321
Transportable Laser Ranging System: 223
Traveling Wave Tube: 247, 249, 271
True, Virgil: 85, 138
Truly, Richard H.: 230, 284
Truszynski, Gerald M.: xi, xviii, xx, 152, 199, 204, 205, 207, 213, 255, 325, 332-333
TRW: 256, 257-258, 260, 271, 273, 276
Tsiranana, Philibert: 54
Tula Peak, New Mexico: 226, 231-232

U

Ultra high frequency: 38, 82, 112, 113, 115, 162-163, 224, 226, 306, 308, 330
Unattended operations (*See* Lights-out operations)
Unified S-band (*See also* S-band): 37, 108-109, 117, 118, 136-137, 139, 144, 146, 147, 159, 162-163, 165, 173, 184, 200, 221, 224, 232, 233, 234, 306, 340; Implementation Sub-panel, 136-139; CSM autotracking antenna, 169; Lunar Module directional antenna limitations, 168, 174
Uninterruptible power supply: 48, 311
United Nations: 54, 128
UNIVAC: 61; 1218 computer, 110
Universal Space Network: xlii, 318, 321, 323
University of Chile: 40, 56, 236, 305, 318
University of New South Wales: 235
University of Tasmania: 56, 220, 234
Uplink Downlink (book): xii

V

V-2 rocket: 2, 8, 11
Vaccaro, Michael J.: 79
Van Allen, James A.: 24
Vanguard, Project: 10, 12, 14-16, 18-19, 23, 24, 31, 32, 33, 35, 39, 263; Computing Center, 21-22, 74; Control Center, 22
Varson, W. F.: 137, 138, 139
Vaught Aircraft Company: 86
Vavra, Paul: 71
Vensel, Joseph: 207
Verwoerd, Hendrik F.: 209, 210
Very Long Baseline Interferometry: 304
Viking, Project: ix, xv, 161, 212
Vision for Space Exploration: 326; Exploration Systems Mission Directorate, 340
Voice communications: 61, 64, 70, 71, 72, 75, 83, 98, 100, 109, 110, 112-113, 120, 121, 128, 129, 131, 151, 158, 159, 171, 184, 185, 191, 192, 194, 224-226, 227, 240, 250, 252, 306, 308, 309, 311
von Braun, Wernher: 2, 15
von Bun, F. O.: 194
Voskhod: 141
Vostok: 105; Vostok 1, 66-67; Vostok 6, 105, 141
Voyager space probe: ix, xv, 161

W

WAC Corporal sounding rocket: 2
Wainright, Lewis: 138
Wake Island: 98
Wallops Flight Facility: 55, 58, 91, 100, 114, 219, 228, 308, 310-312, 335; As the GSFC Suborbital Projects and Operations Directorate, 310; Electrical Systems Branch, 311; Mid-Atlantic Regional Spaceport, 311; Mobile Range Control System, 311; Scout launch vehicle, 310
Wallops Island (*See* Wallops Flight Facility)
Watson, Bill: 200, 305, 318, 320-321, 337-338
Weather Bureau: 5, 121, 122
Webb, James E.: 106, 122, 127, 136, 166, 339
Weingarten, Murray T.: xix, 29-30, 331-332
Wiesner, Jerome: 106
Weitz, Paul J.: 187, 267
Western Electric: 80
Western Union: 256-259, 278, 292, 293; 111 Torn-Tape Relay System, 64; Goetchius, Rod, 80
Whipple, Fred L.: 3-6
White, Edward H. II: 107, 166, 167
White Sands (*See* White Sands Missile Range)
Whites Sands Missile Range: 2, 7-8, 9, 10, 12, 30, 60, 70, 74, 78, 95, 99, 109, 114-115, 146, 147, 163, 219, 232, 252, 278, 281, 282, 283, 284, 338
White, Tom: 96
Williams, Walt: 207
Wing-station: 147-148, 232-233, 308
Winkfield Tracking Station: 42, 56, 220, 222, 232
Womack, Otto: 138
Wood, H. William: xl, 68, 71, 72, 73, 76, 81, 138, 178-179, 180, 193, 201, 330
Woomera, South Australia: xxxi, 4, 19, 21, 23, 44, 45, 47, 48, 54, 56, 84, 95, 114, 128, 163, 203, 208, 212, 234
Working Group on Tracking and Computation: 2
World Administrative Radio Conference: xxii
Wright, G. R.: 5

X

X-band: 11, 312, 314, 320, 338

Y

Yagi antenna: 34, 36, 39, 48, 51, 219-220
Yarragadee, Western Australia: 226
Young, John W: 224, 234
Yven, Clet: 238

Z

Zanzibar Tracking Station: xxxi, 30, 54, 93, 95, 115, 124-128, 333
Zimbabwe: 128

The NASA History Series

Reference Works, NASA SP-4000:

Grimwood, James M. *Project Mercury: A Chronology*. NASA SP-4001, 1963.

Grimwood, James M., and C. Barton Hacker, with Peter J. Vorzimmer. *Project Gemini Technology and Operations: A Chronology*. NASA SP-4002, 1969.

Link, Mae Mills. *Space Medicine in Project Mercury*. NASA SP-4003, 1965.

Astronautics and Aeronautics, 1963: Chronology of Science, Technology, and Policy. NASA SP-4004, 1964.

Astronautics and Aeronautics, 1964: Chronology of Science, Technology, and Policy. NASA SP-4005, 1965.

Astronautics and Aeronautics, 1965: Chronology of Science, Technology, and Policy. NASA SP-4006, 1966.

Astronautics and Aeronautics, 1966: Chronology of Science, Technology, and Policy. NASA SP-4007, 1967.

Astronautics and Aeronautics, 1967: Chronology of Science, Technology, and Policy. NASA SP-4008, 1968.

Ertel, Ivan D., and Mary Louise Morse. *The Apollo Spacecraft: A Chronology, Volume I, Through November 7, 1962*. NASA SP-4009, 1969.

Morse, Mary Louise, and Jean Kernahan Bays. *The Apollo Spacecraft: A Chronology, Volume II, November 8, 1962–September 30, 1964*. NASA SP-4009, 1973.

Brooks, Courtney G., and Ivan D. Ertel. *The Apollo Spacecraft: A Chronology, Volume III, October 1, 1964–January 20, 1966*. NASA SP-4009, 1973.

Ertel, Ivan D., and Roland W. Newkirk, with Courtney G. Brooks. *The Apollo Spacecraft: A Chronology, Volume IV, January 21, 1966–July 13, 1974*. NASA SP-4009, 1978.

Astronautics and Aeronautics, 1968: Chronology of Science, Technology, and Policy. NASA SP-4010, 1969.

Newkirk, Roland W., and Ivan D. Ertel, with Courtney G. Brooks. *Skylab: A Chronology*. NASA SP-4011, 1977.

Van Nimmen, Jane, and Leonard C. Bruno, with Robert L. Rosholt. *NASA Historical Data Book, Volume I: NASA Resources, 1958–1968*. NASA SP-4012, 1976, rep. ed. 1988.

Ezell, Linda Neuman. *NASA Historical Data Book, Volume II: Programs and Projects, 1958–1968*. NASA SP-4012, 1988.

Ezell, Linda Neuman. *NASA Historical Data Book, Volume III: Programs and Projects, 1969–1978*. NASA SP-4012, 1988.

Gawdiak, Ihor Y., with Helen Fedor, compilers. *NASA Historical Data Book, Volume IV: NASA Resources, 1969–1978*. NASA SP-4012, 1994.

Rumerman, Judy A., compiler. *NASA Historical Data Book, 1979–1988: Volume V, NASA Launch Systems, Space Transportation, Human Spaceflight, and Space Science*. NASA SP-4012, 1999.

Rumerman, Judy A., compiler. *NASA Historical Data Book, Volume VI: NASA Space Applications, Aeronautics and Space Research and Technology, Tracking and Data Acquisition/Space Operations, Commercial Programs, and Resources, 1979–1988*. NASA SP-2000-4012, 2000.

Astronautics and Aeronautics, 1969: Chronology of Science, Technology, and Policy. NASA SP-4014, 1970.

Astronautics and Aeronautics, 1970: Chronology of Science, Technology, and Policy. NASA SP-4015, 1972.

Astronautics and Aeronautics, 1971: Chronology of Science, Technology, and Policy. NASA SP-4016, 1972.

Astronautics and Aeronautics, 1972: Chronology of Science, Technology, and Policy. NASA SP-4017, 1974.

Astronautics and Aeronautics, 1973: Chronology of Science, Technology, and Policy. NASA SP-4018, 1975.

Astronautics and Aeronautics, 1974: Chronology of Science, Technology, and Policy. NASA SP-4019, 1977.

Astronautics and Aeronautics, 1975: Chronology of Science, Technology, and Policy. NASA SP-4020, 1979.

Astronautics and Aeronautics, 1976: Chronology of Science, Technology, and Policy. NASA SP-4021, 1984.

Astronautics and Aeronautics, 1977: Chronology of Science, Technology, and Policy. NASA SP-4022, 1986.

Astronautics and Aeronautics, 1978: Chronology of Science, Technology, and Policy. NASA SP-4023, 1986.

Astronautics and Aeronautics, 1979–1984: Chronology of Science, Technology, and Policy. NASA SP-4024, 1988.

Astronautics and Aeronautics, 1985: Chronology of Science, Technology, and Policy. NASA SP-4025, 1990.

Noordung, Hermann. *The Problem of Space Travel: The Rocket Motor*. Edited by Ernst Stuhlinger and J. D. Hunley, with Jennifer Garland. NASA SP-4026, 1995.

Astronautics and Aeronautics, 1986–1990: A Chronology. NASA SP-4027, 1997.

Astronautics and Aeronautics, 1990–1995: A Chronology. NASA SP-2000-4028, 2000.

Management Histories, NASA SP-4100:

Rosholt, Robert L. *An Administrative History of NASA, 1958–1963*. NASA SP-4101, 1966.

Levine, Arnold S. *Managing NASA in the Apollo Era*. NASA SP-4102, 1982.

Roland, Alex. *Model Research: The National Advisory Committee for Aeronautics, 1915–1958*. NASA SP-4103, 1985.

Fries, Sylvia D. *NASA Engineers and the Age of Apollo*. NASA SP-4104, 1992.

Glennan, T. Keith. *The Birth of NASA: The Diary of T. Keith Glennan*. J. D. Hunley, editor. NASA SP-4105, 1993.

Seamans, Robert C., Jr. *Aiming at Targets: The Autobiography of Robert C. Seamans, Jr*. NASA SP-4106, 1996.

Garber, Stephen J., editor. *Looking Backward, Looking Forward: Forty Years of U.S. Human Spaceflight Symposium*. NASA SP-2002-4107, 2002.

Mallick, Donald L. with Peter W. Merlin. *The Smell of Kerosene: A Test Pilot's Odyssey*. NASA SP-4108, 2003.

Iliff, Kenneth W. and Curtis L. Peebles. *From Runway to Orbit: Reflections of a NASA Engineer*. NASA SP-2004-4109, 2004.

Chertok, Boris. *Rockets and People, Volume 1*. NASA SP-2005-4110, 2005.

Laufer, Alexander, Todd Post, and Edward Hoffman. *Shared Voyage: Learning and Unlearning from Remarkable Projects*. NASA SP-2005-4111, 2005.

Dawson, Virginia P. and Mark D. Bowles. *Realizing the Dream of Flight: Biographical Essays in Honor of the Centennial of Flight, 1903-2003*. NASA SP-2005-4112, 2005.

Mudgway, Douglas J. *William H. Pickering: America's Deep Space Pioneer*, NASA SP-2007-4113, 2007.

Project Histories, NASA SP-4200:

Swenson, Loyd S., Jr., James M. Grimwood, and Charles C. Alexander. *This New Ocean: A History of Project Mercury*. NASA SP-4201, 1966; rep. ed. 1998.

Green, Constance McLaughlin, and Milton Lomask. *Vanguard: A History*. NASA SP-4202, 1970; rep. ed. Smithsonian Institution Press, 1971.

Hacker, Barton C., and James M. Grimwood. *On the Shoulders of Titans: A History of Project Gemini*. NASA SP-4203, 1977.

Benson, Charles D., and William Barnaby Faherty. *Moonport: A History of Apollo Launch Facilities and Operations*. NASA SP-4204, 1978.

Brooks, Courtney G., James M. Grimwood, and Loyd S. Swenson, Jr. *Chariots for Apollo: A History of Manned Lunar Spacecraft*. NASA SP-4205, 1979.

Bilstein, Roger E. *Stages to Saturn: A Technological History of the Apollo/Saturn Launch Vehicles*. NASA SP-4206, 1980, rep. ed. 1997.

SP-4207 not published.

Compton, W. David, and Charles D. Benson. *Living and Working in Space: A History of Skylab*. NASA SP-4208, 1983.

Ezell, Edward Clinton, and Linda Neuman Ezell. *The Partnership: A History of the Apollo-Soyuz Test Project*. NASA SP-4209, 1978.

Hall, R. Cargill. *Lunar Impact: A History of Project Ranger*. NASA SP-4210, 1977.

Newell, Homer E. *Beyond the Atmosphere: Early Years of Space Science*. NASA SP-4211, 1980.

Ezell, Edward Clinton, and Linda Neuman Ezell. *On Mars: Exploration of the Red Planet, 1958–1978*. NASA SP-4212, 1984.

Pitts, John A. *The Human Factor: Biomedicine in the Manned Space Program to 1980*. NASA SP-4213, 1985.

Compton, W. David. *Where No Man Has Gone Before: A History of Apollo Lunar Exploration Missions*. NASA SP-4214, 1989.

Naugle, John E. *First Among Equals: The Selection of NASA Space Science Experiments*. NASA SP-4215, 1991.

Wallace, Lane E. *Airborne Trailblazer: Two Decades with NASA Langley's Boeing 737 Flying Laboratory*. NASA SP-4216, 1994.

Butrica, Andrew J., editor. *Beyond the Ionosphere: Fifty Years of Satellite Communication*. NASA SP-4217, 1997.

Butrica, Andrew J. *To See the Unseen: A History of Planetary Radar Astronomy*. NASA SP-4218, 1996.

Mack, Pamela E., editor. *From Engineering Science to Big Science: The NACA and NASA Collier Trophy Research Project Winners*. NASA SP-4219, 1998.

Reed, R. Dale, with Darlene Lister. *Wingless Flight: The Lifting Body Story*. NASA SP-4220, 1997.

Heppenheimer, T. A. *The Space Shuttle Decision: NASA's Search for a Reusable Space Vehicle*. NASA SP-4221, 1999.

Hunley, J. D., editor. *Toward Mach 2: The Douglas D-558 Program*. NASA SP-4222, 1999.

Swanson, Glen E., editor. *"Before this Decade Is Out . . .": Personal Reflections on the Apollo Program*. NASA SP-4223, 1999.

Tomayko, James E. *Computers Take Flight: A History of NASA's Pioneering Digital Fly-by-Wire Project*. NASA SP-2000-4224, 2000.

Morgan, Clay. *Shuttle-Mir: The U.S. and Russia Share History's Highest Stage*. NASA SP-2001-4225, 2001.

Leary, William M. *"We Freeze to Please": A History of NASA's Icing Research Tunnel and the Quest for Flight Safety.* NASA SP-2002-4226, 2002.

Mudgway, Douglas J. *Uplink-Downlink: A History of the Deep Space Network 1957–1997.* NASA SP-2001-4227, 2001.

Dawson, Virginia P. and Mark D. Bowles. *Taming Liquid Hydrogen: The Centaur Upper Stage Rocket, 1958-2002.* NASA SP-2004-4230, 2004.

Meltzer, Michael. *Mission to Jupiter: A History of the Galileo Project.* NASA SP-2007-4231.

Heppenheimer, T. A. *Facing the Heat Barrier: A History of Hypersonics.* NASA SP-2007-4232, 2007.

Center Histories, NASA SP-4300:

Rosenthal, Alfred. *Venture into Space: Early Years of Goddard Space Flight Center.* NASA SP-4301, 1985.

Hartman, Edwin P. *Adventures in Research: A History of Ames Research Center, 1940–1965.* NASA SP-4302, 1970.

Hallion, Richard P. *On the Frontier: Flight Research at Dryden, 1946–1981.* NASA SP-4303, 1984.

Muenger, Elizabeth A. *Searching the Horizon: A History of Ames Research Center, 1940–1976.* NASA SP-4304, 1985.

Hansen, James R. *Engineer in Charge: A History of the Langley Aeronautical Laboratory, 1917–1958.* NASA SP-4305, 1987.

Dawson, Virginia P. *Engines and Innovation: Lewis Laboratory and American Propulsion Technology.* NASA SP-4306, 1991.

Dethloff, Henry C. *"Suddenly Tomorrow Came . . .": A History of the Johnson Space Center.* NASA SP-4307, 1993.

Hansen, James R. *Spaceflight Revolution: NASA Langley Research Center from Sputnik to Apollo.* NASA SP-4308, 1995.

Wallace, Lane E. *Flights of Discovery: 50 Years at the NASA Dryden Flight Research Center.* NASA SP-4309, 1996.

Herring, Mack R. *Way Station to Space: A History of the John C. Stennis Space Center.* NASA SP-4310, 1997.

Wallace, Harold D., Jr. *Wallops Station and the Creation of the American Space Program.* NASA SP-4311, 1997.

Wallace, Lane E. *Dreams, Hopes, Realities: NASA's Goddard Space Flight Center, The First Forty Years.* NASA SP-4312, 1999.

Dunar, Andrew J., and Stephen P. Waring. *Power to Explore: A History of the Marshall Space Flight Center.* NASA SP-4313, 1999.

Bugos, Glenn E. *Atmosphere of Freedom: Sixty Years at the NASA Ames Research Center.* NASA SP-2000-4314, 2000.

Schultz, James. *Crafting Flight: Aircraft Pioneers and the Contributions of the Men and Women of NASA Langley Research Center.* NASA SP-2003-4316, 2003.

General Histories, NASA SP-4400:

Corliss, William R. *NASA Sounding Rockets, 1958–1968: A Historical Summary.* NASA SP-4401, 1971.

Wells, Helen T., Susan H. Whiteley, and Carrie Karegeannes. *Origins of NASA Names.* NASA SP-4402, 1976.

Anderson, Frank W., Jr. *Orders of Magnitude: A History of NACA and NASA, 1915–1980.* NASA SP-4403, 1981.

Sloop, John L. *Liquid Hydrogen as a Propulsion Fuel, 1945–1959*. NASA SP-4404, 1978.

Roland, Alex. *A Spacefaring People: Perspectives on Early Spaceflight*. NASA SP-4405, 1985.

Bilstein, Roger E. *Orders of Magnitude: A History of the NACA and NASA, 1915–1990*. NASA SP-4406, 1989.

Logsdon, John M., editor, with Linda J. Lear, Jannelle Warren-Findley, Ray A. Williamson, and Dwayne A. Day. *Exploring the Unknown: Selected Documents in the History of the U.S. Civil Space Program, Volume I, Organizing for Exploration*. NASA SP-4407, 1995.

Logsdon, John M., editor, with Dwayne A. Day and Roger D. Launius. *Exploring the Unknown: Selected Documents in the History of the U.S. Civil Space Program, Volume II, Relations with Other Organizations*. NASA SP-4407, 1996.

Logsdon, John M., editor, with Roger D. Launius, David H. Onkst, and Stephen J. Garber. *Exploring the Unknown: Selected Documents in the History of the U.S. Civil Space Program, Volume III, Using Space*. NASA SP-4407, 1998.

Logsdon, John M., general editor, with Ray A. Williamson, Roger D. Launius, Russell J. Acker, Stephen J. Garber, and Jonathan L. Friedman. *Exploring the Unknown: Selected Documents in the History of the U.S. Civil Space Program, Volume IV, Accessing Space*. NASA SP-4407, 1999.

Logsdon, John M., general editor, with Amy Paige Snyder, Roger D. Launius, Stephen J. Garber, and Regan Anne Newport. *Exploring the Unknown: Selected Documents in the History of the U.S. Civil Space Program, Volume V, Exploring the Cosmos*. NASA SP-2001-4407, 2001.

Siddiqi, Asif A. *Challenge to Apollo: The Soviet Union and the Space Race, 1945–1974*. NASA SP-2000-4408, 2000.

Hansen, James R., editor. *The Wind and Beyond: Journey into the History of Aerodynamics in America, Volume I, The Ascent of the Airplane*. NASA SP-2003-4409, 2003.

Hogan, Thor. *Mars Wars: The Rise and Fall of the Space Exploration Initiative*. NASA SP-2007-4410, 2007.

Hansen, James R., editor. *The Wind and Beyond: Journey into the History of Aerodynamics in America, Volume II, Reinventing the Airplane*. NASA SP-2007-4409, 2007.

Monographs in Aerospace History, NASA SP-4500:

Launius, Roger D. and Aaron K. Gillette, compilers, *Toward a History of the Space Shuttle: An Annotated Bibliography*. Monograph in Aerospace History, No. 1, 1992.

Launius, Roger D., and J. D. Hunley, compilers, *An Annotated Bibliography of the Apollo Program*. Monograph in Aerospace History, No. 2, 1994.

Launius, Roger D. *Apollo: A Retrospective Analysis*. Monograph in Aerospace History, No. 3, 1994.

Hansen, James R. *Enchanted Rendezvous: John C. Houbolt and the Genesis of the Lunar-Orbit Rendezvous Concept*. Monograph in Aerospace History, No. 4, 1995.

Gorn, Michael H. *Hugh L. Dryden's Career in Aviation and Space*. Monograph in Aerospace History, No. 5, 1996.

Powers, Sheryll Goecke. *Women in Flight Research at NASA Dryden Flight Research Center, from 1946 to 1995*. Monograph in Aerospace History, No. 6, 1997.

Portree, David S. F. and Robert C. Trevino. *Walking to Olympus: An EVA Chronology*. Monograph in Aerospace History, No. 7, 1997.

Logsdon, John M., moderator. *Legislative Origins of the National Aeronautics and Space Act of 1958: Proceedings of an Oral History Workshop*. Monograph in Aerospace History, No. 8, 1998.

Rumerman, Judy A., compiler, *U.S. Human Spaceflight, A Record of Achievement 1961–1998*. Monograph in Aerospace History, No. 9, 1998.

Portree, David S. F. *NASA's Origins and the Dawn of the Space Age*. Monograph in Aerospace History, No. 10, 1998.

Logsdon, John M. *Together in Orbit: The Origins of International Cooperation in the Space Station*. Monograph in Aerospace History, No. 11, 1998.

Phillips, W. Hewitt. *Journey in Aeronautical Research: A Career at NASA Langley Research Center*. Monograph in Aerospace History, No. 12, 1998.

Braslow, Albert L. *A History of Suction-Type Laminar-Flow Control with Emphasis on Flight Research*. Monograph in Aerospace History, No. 13, 1999.

Logsdon, John M., moderator. *Managing the Moon Program: Lessons Learned From Apollo*. Monograph in Aerospace History, No. 14, 1999.

Perminov, V. G. *The Difficult Road to Mars: A Brief History of Mars Exploration in the Soviet Union*. Monograph in Aerospace History, No. 15, 1999.

Tucker, Tom. *Touchdown: The Development of Propulsion Controlled Aircraft at NASA Dryden*. Monograph in Aerospace History, No. 16, 1999.

Maisel, Martin D., Demo J. Giulianetti, and Daniel C. Dugan. *The History of the XV-15 Tilt Rotor Research Aircraft: From Concept to Flight*. NASA SP-2000-4517, 2000.

Jenkins, Dennis R. *Hypersonics Before the Shuttle: A Concise History of the X-15 Research Airplane*. NASA SP-2000-4518, 2000.

Chambers, Joseph R. *Partners in Freedom: Contributions of the Langley Research Center to U.S. Military Aircraft in the 1990s*. NASA SP-2000-4519, 2000.

Waltman, Gene L. *Black Magic and Gremlins: Analog Flight Simulations at NASA's Flight Research Center*. NASA SP-2000-4520, 2000.

Portree, David S. F. *Humans to Mars: Fifty Years of Mission Planning, 1950–2000*. NASA SP-2001-4521, 2001.

Thompson, Milton O., with J. D. Hunley. *Flight Research: Problems Encountered and What They Should Teach Us*. NASA SP-2000-4522, 2000.

Tucker, Tom. *The Eclipse Project*. NASA SP-2000-4523, 2000.

Siddiqi, Asif A. *Deep Space Chronicle: A Chronology of Deep Space and Planetary Probes, 1958–2000*. NASA SP-2002-4524, 2002.

Merlin, Peter W. *Mach 3+: NASA/USAF YF-12 Flight Research, 1969–1979*. NASA SP-2001-4525, 2001.

Anderson, Seth B. *Memoirs of an Aeronautical Engineer—Flight Tests at Ames Research Center: 1940–1970*. NASA SP-2002-4526, 2002.

Renstrom, Arthur G. *Wilbur and Orville Wright: A Bibliography Commemorating the One-Hundredth Anniversary of the First Powered Flight on December 17, 1903*. NASA SP-2002-4527, 2002.

No monograph 28.

Chambers, Joseph R. *Concept to Reality: Contributions of the NASA Langley Research Center to U.S. Civil Aircraft of the 1990s*. SP-2003-4529, 2003.

Peebles, Curtis, editor. *The Spoken Word: Recollections of Dryden History, The Early Years*. SP-2003-4530, 2003.

Jenkins, Dennis R., Tony Landis, and Jay Miller. *American X-Vehicles: An Inventory-X-1 to X-50*. SP-2003-4531, 2003.

Renstrom, Arthur G. *Wilbur and Orville Wright: A Chronology Commemorating the One-Hundredth Anniversary of the First Powered Flight on December 17, 1903*. NASA SP-2003-4532, 2002.

Bowles, Mark D. and Robert S. Arrighi. *NASA's Nuclear Frontier: The Plum Brook Research Reactor.* SP-2004-4533, 2003.

Matranga, Gene J. and C. Wayne Ottinger, Calvin R. Jarvis with D. Christian Gelzer. *Unconventional, Contrary, and Ugly: The Lunar Landing Research Vehicle.* NASA SP-2006-4535.

McCurdy, Howard E. *Low Cost Innovation in Spaceflight: The History of the Near Earth Asteroid Rendezvous (NEAR) Mission.* NASA SP-2005-4536, 2005.

Seamans, Robert C. Jr. *Project Apollo: The Tough Decisions.* NASA SP-2005-4537, 2005.

Lambright, W. Henry. *NASA and the Environment: The Case of Ozone Depletion.* NASA SP-2005-4538, 2005.

Chambers, Joseph R. *Innovation in Flight: Research of the NASA Langley Research Center on Revolutionary Advanced Concepts for Aeronautics.* NASA SP-2005-4539, 2005.

Phillips, W. Hewitt. *Journey Into Space Research: Continuation of a Career at NASA Langley Research Center.* NASA SP-2005-4540, 2005.

Rumerman, Judith A., compiler, with Chris Gamble and Gabriel Okolski, *U. S. Human Spaceflight: A Record of Achievement, 1961-2006.* NASA SP-2007-4541, 2007.

Electronic Media, NASA SP-4600:

Remembering Apollo 11: The 30th Anniversary Data Archive CD-ROM. NASA SP-4601, 1999.

The Mission Transcript Collection: U.S. Human Spaceflight Missions from Mercury Redstone 3 to Apollo 17. NASA SP-2000-4602, 2001.

Shuttle-Mir: The United States and Russia Share History's Highest Stage. NASA SP-2001-4603, 2002.

U.S. Centennial of Flight Commission Presents Born of Dreams—Inspired by Freedom. NASA SP-2004-4604, 2004.

Of Ashes and Atoms: A Documentary on the NASA Plum Brook Reactor Facility. NASA SP-2005-4605, 2005.

Taming Liquid Hydrogen: The Centaur Upper Stage Rocket Interactive CD-ROM. NASA SP-2004-4606, 2004.

Fueling Space Exploration: The History of NASA's Rocket Engine Test Facility DVD. NASA SP-2005-4607, 2005.

Conference Proceedings, NASA SP-4700:

Dick, Steven J., and Keith L. Cowing, editors. *Risk and Exploration: Earth, Sea and the Stars.* NASA SP-2005-4701, 2005.

Dick, Steven J., and Roger D. Launius, editors. *Critical Issues in the History of Spaceflight.* NASA SP-2006-4702, 2006.

Societal Impact, NASA SP-4800:

Dick, Steven J., and Roger D. Launius, editors. *Societal Impact of Spaceflight.* NASA SP-2007-4801, 2007.